Organic Materials for Non-linear Optics

Special Publication No. 69

Organic Materials for Non-linear Optics

The Proceedings of a conference organised by the Applied Solid State Chemistry Group of the Dalton Division of The Royal Society of Chemistry

Oxford, 29th—30th June 1988

Edited by

R.A. Hann
ICI Electronics, Runcorn

D. Bloor
Queen Mary College, London

ROYAL
SOCIETY OF
CHEMISTRY

British Library Cataloguing in Publication Data

Organic materials for non-linear optics
1. Organic compounds. Molecules.
Nonlinear optics
I. Hann, R.A.   II. Bloor, D.
III. Series
535'.2

ISBN 0-85186-806-1

CHEM
seplae

© The Royal Society of Chemistry 1989

Published by The Royal Society of Chemistry,
Burlington House, London W1V 0BN

Printed in Great Britain by
Whitstable Litho Printers Ltd., Whitstable, Kent

Editors' Preface

This publication represents the proceedings of the conference on 'Organic Materials for Non-Linear Optics' held in Oxford from 29-30 June 1988. OMNO 88 (as it became known) was a truly international conference attended by over 160 delegates from academia, industry, and government research establishments. The delegates represented a very wide range of scientific disciplines, and included organic and organometallic chemists, theoretical and experimental physicists and device engineers. Papers presented at the conference covered the theoretical modelling and prediction of non-linear optical (NLO) properties of organic molecules and solids, synthesis and NLO characterisation of organic materials and polymers and application in both bulk and thin film forms. The blend of backgrounds and affiliations emphasises that this area of non-linear optics (NLO) is both interdisciplinary and holds promise for extensive application in the future.

The basis for the development of new materials is a sound theoretical understanding of the origins of optical nonlinearities. The contributions by Munn, Garito and Morley and Pugh show that this is the case for second-order nonlinearities. In this instance an asymmetric molecule structure, with electron donating and accepting groups coupled through a conjugated structure, and a non-centrosymmetric supermolecular organisation are required. Current theory has good predictive capabilities at the molecular level and for crystals of known structure. The models for third-order effects, which are ubiquitous though smaller in magnitude, are less well established. However, progress is being made and theory is being applied with almost as much confidence as for second-order effects.

In spite of the above success in modelling the electronic properties of molecules, the superficially more simple problem of predicting crystal structure from molecular structure remains problematic, as shown by Gavezzotti's paper. There are, however, indications that new approaches may provide leads for future work. Fortunately, as discussed in Sherwood's paper, the practical aspects of growing large high-quality single crystals of organic compounds, known to have large optical nonlinearities, have become very sophisticated. Thus large crystals of excellent optical quality of a number of materials are now available.

Equal sophistication is necessary if these new materials are to be properly characterised and end applications identified. The papers by Zyss and Meredith show how the use of appropriate techniques can yield detailed information on the nonlinear properties of organic molecules and crystals. The large number of papers on small organic molecules shows the continuing interest in this area, and also reveals the range of characterisation techniques used in practise. Second harmonic generation by powdered samples still remains the favoured first screen, but an encouraging number of workers are turning to more sophisticated measurements on single crystals. The range of materials available for investigation is being broadened by the inclusion of organo-metallic molecules, which hold the promise of an extension of the range of available properties.

There is an important trend towards the use of polymers as NLO materials for reasons outlined by Ulrich and exemplified by Prasad and Mohlmann. Clearly the relative ease of fabrication of polymer films and their generally excellent mechanical properties make them much more attractive than single crystals as elements in devices. The ability to induce supermolecular non-centrosymmetry by poling and other techniques is also an important driving force for the use of polymers. Nevertheless, much work remains to be done before polymer films match the performance of single crystals.

An alternative method of forcing a non-centrosymmetric super-molecular structure is to use the Langmuir-Blodgett (LB) technique as discussed by Peterson and Shen. The beauty of the LB technique is its ability to produce ordered mono- and multi- layers with polar molecular arrangement. Thus it provides a powerful method for the realisation of the intrinsic second-order NLO properties of a molecular chromophore in a solid film at the expense of a modest overhead in synthetic chemistry. The main limitation on the LB technique is the difficulty of building up thick multilayers of sufficient quality at realistic speeds.

The ultimate application of NLO materials will be in devices for controlling the flow of optical data in telecommunications, optical computing etc., Stegeman and Lytel discuss some of the devices that have been made so far and outline the performance that organic materials will have to achieve in order to become widely accepted as replacements for or additions to the established inorganic materials such as Lithium Niobate. Progress in this direction is encouraging and bodes well for the

vi

emergence of practical devices in the near future. When making comparisons between organics and other materials it must be emphasised that NLO coefficients on their own are often an insufficient measure of performance for a particular application. An appropriate figure of merit must be used in order to characterise the inevitable trade-off between properties. It is encouraging that workers such as Allen are starting to quote these figures of merit for their materials.

OMNO 88 provided a comprehensive, international survey of the current status of studies of organic NLO materials. We hope that this volume will make this information available to a wider audience and encourage more scientists to become involved in this rapidly developing area of scientific and technical endeavour. With the present upsurge of interest one can confidently expect the discovery of new materials with improved NLO properties, the development of existing materials and the successful incorporation of these materials into a range of active optical devices. Plans are already in hand for an OMNO 90 meeting in Oxford in the autumn of 1990 at which we can confidently expect such advances to be reported.

R.A. Hann
D. Bloor

Contents

Crystals and Crystal Structure

Characterization of Non-linear Optical Properties

Small Organic Molecules

Theory of Non-linear Properties

Theory of Molecular Opto-electronics. VI. Comparisons between Nitroanilines

M. Hurst and R.W. Munn*

DEPARTMENT OF CHEMISTRY AND CENTRE FOR ELECTRONIC MATERIALS, UMIST, MANCHESTER M60 1QD, UK

1 INTRODUCTION

Molecular materials are attracting considerable interest for potential applications in nonlinear optics.[1] Evidence for this interest is provided by growth in publications, specialist conferences, and industrial participation. Reasons for the interest include the high nonlinear response and optical damage resistance observed in some molecular materials, the prospect of chemical modification to optimize the material properties, and the possibility of theoretical prediction to guide synthetic programmes seeking improved materials. To a large extent, these reasons exemplify the advantages foreseen for the general area of molecular electronics.[2] However, molecular opto-electronics is a field where theoretical prediction is particularly promising, because in molecular materials the molecules largely retain their separate identities. Properties of the materials can then be expressed in terms of those of the molecules, their arrangement and the interactions between them.[3]

Aspects of theoretical prediction and interpretation are illustrated here by comparing meta-nitroaniline (mNA) and para-nitroaniline (pNA). These isomeric molecules are closely similar, but pNA crystallizes in a centro-symmetric structure[4] while mNA does not,[5] so that only mNA has a nonzero second-order susceptibility $\chi^{(2)}$. Screening crystal structure databases shows that only about 25% of structures lack a centre of symmetry,[6] an observation which has been attributed to a preference among polar molecules for adopting antiferroelectric structures. This idea is explored for the nitroanilines by calculating the dipole energy taking full account of dielectric screening.

3

All crystals have a nonzero third-order susceptibility $\chi^{(3)}$, whether or not they have a centre of symmetry. It is known that pNA has larger hyperpolarizabilities than mNA but not how these interact with the different crystal structures. This is investigated by calculations of $\chi^{(3)}$ using procedures developed previously.

2 COMPARISONS BETWEEN NITROANILINES

2.1 Linear response

The linear polarizability is required to calculate the local electric field which contributes to electric susceptibilities. This polarizability should be the effective polarizability appropriate to the crystal environment. It is therefore calculated from the crystal linear susceptibility tensor $\chi^{(1)}$, via[7]

$$a_{k}^{-1} = Z[\chi^{(1)}]^{-1} + \sum_{k'} L_{kk'} \quad , \tag{1}$$

Here $a_k = \alpha_k/\varepsilon_o v$, where α_k is the polarizability of molecule k and v is the volume of the primitive unit cell, which contains Z molecules. The $L_{kk'}$ are Lorentz-factor tensors or lattice dipole sums conveniently evaluated by the Ewald method[8] treating each molecule as a set of submolecules.

For mNA, the linear response has already been analysed.[7] We follow the same procedures to obtain results for pNA for comparison. Thus we treat pNA as three submolecules centred on the aromatic ring and each nitrogen atom. The crystal structure[4] belongs to the monoclinic space group $P2_1/n$ with $Z = 4$ and $a = 12.337$ Å, $b = 6.037$ Å, $c = 8.597$ Å, and $\beta = 91.42°$. Calculated Lorentz-factor tensors averaged over submolecules are given in Table 1.

Table 1. Lorentz-factor tensor components $L_{\alpha\beta}(kk')$ averaged over submolecules for pNA, expressed in the crystal abc' axis system.

kk'	αβ					
	aa	ab	ac'	bb	bc'	c'c'
11	-0.006	0.164	0.004	0.773	-0.107	0.233
12	0.190	0	0.487	0.072	0	0.739
13	0.586	0.107	0.522	-0.078	0.519	0.492
14	0.528	0	-0.315	0.338	0	0.134

The effective polarizability is obtained from the optical susceptibility. The principal refractive indices at a wavelength of 589 nm are 1.756 along the $\underset{\sim}{b}$ axis and 1.525 and 1.788 in the $\underline{ac}$ plane,[9] but the orientation of the principal axes in the $\underline{ac}$ plane is not specified However, this can be deduced from the quoted[9] molecular refractivities in the molecular axes. The molecular long $\underset{\sim}{L}$ axis is taken from the centre of the ring to the nitrogen of the nitro group, the medium $\underset{\sim}{M}$ axis perpendicular to $\underset{\sim}{L}$ in the mean molecular plane, and the normal $\underset{\sim}{N}$ axis perpendicular to the plane. Transformation of the molecular refractivities to molar refractivities and diagonalization gives back the quoted principal refractivities within 3% and shows that the principal axes make an angle of 41° with the crystal axes in the $\underline{ac}$ plane. The resulting susceptibility components in the crystal axes are $\chi_{aa} = 1.704$, $\chi_{ac'} = -0.432$, $\chi_{bb} = 2.084$ and $\chi_{c'c'} = 1.819$.

The calculated effective polarizability components for pNA, expressed in the molecular axes, are given in Table 2 together with those for mNA at a wavelength of 532 nm.[7] The results for pNA are closely similar to those for mNA, though it should be remembered that the $\underset{\sim}{L}$ and $\underset{\sim}{M}$ axes are not strictly comparable in the two molecules. The mean polarizabilities differ by some 6%, consistent with a simple group additivity scheme. Perhaps surprisingly, considering the greater prospects for electron transfer, pNA is the less polarizable. This may be because the pNA values refer to a longer wavelength than the mNA ones while the lowest $\pi-\pi^*$ transition in pNA lies at a shorter wavelength than that in mNA,[10] so that the energy denominator enhances the mNA polarizability somewhat; by a wavelength of 1064 nm the mean polarizability of mNA has fallen to less than that calculated here for pNA.[7]

Table 2. Polarizability volumes, $V_{AB} = (\alpha_{AB}/4\pi\varepsilon_o)/10^{-30}$ m^3, for pNA and mNA, expressed in the molecular $\underset{\sim}{L}, \underset{\sim}{M}, \underset{\sim}{N}$ axis system

Species	AB					
	LL	LM	LN	MM	MN	NN
pNA	19.9	−0.9	1.2	16.3	1.7	10.5
mNA	20.9	0.4	−0.7	15.9	1.7	13.1

Given the effective polarizability, the local-field tensor relating the local field to the macroscopic field is given by

$$\underset{\approx}{d}_k = \sum_{k'} (\underset{\approx}{a}^{-1} - \underset{\approx}{L})^{-1}_{kk'} \equiv \sum_{k'} \underset{\approx}{D}_{kk'} . \tag{2}$$

Here $\underset{\approx}{a}$ is the $3Z \times 3Z$ matrix having the $a_k \delta_{kk'}$ as 3×3 submatrices and $\underset{\approx}{L}$ is the $3Z \times 3Z$ matrix having the $\underset{\approx}{L}_{kk'}$ as 3×3 submatrices. The calculated local field components for pNA, expressed in the crystal axes, are given in Table 3 (Tensor method) together with those for mNA at 532 nm.[7] The two tensors are broadly similar, with the components for mNA generally larger than those for pNA, reflecting the higher polarizability. Table 3 also shows the local fields given by the anisotropic Lorentz approximation $\underset{\approx}{d}^{AL} = \underset{\approx}{1} + \underset{\approx}{\chi}^{(1)}/3$. Unlike those obtained by the tensor method, these tensors are necessarily symmetric, and diagonal in the principal dielectric axes. They also give different orderings of the diagonal components:[11] in pNA, d_{bb} and $d_{c'c'}$ are reversed, while in mNA, d_{cc} becomes the smallest component instead of the largest. Such differences may have significant effects on calculated nonlinear susceptibilities, depending on the nature of the molecular hyperpolarizabilities.

Table 3. Local-field components $d_{\alpha\beta}$ for pNA and mNA, expressed in the crystal abc' axis system for pNA and the abc axis system for mNA.

Species	$\alpha\beta$								
	11	12	13	21	22	23	31	32	33
(i) Tensor method									
pNA	1.478	0.141	0.177	0.110	1.576	−0.006	0.125	0.027	1.651
mNA	1.682	−0.194	−0.253	−0.200	1.563	0.037	−0.294	0.041	1.918
(ii) Anisotropic Lorentz approximation									
pNA	1.568	0	−0.144	0	1.695	0	−0.144	0	1.606
mNA	1.759	0	0	0	1.735	0	0	0	1.653

2.2 Screened dipole energy

Both pNA and mNA have sizeable permanent dipole moments $\underset{\sim}{\mu}$. In the crystal, these dipole moments produce electric fields which polarize the molecules so that they acquire total dipole moments p. The energy of the crystal of polarized molecules relative to the unpolarized molecules at infinite separation is the screened dipole energy W. This is the energy of interaction of the permanent dipole moments reduced by the polarization energy; it is the polarization, incorporating the dielectric response of the crystal, which screens the permanent dipole interactions.

The screened dipole energy is given by[12,13]

$$4\pi\varepsilon_o W = -\tfrac{1}{2} \sum_{nn'n''} \underset{\sim}{\mu}_n \cdot \underset{\approx}{\varepsilon}^{-1}_{nn'} \cdot \underset{\approx}{T}_{n'n''} \underset{\sim}{\mu}_{n''} \tag{3}$$

where the sum runs over all NZ molecules labelled $n (\equiv \ell k)$ in the crystal, where ℓ labels the unit cell. The quantity $\underset{\sim}{T}$ is the dipole tensor given by

$$\underset{\approx}{T}_{nn'} = \lim_{\underset{\sim}{r} \to \underset{\sim}{r}_{n'}} \underset{\sim}{\nabla}\underset{\sim}{\nabla}(1/|\underset{\sim}{r}_n - \underset{\sim}{r}|), \tag{4}$$

where $\underset{\sim}{r}_n$ is the position of molecule n, and $\underset{\approx}{\varepsilon}$ is the inverse dielectric function given by

$$\underset{\approx}{\varepsilon}^{-1}_{nn'} = (\underset{\approx}{J} - \underset{\approx}{T} \cdot a)^{-1}_{nn'}, \tag{5}$$

where the matrices on the right-hand side are 3NZx3NZ matrices with elements $\underset{\approx}{1}\delta_{nn'}$, $\underset{\approx}{T}_{nn'}$ and $\underset{\approx}{a}_n\delta_{nn'}$ respectively. It can be seen from (3) how $\underset{\approx}{\varepsilon}^{-1}$ screens the bare field due to the dipoles.

Because of translational symmetry it proves possible to evaluate these expressions via Fourier transformation. The dipole moment $\underset{\sim}{\mu}_n$ depends only on k and not on ℓ, while $\underset{\approx}{T}_{n'n''}$ depends only on the difference $\ell'-\ell''$ and not on ℓ' and ℓ'' separately, so that the same result holds for $\underset{\approx}{\varepsilon}^{-1}_{n'n''}$. If we introduce the transformed dipole tensor $\underset{\sim}{t}(\underset{\sim}{y})$ via

$$(v/4\pi)\underset{\approx}{T}_{nn'} = N^{-1} \sum_{\underset{\sim}{y}} \exp[-2\pi i \underset{\sim}{y} \cdot (\underset{\sim}{r}_n - \underset{\sim}{r}_{n'})]\underset{\approx}{t}_{kk'}(\underset{\sim}{y}) \tag{6}$$

and the transformed inverse dielectric function given by[14]

$$\underset{\approx}{\varepsilon}_{kk'}^{-1} (\underline{y}) = [\underset{\approx}{I} - \underset{\approx}{t}(\underline{y}) \cdot \underset{\approx}{a}]_{kk'}^{-1} , \tag{7}$$

where $3Zx3Z$ matrices on the right-hand side have elements $\underset{\approx}{1}\delta_{kk'}$, $\underset{\approx}{t}_{kk'}(\underline{y})$ and $\underset{\approx}{a}_k\delta_{kk'}$ respectively, then we can write

$$W/N = -(1/2\varepsilon_o v) \underset{kk'k''}{\Sigma} \underset{\underline{y}\to 0}{\lim} \underline{\mu}_k \cdot \underset{\approx}{\varepsilon}_{kk'}^{-1}(\underline{y}) \cdot \underset{\approx}{t}_{k'k''}(\underline{y}) \cdot \underline{\mu}_{k''} . \tag{8}$$

It is not sufficient merely to set $\underline{y}=0$ in this expression, because as $\underline{y}\to 0$

$$\underset{\approx}{t}_{k'k''}(\underline{y}) \longrightarrow \underset{\approx}{L}_{k'k''} - \underline{y}\underline{y}/y^2 \tag{9}$$

where the last term is non-regular, its value depending on the manner in which $\underline{y}\to 0$. When (9) is substituted in (8) the sum over k'' contains the non-regular part multiplied by $\Sigma \underline{\mu}_{k''}$. This gives the macroscopic field due to the k'' net dipole moment of the unit cell. This field is clearly zero in a non-polar crystal such as pNA, while in a polar crystal such as mNA surface charges accumulate to annul the net dipole and hence the internal field; this can be proved mathematically for an ionic crystal[15] and is shown experimentally by the absence of a macroscopic dipole moment. Similar arguments show that the irregular part of $\underset{\approx}{\varepsilon}^{-1}(\underline{y})$ as $\underline{y}\to 0$ makes no contribution to W, so that the limit in (8) is derived by replacing $\underset{\approx}{t}$ by $\underset{\approx}{L}$ everywhere, with the result

$$W/N = -(1/2\varepsilon_o v) \underset{kk'k''}{\Sigma} \underline{\mu}_k \cdot \underset{\approx}{D}_{kk'} \cdot \underset{\approx}{L}_{k'k''} \cdot \underline{\mu}_{k''} , \tag{10}$$

where $\underset{\approx}{D}_{kk'}$ is the local-field contribution defined by (2). Thus the evaluation of the dipole energy requires quantities already computed, apart from the dipole moments themselves.

We take the dipole moments from theoretical CNDO calculations[10] which are also the source of first hyperpolarizability values used later. With $\mu = 7.95$ D $(26.5 \times 10^{-30}$ C m) for pNA we obtain $W = -31$ kJ mol^{-1}, while with $\mu = 6.64$ D $(22.1 \times 10^{-30}$ C m) for mNA we obtain $W = -36$ kJ mol^{-1}. Both energies are comparable with those calculated in the HCN crystal.[16] The stabilizing (negative) dipole energy for pNA is in accord with the expectation that the dipole energy tends to favour a centrosymmetric structure. However, the dipole energy is also stabilizing for mNA in its non-centrosymmetric structure, and somewhat greater in magnitude. This conflicts with the idea that the dipole energy dominates the structure, but accords with the observation[17] that the dipole energy is typically not large enough or sufficiently strongly dependent on crystal structure to change the structure from that determined by van der Waals forces for close packing.

We have attempted to analyse the dominant contributions to W for pNA and mNA by calculating the unscreened dipole energy for a few of the near neighbour molecules in each crystal. In each case both positive and negative contributions are found. However, the calculations show that the nearest-neighbour molecules actually give positive contributions in both crystals. This illustrates the importance of the long-range contribution to W, which in turn helps to explain why W need not depend strongly on details of the crystal structure.

The dipole energy can be written in the form[12]

$$W/N = -\tfrac{1}{2} \sum_k \mu_k \cdot F^p_k \; , \tag{11}$$

where F^p_k is the local field at molecule k due to the permanent dipole moments. Comparison with (10) yields

$$F^p_k = (1/\varepsilon_o v) \sum_{k'k''} D_{kk'} \cdot L_{k'k''} \cdot \mu_{k''} \; , \tag{12}$$

from which we calculate in the respective crystal axis systems $F^p/10^9$ V m^{-1} = (3.1, 0.3, 2.7) for pNA and (0.3, 2.4, 5.4) for mNA. Such high fields emphasize the need to evaluate molecular response coefficients in the crystal field rather than at zero field.[18] However, in practice zero-field calculations[10] of β suffice to yield adequate predictions of $\chi^{(2)}$ for mNA.[7] It should also be noted that the effective polarizability deduced from crystal refractive indices describes the response to a field imposed in addition to F^p (which is just as "permanent" as the dipole moment μ) and hence is the polarizability at a field F^p rather than at zero field.[18]

2.3 Nonlinear response

The quadratic nonlinear susceptibility $\chi^{(2)}$ has been calculated and analysed for mNA elsewhere.[7] That for pNA is zero by symmetry, so no comparison is feasible.

We therefore examine the cubic nonlinear susceptibility $\chi^{(3)}$. This consists of the sum of a "direct" contribution $\chi^{(d)}$ proportional to the second hyperpolarizability γ and a "cascading" contribution $\chi^{(c)}$ proportional to the square of the first hyperpolarizability β.[19,20] We can therefore compare not only the total $\chi^{(3)}$ for pNA and mNA but also the relative contributions of $\chi^{(d)}$ and $\chi^{(c)}$ in the two cases.

The direct contribution is given by[20-22]

$$\underset{\approx}{\chi}^{(d)} = \sum_k \underset{\approx}{d}_k^T \cdot \underset{\approx}{c}_k : \underset{\approx}{d}_k \underset{\approx}{d}_k \underset{\approx}{d}_k \ , \tag{13}$$

where the superscript T denotes the transpose and $\underset{\approx}{c} = \gamma/\varepsilon_0 v$. Successive local fields from left to right are understood to refer to frequencies ω_0, ω_1, ω_2 and ω_3, with $\omega_0 = \omega_1 + \omega_2 + \omega_3$.
The cascading contribution is given by[20-22]

$$\underset{\approx}{\chi}^{(c)} = 2 \sum_{kk'k''} \underset{\approx}{d}_k^T \cdot \underset{\approx}{b}_k : \underset{\approx}{d}_k \underset{\approx}{D}_{kk'} \cdot \underset{\approx}{L}_{k'k''} \cdot \underset{\approx}{b}_{k''} : \underset{\approx}{d}_{k''} \underset{\approx}{d}_{k''} \ , \tag{14}$$

where $\underset{\approx}{b} = \beta/\varepsilon_0 v$. Successive local fields are understood to refer to frequencies ω_0, ω_1, ω_4, ω_2 and ω_3, where $\omega_4 = \omega_2 + \omega_3$ and $\omega_0 = \omega_1 + \omega_4$ so that again $\omega_0 = \omega_1 + \omega_2 + \omega_3$. As in (12), the factor $\sum_{k'} \underset{\approx}{D}_{kk'} \cdot \underset{\approx}{L}_{k'k''}$ gives the local field at molecule k due to polarization at molecule k", but now the polarization is quadratic and induced rather than permanent.

All the quantities required to calculate $\chi^{(3)}$ have already been obtained except for the hyperpolarizabilities. For $\underset{\sim}{\beta}$ there are CNDO calculations[10] as used previously in our calculations[7] for $\chi^{(2)}$ for mNA; values are available for the static $\underset{\sim}{\beta}(0;0,0)$ and that for SHG, $\underset{\sim}{\beta}(-2\omega;\omega,\omega)$ at $\omega = 283$ THz ($\lambda = 1.064$ μm). For $\underset{\sim}{\gamma}$ there are values from four-wave mixing experiments in solution,[23] which have been combined with EFISH results to obtain values for $\underset{\sim}{\beta}(-\omega;\omega,0)$. These measurements do not yield the individual components of $\underset{\sim}{\gamma}$ and $\underset{\sim}{\beta}$ but rather the scalar part of $\underset{\sim}{\gamma}$ depending on $\gamma_{\alpha\alpha\beta\beta}$ and the vector part of $\underset{\sim}{\beta}$ depending on β_{LBB}, where repeated subscripts are understood to be summed and $\underset{\sim}{L}$ is the molecular axis parallel to the molecular dipole moment.[24] We have therefore assumed that each molecule can be treated as roughly one-dimensional, so that β_{LLL} and γ_{LLLL} are the only nonzero components. Though crude, this assumption is supported by the CNDO calculations of β, which show that β_{LLL} is a factor of 5 or more larger~than other components. It should allow the ratio between corresponding components of $\chi^{(d)}$ and $\chi^{(c)}$ to be obtained more reliably than the individual components. Since the full β tensor is available for two input frequencies, we have used them to calculate the components of $\chi^{(c)}$. This illustrates the effect of dispersion and allows~ a more rigorous comparison of this contribution for pNA and mNA, though it should be noted that the local fields are not available at all relevant frequencies and so all refer to ω. The results are given in Table 4.

Table 4. Cubic susceptibility components $\chi_{\alpha\beta\gamma\delta}/(\text{pm } V^{-1})^2$ $(=\chi_{\alpha\beta\gamma\delta}/8.9858\times10^{-16}$ esu) for pNA and mNA, expressed in the crystal axis system. Where two (or three) values are given these represent $\chi_{\alpha\beta\gamma\delta}/\chi_{\delta\gamma\beta\alpha}/(\chi_{\beta\alpha\gamma\delta})$.

| $\alpha\beta\gamma\delta$ | Tensor | | One dimensional β and γ | | | |
	$\chi^{(c)}$	$\chi^{(c)}(\omega)$	$\chi^{(c)}$	$\chi^{(d)}$	Total	% c:d
			(a) pNA			
1111	134	400	-10	4370	4360	-0.2
2222	72	157	910	12610	13520	7.2
3333	5	10	100	2360	2460	4.2
1122	59	157/199	100	1440	1540	6.9
1212	-64	-157/-92	50	"	1490	3.5
2233	-13	-39/-24	310	4970	5280	6.2
2323	33	72/59	180	"	5150	3.6
1133	-25	-72/-54	30	1600	1630	1.9
1313	21	44/33	"	"	"	"
1333	-8	-19/-12/-15	-40	-1550	-1590	2.6
3111	30	68/72/88	-10	-2420	-2430	0.4
1232	-35	-58/-118	-90	-2370	-2460	3.8
3212	"	-89/-86	"	"	"	"
1322	38	83/77/61	-140	"	-2510	5.9
			(b) mNA			
1111	-2	-7	0	3	3	1.6
2222	35	93	40	2001	2041	2.0
3333	27	104	8	390	398	2.1
1122	-7	-12/-32	1	57	58	1.9
1212	-3	-7/-6	5	"	62	8.8
2233	31	91/107	18	815	833	2.2
2323	40	114/143	102	"	917	12.5
1133	-5	-9/-37	1	22	23	2.3
1313	12	27/35	2	"	24	9.1

Before discussing the numerical values, we consider some symmetry matters. Components $\chi_{\alpha\beta\gamma\delta}$ are symmetric under the interchange ($\gamma \rightleftharpoons \delta$) and hence are not both quoted; otherwise, elements not quoted are zero. The tensor β has Kleinman symmetry at zero frequency but not at ω, so that $\chi^{(c)}(\omega)$ has more independent elements than $\chi^{(c)}(0)$, which itself lacks full Kleinman symmetry.[22] In the one-dimensional approximation, β and γ both have Kleinman symmetry, so that $\chi^{(d)}$ also has Kleinman symmetry and hence the fewest independent nonzero elements, while the total $\chi^{(3)}$ has the symmetry of $\chi^{(c)}$. Finally we note that there are more nonzero elements for pNA than for mNA because of its lower symmetry, which only requires $\alpha\beta\gamma\delta$ to contain 1 and 3 an even number of times taken together rather than an even number of times each.

Among the cascading contributions, a significant number are negative in sign. Between zero frequency and ω, the cascading contribution increases in magnitude by a factor of typically 2-4 in each crystal, the sign remaining unchanged.

In the one-dimensional approximation, the largest cascading contribution in pNA is some **23** times larger than in mNA, while the corresponding direct contribution is some 6 times larger. In all cases but one the cascading contribution increases the magnitude of $\chi^{(3)}$. The magnitude of β in pNA is some 6 times larger than in mNA and that of γ some 4 times larger.[23] Given that $\chi^{(c)} \sim \beta^2$ while $\chi^{(d)} \sim \gamma$, scaling from mNA would predict increases by factors of 36 and 4 in pNA. We therefore see that the local fields and molecular orientations in pNA are less than those in mNA in converting β but more effective in converting γ into $\chi^{(3)}$. The importance of $\chi^{(c)}$ relative to $\chi^{(d)}$ is generally higher in pNA, but never very high. Values of $\chi^{(c)}$ comparable with those of $\chi^{(d)}$ were calculated previously.[19,22]

Comparison of the cascading contributions obtained from the tensor β and from the one-dimensional model for β shows that the latter can be much larger in magnitude and may differ in sign. Tabulations[10,25] show that CNDO calculations usually underestimate the vector part of β obtained experimentally, and that this discrepancy is much larger for the values used here for pNA than for mNA, in general accord with the pattern of differences in table 4. The one-dimensional model assigns all the vector part of β to β_{LLL}, and this will affect the magnitude and sign of the $\chi^{(c)}$ components of $\chi^{(c)}$ depending on other components of β. In pNA there are numerous discrepancies of sign between the two approaches, while in mNA only the tensor β gives any negative components of $\chi^{(c)}$.

One might ask which approach to calculating $\chi^{(c)}$ is more correct. The tensor β is calculated for the isolated molecule, whereas the one-dimensional β is an approximation to information on β obtained from solution studies, so that neither refers to the crystal environment. For mNA the tensor β gives values of $\chi^{(2)}$ in satisfactory agreement with the measured values.[7] The magnitude of β is not greatly different in the one-dimensional model, which yields χ_{ccc} a factor of 2 or 3 too small, with χ_{aac} somewhat too large and χ_{bbc} about right. Thus the one-dimensional model for β does not reproduce the detailed behaviour of $\chi^{(2)}$ in mNA as well as the tensor β, as might have been expected. Presumably a similar conclusion about the predicted pattern of components can be drawn concerning $\chi^{(c)}$.

3 DISCUSSION

In many ways pNA and mNA are quite closely similar. Their effective polarizabilities are much the same, but that for mNA is lightly higher. This leads to local fields which are much the same, with that for mNA slightly higher.

The obvious difference between pNA and mNA lies in their crystal structures. We find that this difference is not driven by dipolar interactions, since the screened dipole energies are stabilizing in each crystal. This leaves packing as the driving force towards the different structures. This conclusion is supported by the well-known case of 2-methyl-4-nitroaniline (MNA). As far as its dipole moment is concerned, MNA resembles pNA, but the extra methyl group gives it an "awkward" shape like mNA. As a result, it crystallizes in a non-centrosymmetric structure with a high $\chi^{(2)}$ activity.[27,28] Calculations to see whether the screened dipole energy is stabilizing in MNA as well as in pNA and mNA would be interesting, but we have been unable to find the complete refractive index data needed as input.

It is clear that the dipole energy may affect the structure, so that efforts to keep it small can be beneficial. An example is provided by 3-methyl-4-nitro-pyridine-N-oxide (POM), which has a calculated ground-state dipole moment[10] of 0.48 D (1.58×10^{-30} C m) but a large movement of charge in the first excited state to give a high dipole moment and hyperpolarizability. This molecule crystallizes in a noncentrosymmetric structure

for which we calculate a screened dipole energy of about
+0.1 kJ mol^{-1}. Thus the dipolar forces are destabilizing
but too small to matter. The role of the methyl group in
again producing a shape which favours a noncentrosymmetric
structure is shown by the fact that the compound without
the methyl group crystallizes with a centre of symmetry.[23]

A final comment on the role of the dipole energy is
determining structure in potential $\chi^{(2)}$ materials arise
from the observation that molecules with a high β are
likely also to have a high α. This means that the dipolar
interactions will be rather effectively screened, so that
their magnitude and hence their likely significance will
be reduced.

Since all structures are $\chi^{(3)}$ active, the effect of
crystal structure has not been intensively studied in the
same way as for $\chi^{(2)}$. We have shown by our simplified
calculations that the centrosymmetric structure which
precludes $\chi^{(2)}$ activity in pNA seems to reduce the
cascading but increase the direct contribution to $\chi^{(3)}$
activity, each by a factor of about 1.5 compared with mNA.
More algebraic and numerical work is required in order to
establish whether such effects are generally to be expected
in centrosymmetric crystals. If so, efforts to produce
noncentrosymmetric crystals of molecules with high β to
optimize $\chi^{(2)}$ may tend not to optimize $\chi^{(3)}$.

In terms of the role of theory in molecular opto-
electronics, the present work makes various contributions.
Analysis of the refractive indices for pNA yields the
effective polarizability, which is a molecular property
modified by molecular interactions. Calculations of
molecular interactions yield the local fields and the
screened dipole energy, which is a factor determining the
molecular arrangement. Together, the molecular properties,
molecular arrangement and molecular interactions yield
the crystal opto-electronic property $\chi^{(3)}$. Although pNA
and mNA are well-known materials, we have been able to
provide a new and more detailed analysis and interpretation
of their properties relevant to opto-electronics. These
in turn can contribute to improved strategies for the
design of novel materials.

Acknowledgments

This work was supported by the UK Science and
Engineering Research Council under the Joint Opto-Electronics Research Scheme. We are grateful to Dr J O Morley
for providing details of the work in Reference 10.

REFERENCES

1. 'Nonlinear Optical Properties of Organic Molecules and Crystals',
    2 vols, D.S. Chemla and J. Zyss, eds. (Academic Press, Orlando,
    1987).
2. R.W. Munn, in 'The Chemistry of the Semiconductor Industry',
    S.J. Moss and A. Ledwith, eds (Blackie, Glasgow, 1987), p292.
3. R.W. Munn, J. Mol. Electron., 1988, 4, 31.
4. M. Colapietro, A. Domenicano, C. Marciante and G. Portalone,
    Z. Naturforsch., 1982, B37, 1309.
5. A.C. Skapski and J.L. Stevenson, J.Chem.Soc.Perkin Trans. 2,
    1973, 1197.
6. J.F. Nicoud and R.J. Twieg, in ref.1, vol. 1, p.227.
7. M. Hurst and R.W. Munn, J. Mol. Electron., 1986, 2, 139.
8. P.G. Cummins, D.A. Dunmur, R.W. Munn and R.J. Newham,
    Acta Cryst. A, 1976, 32, 847.
9. M.A. Lasheen and I.H. Ibrahim, Acta Cryst. A, 1975, 31, 136.
10. V.J. Docherty, D. Pugh and J.O. Morley, JCS Faraday Trans. 2,
    1985, 81, 1179.
11. J.H. Meyling, P.J. Bounds and R.W. Munn, Chem. Phys. Letters,
    1977, 51, 234.
12. R.W. Munn, Chem. Phys., 1983, 76, 243.
13. R. Frech, Phys. Rev. B, 1985, 32, 6832.
14. R.W. Munn and T. Luty, Chem. Phys., 1979, 38, 413.
15. E.R. Smith, Proc.Roy.Soc. Lond. A, 1982, 381, 241.
16. A.I.M. Rae, Mol. Phys., 1969, 16, 257.
17. A.I. Kitaigorodsky, 'Molecular Crystals and Molecules' (Academic,
    New York, 1973).
18. M. Hurst and R.W. Munn, J. Mol. Electron., 1986, 2, 43.
19. G.R. Meredith, in 'Nonlinear Optics: Materials and Devices',
    C. Flytzanis and J.L. Oudar, eds (Springer, Berlin, 1986) p.116.
20. M. Hurst and R.W. Munn, J. Mol. Electron., 1986, 2, 35.
21. G.R. Meredith, B. Buchalter and C. Hanzlik, J. Chem. Phys.,
    1983, 78, 1533.
22. M. Hurst and R.W. Munn, J. Mol. Electron., 1987, 3, 75.
23. J.L. Oudar and D.S. Chemla, J. Chem. Phys., 1977, 66, 2664.
24. J. Zyss and D. Chemla, in ref. 2, vol. 1, p.23.
25. D. Li, T.J. Marks and M.A. Ratner, Chem.Phys. Letters, 1986,
    131, 370.
26. A. Gavezzotti, this conference.
27. B.F. Levine, C.G. Bethea, C.D. Thurmond, R.T. Lynch and
    J.L. Bernstein, J. Appl. Phys., 1979, 50, 2523.
28. G.F. Lipscomb, A.F. Garito and R.S. Narang, J. Chem. Phys.,
    1981, 75, 1509.

Enhancement of Non-linear Optical Properties of Conjugated Linear Chains through Lowered Symmetry

A.F. Garito*, J.R. Heflin, K.Y. Wong, and O. Zamani-Khamiri

DEPARTMENT OF PHYSICS AND LABORATORY FOR RESEARCH ON THE STRUCTURE OF MATTER, UNIVERSITY OF PENNSYLVANIA, PHILADELPHIA, PA19104, USA

I. INTRODUCTION

Reduced dimensionality and quantum confinement of electrons have been subjects of increasing interest in fundamental condensed matter physics. Much effort has been devoted to physical studies of materials designed and prepared at atomic and molecular levels having novel structures confined at nanometer length scales. In the field of nonlinear optics, experimental and theoretical studies are centered on semiconductor and polymer structures as nonlinear dynamical systems that exhibit unusual nonlinear optical and electrooptical properties.

Quantum confinement of electrons is an important feature in the dynamics of material structures having effective length scales L from one to several tens of nanometers. Atoms, with $L < 10^{-1}$ nm, naturally represent the extreme microscopic quantum limit in which electronic excitations are completely confined in all three spatial dimensions. The opposite macroscopic limit occurs in regular bulk structures where $L > 10$ nm. Examples of confined structures include microfabricated semiconducting quantum well heterostructures such as AlGaAs/GaAs/AlGaAs in which one dimension ($L < 20$ nm) is confined and quasi-one-dimensional conductors such as TTF-TCNQ and $NbSe_3$ in which two dimensions ($L < 1$ nm) are confined.

In common three dimensional bulk solids, the many-electron system behaves as a Fermi gas of weakly interacting particles. In the weak coupling regime, the electron motions are weakly correlated and well described by single-particle theory in the effective mass approximation. However, as the spatial dimensionality of the many-electron system is effectively lowered and the motion of the electrons is confined, the effects of Coulomb interactions are markedly enhanced, and electron motion becomes strongly correlated. The proper description is then that of a strongly interacting gas in reduced dimensions. In this strong coupling regime, many-body correlations and fluctuations are extremely subtle. Many different broken-symmetry states are possible, and classes of quantum confined systems can exhibit a large variety of unusual electronic and structural properties[1].

In the case of transport studies, examples of recent interest include fractional quantum Hall states in two dimensions and charge density waves in one dimension. There is an equally important role for low-dimensional structures in recent nonlinear optic and electrooptic studies where a unifying concept is emerging that reduced dimensionality and quantum confinement place special constraints on electronic excitations that result in unusually large nonlinear optical responses. In the one-dimensional case of conjugated organic linear chains, strong electron correlation behavior in virtual two-photon states appears to be the origin of unusually large, nonresonant third order nonlinear optical susceptibilities $\chi^{(3)}(-\omega_4;\omega_1,\omega_2,\omega_3)$ widely observed for these and related polymer structures. Our recent efforts have centered on providing fundamental understanding of the origin of the unique second[2-6] and third[7-10] order nonlinear optical properties of these quasi-one-dimensional conjugated structures. In two dimensional AlGaAs/GaAs/AlGaAs quantum well heterostructures[11], confined resonant excitons contribute to a large intensity dependent refractive index n_2.

16

The initial subjects of our investigations have been the simplest conjugated linear chains known as polyenes, which are hydrocarbon chains in which each carbon site is covalently bonded to its two nearest neighbor carbons and hydrogen. The remaining valence electron of each carbon atom contributes to a delocalized, strongly correlated π-electron distribution along the carbon chain. The ground state of this system is a spin-singlet, broken-symmetry state in which the carbon lattice possesses a single-bond/double-bond alternation. For descriptions of these systems[12], important evidence for the inadequacy of independent particle theories such as Huckel theory has been provided by the experimental discovery[13] that below the one-photon 1^1B_u π-electronic state exists a two-photon 2^1A_g state in direct contradiction with these simple theories. It was then demonstrated[14 - 18] that a proper account of the electron correlation resulting from electron-electron interactions was required in order to obtain the correct ordering of these two states.

In this paper, we discuss the microscopic origin and features of the third order susceptibility $\gamma_{ijkl}(-\omega_4;\omega_1,\omega_2,\omega_3)$ for both centrosymmetric and noncentrosymmetric conjugated linear chains. We begin with a review of recent results for centrosymmetric polyene chains in which it was found that the third order response of these structures is properly described by a theory which takes explicit account of electron correlation. Highly correlated π-electron virtual excitations to high-lying two-photon ^1A_g states make important contributions to $\gamma_{ijkl}(-\omega_4;\omega_1,\omega_2,\omega_3)$. We will review results for $\gamma_{ijkl}(-\omega_4;\omega_1,\omega_2,\omega_3)$ both as a function of structural conformation and of chain length. Finally, we present recently obtained results for noncentrosymmetric polyenes with donor and acceptor substituents. It is shown that the resultant lowering of symmetry allows a new type of virtual process which is predicted to drastically enhance $\gamma_{xxxx}(-\omega_4;\omega_1,\omega_2,\omega_3)$.

II. MICROSCOPIC ORIGIN OF $\gamma_{ijkl}(-\omega_4;\omega_1,\omega_2,\omega_3)$ FOR CONJUGATED LINEAR CHAINS.

We have recently presented[8-10] a many-electron theory of $\gamma_{ijkl}(-\omega_4;\omega_1,\omega_2,\omega_3)$ for a large range of chain lengths and two structural conformations of polyenes, and our results are in good agreement with available experimental data. Using the direct summation over states expression for $\gamma_{ijkl}(-\omega_4;\omega_1,\omega_2,\omega_3)$ derived from time-dependent perturbation theory[19,20], we identified the most significant transitions in the virtual excitation process and calculated the dispersion of the third harmonic susceptibility $\gamma_{ijkl}(-3\omega;\omega,\omega,\omega)$ and the dc-induced second harmonic susceptibility $\gamma_{ijkl}(-2\omega;\omega,\omega,0)$ for various chain lengths of the all-*trans* and *cis-transoid* conformations(hereafter referred to as *trans* and *cis*, respectively) of polyenes (Figure 1).

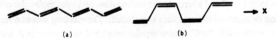

(a)　　　　　　　(b)

Figure 1. Schematic diagrams of the molecular structures of (a) *trans*-octatetraene and (b) *cis*-octatetraene.

The microscopic third harmonic susceptibility $\gamma_{ijkl}(-3\omega;\omega,\omega,\omega)$ is defined by the expression

$$p_i^{3\omega} = \gamma_{ijkl}(-3\omega;\omega,\omega,\omega)\, E_j^\omega E_k^\omega E_l^\omega \qquad (1)$$

where $p_i^{3\omega}$ is a component of the molecular polarization induced at a frequency of 3ω in response to the cube of an electromagnetic field E^ω oscillating at a frequency of ω. From time-dependent, quantum electrodynamic perturbation theory making use of the Bogoliubov-Mitropolsky method of averages one obtains the analytic expression

$$\gamma_{ijkl}(-3\omega;\omega,\omega,\omega) = \frac{1}{3!}\left(\frac{e^4}{4\hbar^3}\right)\Bigg[\sum_{n_1n_2n_3}{}'\Bigg\{\frac{P_{jkl}[r^i_{gn_3}\,\bar{r}^j_{n_3n_2}\,\bar{r}^k_{n_2n_1}\,r^l_{n_1g}]}{(\omega_{n_3g}-3\omega)(\omega_{n_2g}-2\omega)(\omega_{n_1g}-\omega)}$$

$$+\frac{P_{jkl}[r^j_{gn_3}\,\bar{r}^i_{n_3n_2}\,\bar{r}^k_{n_2n_1}\,r^l_{n_1g}]}{(\omega_{n_3g}+\omega)(\omega_{n_2g}-2\omega)(\omega_{n_1g}-\omega)} \;+\; \frac{P_{jkl}[r^j_{gn_3}\,\bar{r}^k_{n_3n_2}\,\bar{r}^i_{n_2n_1}\,r^l_{n_1g}]}{(\omega_{n_3g}+\omega)(\omega_{n_2g}+2\omega)(\omega_{n_1g}-\omega)}$$

$$+\frac{P_{jkl}[r^j_{gn_3}\,\bar{r}^k_{n_3n_2}\,\bar{r}^l_{n_2n_1}\,r^i_{n_1g}]}{(\omega_{n_3g}+\omega)(\omega_{n_2g}+2\omega)(\omega_{n_1g}+3\omega)}\Bigg\}$$

$$-\sum_{n_1n_2}{}'\Bigg\{\frac{P_{jkl}[r^i_{gn_2}\,r^j_{n_2g}\,r^k_{gn_1}\,r^l_{n_1g}]}{(\omega_{n_2g}-3\omega)(\omega_{n_2g}-\omega)(\omega_{n_1g}-\omega)}$$

$$+\frac{P_{jkl}[r^j_{gn_2}\,r^i_{n_2g}\,r^k_{gn_1}\,r^l_{n_1g}]}{(\omega_{n_2g}-\omega)(\omega_{n_1g}+\omega)(\omega_{n_1g}-\omega)} \;+\; \frac{P_{jkl}[r^j_{gn_2}\,r^k_{n_2g}\,r^i_{gn_1}\,r^l_{n_1g}]}{(\omega_{n_2g}+3\omega)(\omega_{n_2g}+\omega)(\omega_{n_1g}+\omega)}$$

$$+\frac{P_{jkl}[r^j_{gn_2}\,r^k_{n_2g}\,r^l_{gn_1}\,r^i_{n_1g}]}{(\omega_{n_2g}+\omega)(\omega_{n_1g}-\omega)(\omega_{n_1g}+\omega)}\Bigg\}\Bigg] \tag{2}$$

where $r^i_{n_1n_2}$ is the matrix element $\langle n_1|r^i|n_2\rangle$, $\hbar\omega_{n_1g}$ is the excitation energy of state n_1, and P_{jkl} denotes the sum over all permutations of j, k, and l ensuring that γ_{ijkl} is independent of the ordering of these three indices. The convention has been chosen that the electric fields are represented as E^ω $\sin(\omega t - \mathbf{k}\cdot\mathbf{r})$. A similar expression can be derived for the dc-induced second harmonic generation susceptibility $\gamma_{ijkl}(-2\omega;\omega,\omega,0)$ as well as for the other fundamental third order optical processes.

The individual terms of Eq. (2) were directly evaluated from the singlet state excitation energies and transition dipole moments obtained by configuration interaction methods in which all singly (SCI) and doubly (DCI) excited π-electron configurations were included in order to describe electron correlations properly. The CI π-electron basis sets were obtained by an all-valence electron self-consistent field (SCF) molecular orbital (MO) method in the standard, rigid lattice Complete Neglect of Differential Overlap/Spectroscopic (CNDO/S) approximation. Although the self-consistent calculation of the ground state includes all of the valence-shell electrons for each atom in the molecule, only the π-electron orbitals are needed in the SDCI calculation since the low-lying excitations are $\pi \rightarrow \pi^*$ transitions and for conjugated systems the π-electron contributions to $\gamma_{ijkl}(-\omega_4;\omega_1,\omega_2,\omega_3)$ dominate those from the σ-electrons. Bond alternation is treated directly in the molecular coordinates, and the hopping interaction between all pairs of sites is included. The electron-electron repulsion is accounted for via the Ohno potential with the Coulomb repulsion integral between sites A and B given by

$$\gamma_{AB} = 14.397 \text{ eV}\cdot\text{Å}/\{[(28.794 \text{ eV}\cdot\text{Å})/(\gamma_{AA}+\gamma_{BB})]^2 + [R_{AB}(\text{Å})]^2\}^{1/2} \tag{3}$$

where γ_{AA} and γ_{BB} are the intra-atomic repulsion integrals and R_{AB} is the interatomic distance. The values calculated at the SDCI level of the excitation energies for both the one-photon allowed 1^1B_u state and the two-photon allowed 2^1A_g state are in good agreement with experimental values and with previously reported theoretical results[12,14,16].

For all chain lengths and conformations studied, the γ_{xxxx} component of $\gamma_{ijkl}(-3\omega;\omega,\omega,\omega)$ with all fields along the direction of conjugation, is far larger than the others, as expected. This is a direct consequence of the one-dimensional delocalization of the π-electrons along the chain. For the specific case of *trans*-octatetraene (*trans*-OT) with number of carbon sites N=8, the independent tensor components of $\gamma_{ijkl}(-3\omega;\omega,\omega,\omega)$ at a nonresonant fundamental photon energy of 0.65 eV (λ=1.907 μm) are γ_{xxxx}=15.5, γ_{xyyx}=0.6, γ_{yxxy}=0.5, and γ_{yyyy}=0.2x10^{-36} esu.

For centrosymmetric conjugated chains, the π-electron states have definite parity of A_g or B_u, and the one-photon transition moment vanishes between states of like parity. Since the ground state is always ^1A_g, it is evident from Eq. (2) that the π-electron states in a third order process must be connected in the series g$\rightarrow$ ^1B_u $\rightarrow$ ^1A_g $\rightarrow$ ^1B_u $\rightarrow$ g. Virtual transitions to both one-photon ^1B_u and two-photon ^1A_g states are necessarily involved. In the summation over intermediate states for *trans*-OT there are two major terms which constitute 70% of γ_{xxxx}. In both of these terms, the only ^1B_u state involved is the dominant low-lying one-photon $1\,^1B_u$ π-electron excited state. In addition to its low energy, the importance of this state lies in the large value of the x-component of its transition dipole moment with the ground state of 7.8 D. This is the largest of all transition moments involving the ground state. One of the two major terms comes from the double sum of Eq. (2) with both of the intermediate states being the $1\,^1B_u$. In the case of the double sum, the middle intermediate state is always the ground state. This term makes a negative contribution to γ_{xxxx} below resonance since both the numerator and denominator are positive but the double sum has an overall negative contribution. The other major term comes from the triple sum with the $6\,^1A_g$ state as the middle intermediate. This state, calculated at 7.2 eV, has a large transition moment with $1\,^1B_u$ of 13.2 D and is much more significant than the $2\,^1A_g$ state which has a corresponding transition moment of only 2.8 D. This term makes a positive contribution to γ_{xxxx} and is larger than the first leading to an overall positive value for γ_{xxxx}.

Figure 2 displays the calculated dispersion curve for $\gamma_{xxxx}(-3\omega;\omega,\omega,\omega)$ of trans-OT as a function of the input photon energy. The first resonance located at 1.47 eV (λ = 0.84 μm) and indicated by the vertical dash in the figure, is due to the 3ω resonance of the $1\,^1B_u$ state. The second singularity located at 2.08 eV (λ = 0.60 μm), is from the 2ω resonance of the $2\,^1A_g$ state. As seen in Eq. (2), the ^1B_u states will have both 3ω and ω resonances in third harmonic generation, whereas the ^1A_g states will have only 2ω resonances. It should also be noted that in real systems, natural broadening of electronic states will prevent divergence at the resonances.

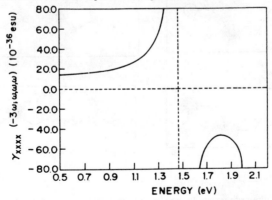

Figure 2. Dispersion of $\gamma_{xxxx}(-3\omega;\omega,\omega,\omega)$ for *trans*-OT. The vertical dash locates the 3ω resonance to the $1\,^1B_u$ state.

III. CONFORMATION DEPENDENCE OF γ_{ijkl}

The results for the *cis* conformations are in direct analogy to those for *trans*. For the range of chain lengths we have considered, we find that the transition energies of the 1^1B_u and 2^1A_g states of the *cis* conformations are slightly red-shifted from the values for *trans* by 0.02 to 0.10 eV with the shift monotonically increasing with increased chain length. Just as for *trans*, the dominant tensor component of $\gamma_{ijkl}(-3\omega;\omega,\omega,\omega)$ is γ_{xxxx}, and the most significant virtual transitions and the dispersion are essentially the same as discussed above. The mechanism for $\gamma_{ijkl}(-3\omega;\omega,\omega,\omega)$ is still a symmetry-dictated virtual excitation process involving strongly correlated π-electron states.

The similarity between the two conformations is further emphasized in the transition density matrix contour diagrams which graphically illustrate the electron redistribution upon virtual excitation. The transition density matrix $\rho_{nn'}$ is defined through the expression

$$<\mu_{nn'}> = - e \int r\rho_{nn'}(r)dr \qquad (4)$$

with

$$\rho_{nn'}(r_1) = \int \psi_n^*(r_1,r_2,...r_M)\, \psi_{n'}(r_1,r_2,...r_M)dr_2...dr_M \qquad (5)$$

where M is the number of valence electrons included in the molecular wavefunction. Contour diagrams for $\rho_{nn'}$ of the ground and 6^1A_g states with the 1^1B_u state for both the *cis* and *trans* forms of OT are compared in Figure 3 where solid and dashed lines correspond to increased and decreased charge density. The (a) *cis* and (b) *trans* virtual $g \rightarrow 1^1B_u$ transition results in a somewhat modulated redistribution of charge with transition moment x-components of 7.9 and 7.8 D, respectively. For the $1^1B_u \rightarrow 6^1A_g$ virtual transition, however, there is a resultant highly separated charge distribution in the (c) *cis* and (d) *trans* conformations. The corresponding transition moments are 12.0 and 13.2 D for the *cis* and *trans* cases, respectively.

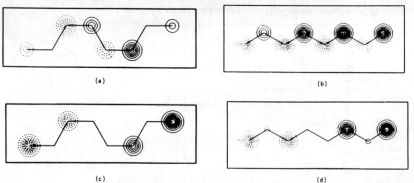

Figure 3. Transition density matrix contour diagrams for (a) *cis* and (b) *trans* ground states with the 1^1B_u state and (c) *cis* and (d) *trans* 6^1A_g states with the 1^1B_u state of OT.

Gas phase third order susceptibility measurements of $\gamma_g(-2\omega;\omega,\omega,0)$ for polyenes have been obtained[21] using dc-induced second harmonic generation (DCSHG). At the nonresonant fundamental input of 1.787 eV ($\lambda = 0.694$ μm), the values for butadiene (BD) with N=4 and hexatriene (HT) with N = 6 are[22] 3.45 ± 0.20 and 11.30 ± 1.05 x 10^{-36} esu, respectively. Although the BD gas was more than 99% *trans*-BD, the HT was believed to contain as much as 40% *cis*-HT[21,23]. Based on the results of our calculations described above and the appropriate expression for $\gamma_g(-2\omega;\omega,\omega,0)$ with isotropic averaging

$$\gamma_g = \frac{1}{5}[\sum_i \gamma_{iiii} + \frac{1}{3} \sum_{i \neq j} (\gamma_{iijj}+\gamma_{ijij}+\gamma_{ijji})] \qquad (6)$$

where the indices i and j represent the Cartesian coordinates x, y, and z, we calculate the π-electron contributions at the same frequency to be 2.1, 11.5, and 9.1 x 10^{-36} esu for BD, *trans*-HT, and *cis*-HT, respectively. Although the σ-electron contribution to γ_g is negligible for longer chains since it increases much more slowly with respect to chain length than the π-electron contribution, it is more signficant for shorter chains and should be included in the cases of BD and HT. After adding in respective σ-contributions of 1.5 and 2.4 x 10^{-36} esu as estimated in Ref. 21, we obtain values for $\gamma_g(-2\omega;\omega,\omega,0)$ of 3.6 and 12.9 x 10^{-36} esu for BD and HT, respectively, in agreement with experiment both in sign and magnitude.

IV. CHAIN LENGTH DEPENDENCE OF γ_{ijkl}

Both the *trans* and *cis* conformations exhibit power law dependences of γ_{xxxx} on the number of carbon sites N with large exponents. For *trans*, the exponent is 5.4±0.2, and for *cis* it is 4.7±0.2. In addition to this smaller exponent for *cis*, for chains with equal numbers of sites, the value of $\gamma_{xxxx}(-3\omega;\omega,\omega,\omega)$ at a fixed frequency is in all cases smaller for the *cis* chain than *trans*. However, when γ_{xxxx} at 0.65 eV is plotted against the actual length of the chain L rather than N, as is done on the lower scale of Figure 4, it becomes clear that the smaller values for *cis* result simply from the shorter chain length. The length L is here defined as the distance along the x-direction between the two end carbon sites. The calculated values for $\gamma_{xxxx}(-3\omega;\omega,\omega,\omega)$ are thus unified by the general result

$$\gamma_{xxxx} \propto L^{4.6\pm0.2}. \qquad (7)$$

Because of the different geometry, a given *cis* chain is always shorter than its corresponding *trans* form. The difference in $\gamma_{xxxx}(-3\omega;\omega,\omega,\omega)$ values for the two is simply due to this fact, and γ_{xxxx} is, therefore, much more sensitive to the physical length of the chain than the conformation.

The very rapid growth that is observed in the nonresonant $\gamma_{xxxx}(-3\omega;\omega,\omega,\omega)$ with increased chain length is a result of several features common to both conformations that emerge from comparison of polyenes of different length. First, the lowest optical excitation energy decreases proprotionally to the inverse of the chain length with a lowering from 5.9 eV in butadiene to 3.7 eV in the case of dodecahexaene (N=12). Second, the magnitudes of transition dipole moments along the chain axis increase steadily with the chain length. Third, while for OT and the shorter chains the nonlinear susceptibility is almost entirely composed of the contributions from only a few states, longer chains have significant contributions from an increasingly larger number of both ^1B_u and ^1A_g states.

From our calculated power law dependence, we can draw several important results for polymers. A typical value of the nonresonant macroscopic third order susceptibility $\chi^{(3)}(-3\omega;\omega,\omega,\omega)$ observed for polymers is 10^{-10} esu[24,25]. For an isotropic distribution of chains considered as independent sources of nonlinear response with a single dominant tensor component $\gamma_{xxxx}(-3\omega;\omega,\omega,\omega)$ we have $\chi^{(3)}=(1/5)N(f^\omega)^3f^{3\omega}\,\gamma_{xxxx}$, where N is the number density of chains and $f^\omega = (2+n_\omega^2)/3$ is the Lorentz-Lorenz local field factor. Using typical values of $N=10^{20}$ molecules/cm^3 and 1.8 for the refractive index, we derive a γ_{xxxx} of roughly 2×10^{-31} esu. From Eqn. (7), this value would correspond to a chain of N≈50 carbon sites, or a length of approximately 60 Å. Since these polymers consist of much longer chains, we infer that γ_{xxxx} must deviate from the power law dependence and begin to saturate at some length shorter than 60 Å. This then suggests that large values of γ, and correspondingly $\chi^{(3)}$, require only chains of intermediate length of order 100 Å.

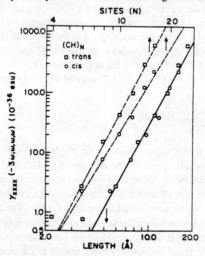

Figure 4. Log-log plot of $\gamma_{xxxx}(-3\omega;\omega,\omega,\omega)$ at 0.65 eV *versus the number* of carbon sites N (upper axis) and the length L (lower axis) for *cis* and *trans*-polyenes. (Ref. 8)

V. EFFECT OF LOWERED SYMMETRY ON γ_{ijkl}

In the previous sections, we have considered the third order nonlinear optical properties of centrosymmetric linear chains and have reviewed the important role of definite parity selection rules in the third order virtual excitation processes. We will demonstrate in this section that lowering the symmetry to a noncentrosymmetric structure can act as a mechanism for the enhancement of nonresonant $\gamma_{ijkl}(-\omega_4;\omega_1,\omega_2,\omega_3)$. As seen in Eq. (2), the Bogoliubov-Mitropolsky formalism admits new types of virtual excitation processes otherwise forbidden under centrosymmetric conditions.

The symmetry is lowered by heteroatomic substitution on the linear chain. A principal noncentrosymmetric analog to OT is 1,1-dicyano-8-N,N-dimethylamino-1,3,5,7-octatetraene (NOT) having a dicyano acceptor group on one end and a dimethylamino donor group on the other as shown in Figure 5. This compound has been synthesized in our laboratory, and its linear and nonlinear optical properties are currently under experimental investigation. Furthermore, the earlier detailed discussion of unsubstituted OT will allow direct understanding of the effect of lowered symmetry on $\gamma_{ijkl}(-\omega_4;\omega_1,\omega_2,\omega_3)$. Results for the second order optical properties and length dependences of β_{ijk} and γ_{ijkl} will be reported in future papers.

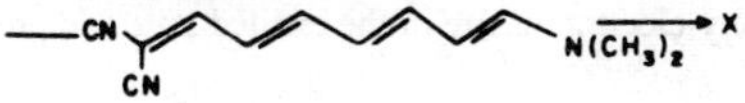

Figure 5. Schematic diagram of the molecular structure of 1,1-dicyano,8-N,N-dimethylamino-1,3,5,7-octatetraene (NOT).

The calculation of the electronic states and nonlinear otpical properties of NOT involved all single and double excited configurations of the six occupied and six unoccupied π-electron molecular orbitals. This leads to 703 configurations in the CI matrix which is then diagonalized to produce 703 singlet π-electron states. The complete calculation including calculation of all transition dipole moments and evaluation of Eq. (2) for γ_{ijkl} requires $5\frac{1}{2}$ CPU hours on a CRAY-X/MP. The calculated excitation energies and oscillator strengths of NOT are given in Table I. The dominant excitation is that from the lowest energy π-electron singlet excited state located at 3.03 eV. This is 1.4 eV lower than the energy of the dominant one-photon 1^1B_u state of OT. There is a secondary peak predicted in the optical absorption spectrum at 3.69 eV which actually corresponds to the 2^1A_g state of OT. Because of the lowered symmetry of NOT, there are no one-photon selection rules as there are in OT. Instead, all of the π-electron states of NOT possess A' symmetry and all are allowed one-photon excitations from the ground state. Thus, in addition to the lowering in energy of the analog to the 1^1B_u state, the symmetry lowering has two interesting effects on linear optical properties. The analog of the 2^1A_g state becomes a one-photon allowed transition which turns out to have a sizeable oscillator strength, and the ordering of the analogs of the 2^1A_g and 1^1B_u states is inverted. We wish to emphasize, however, that although the existence of the 2^1A_g below the 1^1B_u provided the first

TABLE I. Calculated States of NOT

State	Energy (eV)	Oscillator Strength
$2^1A'$	3.03	0.88
$3^1A'$	3.69	0.26
$4^1A'$	4.21	0.00
$5^1A'$	4.55	0.02
$6^1A'$	4.87	0.01
$7^1A'$	5.24	0.24
$8^1A'$	5.35	0.00
$9^1A'$	5.88	0.00
$10^1A'$	5.98	0.00
$11^1A'$	6.21	0.05

definitive evidence of the importance of electron correlation in polyenes, the inverted order in the substituted chain is not due to any less correlation. The $3^1A'$ state of NOT, which is the 2^1A_g analog, is still composed of 48% double excited configurations. The double excited configurations, of course, are those that represent the many-electron nature of the excited state and provide the electron correlation in the CI formalism.

The principle symmetry constraint in the case of centrosymmetric structures that the intermediate states must alternate between one-photon states and two-photon states is lifted upon symmetry lowering. Matrix elements of the form $< n \mid r^i \mid n >$ are no longer symmetry-forbidden and can have an important role in γ_{ijkl}. Diagonal transitions of this form are best illustrated in the difference density matrix $\Delta\rho_n$ where

$$\Delta\rho_n(r) = \rho_n(r) - \rho_g(r) \qquad (8)$$

and

$$<\Delta\mu_n> = -e \int r \, \Delta\rho_n(r) dr. \qquad (9)$$

The function $\rho_n(r)$ is given by Eq. (5) with n equal to n'. The contour diagram of $\Delta\rho_{2^1A'}(r)$ is shown in Figure 6 where the solid and dashed lines correspond to increased and decreased electron density, respectively. There is a large redistribution of electron density along the dipolar x-axis leading to a large dipole moment difference $\Delta\mu_{2^1A'}^x$, of 12.6 D. The sign for $\Delta\mu_{2^1A'}^x$ is seen to be positive as electron density is decreased in the region of the electron donor and increased in the region of the electron acceptor group. This is consistent with the experimentally observed shift to lower energies of the first optical absorption peak in increasingly polar solvents. The magnitude of $\Delta\mu_{2^1A'}^x$ is relatively large and leads to important terms in γ_{ijkl} that involve the matrix element $< 2^1A' \mid x \mid 2^1A' >$. There are no analogous terms in γ_{ijkl} in the case of centrosymmetric linear chains since the dipole moments of the ground state and all excited states are zero by symmetry. Contour diagrams for $\rho_{nn'}$ between the $2^1A'$ state and the $1^1A'$, $3^1A'$, and $7^1A'$ states are similar to the OT diagrams between the 1^1B_u state and the 1^1A_g, 2^1A_g and 6^1A_g states, but these transitions are dominated in NOT by the unusually large terms involving $<\Delta\mu_{2^1A'}>$.

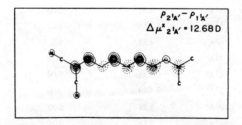

$$P_{2^1A'} - P_{1^1A'}$$
$$\Delta\mu^x_{2^1A'} = 12.68\,D$$

Figure 6. Difference density matrix contour diagram for the state $2^1A'$ with the ground state of NOT.

For the third harmonic susceptibility $\gamma_{ijkl}(-3\omega;\omega,\omega,\omega)$, it is found, once again, that the γ_{xxxx} component is by far the largest. At 0.65 eV, the independent tensor components are $\gamma_{xxxx} = 173.2$, $\gamma_{xxyy} = 1.6$, $\gamma_{yyxx} = 1.0$, and $\gamma_{yyyy} = 0.1 \times 10^{-36}$ esu. The calculated dispersion curve of $\gamma_{xxxx}(-3\omega;\omega,\omega,\omega)$ in Figure 7 smoothly increases to the first resonance occuring at 1.0 eV which is the 3ω resonance of the $2^1A'$ state. Because of the lowered symmetry of NOT, the 3ω and 2ω resonance selection rules for the centrosymmetric polyenes are no longer applicable. Thus, every excited state has allowed 3ω, 2ω and ω resonances, and the dispersive behavior of $\gamma_{xxxx}(-3\omega;\omega,\omega,\omega)$ exhibits all of these many resonances at frequencies beyond the first resonance.

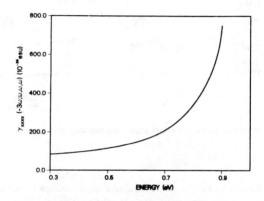

Figure 7. Dispersion of $\gamma_{xxxx}(-3\omega;\omega,\omega,\omega)$ for NOT.

As described earlier, for centrosymmetric structures there are two important types of virtual excitation processes that dominate $\gamma_{xxxx}(-\omega_4;\omega_1,\omega_2,\omega_3)$. We have found for the noncentrosymmetric chains a third type of process is allowed and, in fact, makes a larger contribution to γ_{xxxx} than the other two. These three types of virtual excitation processes in NOT are illustrated in Figure 8. Types I and II are analogous to the dominant processes for centrosymmetric OT. For NOT, the $2^1A'$ state plays the role of the 1^1B_u state of OT because it has the largest transition dipole moment (8.6 D) with the ground state. The type I term is a result of the double summation in the Bogoliubov-Mitropolsky formalism and has the ground state as the middle intermediate state. The largest term of this type is the one with $2^1A'$ as the first and last intermediate state because of its large transition moment with the ground state. The type I term illustrated in Figure 8 makes a negative contribution to γ_{xxxx} because although the numerator and denominator are both positive, the double sum makes a negative contribution to γ_{xxxx}. In the type II process, there is a high-lying middle intermediate state that has a large transition moment with $2^1A'$. While for OT this state was the 6^1A_g, both the $7^1A'$ and the $11^1A'$ states of NOT occupy this role. The type II terms make a positive contribution to γ_{xxxx} because the numerator is effectively the square of two matrix elements and the denominator is positive when below all resonances.

Most importantly, however, for noncentrosymmetric structures, there is a new type of process which is allowed, denoted as type III. For NOT, this is the dominant type of term contributing to γ_{xxxx}. Type III terms involve a diagonal matrix element and are therefore forbidden in centrosymmetric structures which cannot possess a permanent dipole moment. The important quantity in this term is the dipole moment difference between an excited state and the ground state. For the $2^1A'$ state, the value of 12.6 D leads to a very large term in the triple sum in which all three intermediate states are the $2^1A'$. Since the numerator and denominator are both positive, the contribution of this term to γ_{xxxx} is positive. The lowered symmetry of NOT, as compared to OT, produces a new type of virtual excitation process which dominates γ_{xxxx} and causes the value of γ_{xxxx} to be an order of magnitude larger for NOT compared to OT. For example, the calculated nonresonant values of $\gamma_{xxxx}(-3\omega;\omega,\omega,\omega)$ at 0.65 eV for NOT and OT are 173 x 10^{-36} and 15.5 x 10^{-36} esu, respectively.

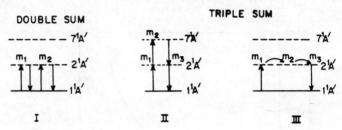

Figure 8. Schematic illustration of the three important contributing terms to γ_{xxxx}.

VI. CONCLUSION

In summary, we first reviewed a symmetry-controlled electron correlation mechanism for $\gamma_{ijkl}(-3\omega;\omega,\omega,\omega)$ of centrosymmetric polyenes ranging in chain length from N = 4 to 16 in both *trans* and *cis* conformations. After showing on general symmetry grounds that two-photon allowed states as well as one-photon allowed states must play a role in γ_{ijkl} for centrosymmetric structures, we summarized the detailed significant contributing terms referred to as Type I and II, especially the dominant Type II terms involving highly correlated two photon ^1A_g states. Good agreement is obtained between our calculations and available experimental gas phase measurements. For both *trans* and *cis* conformations over the range of chain lengths we considered, γ_{xxxx} exhibits a power law dependence on the chain length L with an exponent of 4.6±0.2. In addition, from the power law behavior a principal result suggests that the large values of $\chi^{(3)}$ widely observed for conjugated polymers should be attainable with chains of intermediate length of order 100Å and that infinite conjugated polymer chains, therefore, may not be required. Finally, for chains in which the symmetry is lowered by hetero-atomic substitution of electron donor and acceptor groups, $\gamma_{xxxx}(-3\omega;\omega,\omega,\omega)$ is found to be enhanced by at least an order of magnitude due to a new type of virtual excitation process (Type III) which becomes allowed upon breaking the centrosymmetry. The important quantity in this type of dominant term is the dipole moment difference between a principal excited state and the ground state.

ACKNOWLEDGEMENTS

This research was generously supported by AFOSR and DARPA, F49620-85-C-0105 and NSF/MRL, DMR-85-19059. The calculations were performed on the CRAY X-MP of the Pittsburgh Supercomputing Center.

REFERENCES

1. J. Solyom, Adv. in Phys. **28**, 201(1979).
2. S.J. Lalama and A.F. Garito, Phys. Rev. A**20**, 1179 (1979).
3. C.C. Teng and A.F. Garito, Phys. Rev. Lett. **50**, 350 (1983); Phys. Rev. B**28**, 6766 (1983).
4. A.F. Garito, K.D. Singer and C.C. Teng, in *Nonlinear Optical Properties of Organic and Polymeric Materials*, edited by D.J. Williams, ACS Symp. Series, Vol. 233 (American Chemical Society, Washington, DC, 1983), Chap. 1.
5. A.F. Garito, Y.M. Cai, H.T. Man and O. Zamani-Khamiri, in *Crystallographically Ordered Polymers*, edited by D.J. Sandman, ACS Symp. Series, Vol. 337 (American Chemical Society, Washington, DC, 1987), Chap. 14.
6. A.F. Garito, K.Y. Wong and O. Zamani-Khamiri, in *Nonlinear Optical and Electroactive Polymers*, edited by D. Ulrich and P. Prasad (Plenum, New York, 1987).
7. A.F. Garito, C.C. Teng, K.Y. Wong and O. Zamani-Khamiri, Mol. Cryst. Liq. Cryst. **106**, 219 (1984).
8. J.R. Heflin, K.Y. Wong, O. Zamani-Khamiri and A.F. Garito, Phys. Rev. B (in press).
9. J.R. Heflin, K.Y. Wong, O. Zamani-Khamiri, and A.F. Garito, J. Opt. Soc. Am. B **4**, 136 (1987); in *Photoresponsive Materials*, edited by S. Tazuke (Mater. Res. Soc. Proc.,1988).
10. A. F. Garito, J.R. Heflin, K.Y. Wong, and O. Zamani-Khamiri, in *Nonlinear Optical Properties of Polymers*, edited by A.J. Heeger, D. Ulrich, and J. Orenstein (Mater. Res. Soc. Proc. **109**, Pittsbrugh, PA, 1988), pp. 91-102; Proceedings SPIE 825(1988); Mol. Cryst. Liq. Cryst. (in press).
11. See, for example, D.S. Chemla and D.A.B. Miller, J. Opt. Soc. Am. B **2**, 1155(1985) and references therein.
12. See, for example, B.S. Hudson, B.E. Kohler, and K. Schulten, in *Excited States*, Vol. 6, edited by E.C. Lim (Academic Press, New York,1982), p. 1 and references therein.
13. B.S. Hudson and B.E. Kohler, J. Chem. Phys. **59**, 4984 (1973).
14. K. Schulten, I. Ohmine, and M. Karplus, J. Chem. Phys. **64**, 4422 (1976).
15. A.A. Ovchinnikov, I.I. Ukrainski, and G.V. Kuentsel, Soviet Phys. Uspekhi **15**, 575 (1973) and references therein.
16. K. Schulten, U. Dinur and B. Honig, J. Chem. Phys. **73**, 3927 (1980).
17. Z.G. Soos and S. Ramasesha, Phys. Rev. B**29**, 5410 (1984).
18. P. Tavan and K. Schulten, Phys. Rev. B**36**, 4337 (1987).
19. N.N. Bogoliubov and Y.A. Mitropolsky, *Asymptotic Methods in the Theory of Nonlinear Oscillations*(Gordon and Breach,1961) translated from Russian.
20. B.J. Orr and J.F. Ward, Mol. Phys. **20**, 513 (1971).
21. J.F. Ward and D.S. Elliott, J. Chem. Phys. **69**, 5438 (1978).
22. We have converted Ward and Elliot's $\chi^{(3)}$ to our notation by
$$\gamma_g(-2\omega;\omega,\omega,0) = \frac{3}{2}\chi^{(3)}.$$
23. R.M. Gavin, Jr., S. Risemberg and S.A. Rice, J. Chem. Phys. **58**, 3160 (1983).
24. C. Sauteret et al, Phys. Rev. Lett. **36**, 956 (1976).
25. F. Kajzar and J. Messier, Polymer J. **19**, 275 (1987).

Semi-empirical Calculations of Molecular Hyperpolarizabilities

J.O. Morley

ICI, FINE CHEMICALS RESEARCH, BLACKLEY, MANCHESTER, M9 3DA, UK

D. Pugh*

DEPARTMENT OF PURE AND APPLIED CHEMISTRY, UNIVERSITY OF STRATHCLYDE, GLASGOW G1 1XL, UK

1 INTRODUCTION

The qualitative interpretation of linear and non-linear optical processes in organic materials is almost always derived from the formulism of the time dependent perturbation theory expansion over the excited electronic states of the system. This approach is often referred to as the Sum Over States (SOS) method. The formulae for the linear polarizability tensor, α_{ij}, and the first hyperpolarizability tensor, β_{ijk}, for second harmonic generation are given in equations (1) and (2). (See, for example references 1, 2 and 3).

$$\alpha_{ij} = (e^2/\hbar) \sum r_{gn}{}^i r_{ng}{}^j [(\omega_{ng} - \omega)^{-1} + (\omega_{ng} + \omega)^{-1}] \qquad (1)$$

$$\beta_{ijk} = -(e^3/4\hbar^2) \sum_n \sum_m$$

$$[r_{gm}{}^i r_{mn}{}^j r_{ng}{}^k \{ (\omega_{mg} - \omega)^{-1} (\omega_{ng} + \omega)^{-1} + (\omega_{mg} + \omega)^{-1} (\omega_{ng} - \omega)^{-1} \}$$

$$+ r_{gm}{}^i r_{mn}{}^j r_{ng}{}^k \{ (\omega_{mg} + 2\omega)^{-1} (\omega_{ng} + \omega)^{-1} + (\omega_{mg} - 2\omega)^{-1} (\omega_{ng} - \omega)^{-1} \}$$

$$+ r_{gm}{}^j r_{mn}{}^k r_{ng}{}^i \{ (\omega_{mg} - \omega)^{-1} (\omega_{ng} - 2\omega)^{-1} + (\omega_{mg} + \omega)^{-1} (\omega_{ng} + 2\omega)^{-1} \}]$$

$$(2)$$

The linear tensor can always be referred to the principal axis system at the frequency concerned and the principal values are then given by,

$$\alpha_i = (e^2/\hbar) \sum_n |r_{gn}{}^i|^2 [(\omega_{gn} + \omega)^{-1} + (\omega_{gn} - \omega)^{-1}] \qquad (3)$$

In a general derivation of the above formulae, the perturbed wave-function, produced by the action of the

28

applied fields, is expanded in terms of the complete set of functions comprised of the ground state and all the molecular excited states. If it is assumed that the dominant effects are electronic, the expansion can be restricted to electronic states, calculated for a fixed nuclear configuration. The presence of the ground state in the expansion is inconvenient and leads to apparent divergences in some cases. In the case of the second order tensor, the ground state can be removed by formally transforming to the electronic charge centroid of the molecule[2]. This transformation requires that the diagonal elements $\langle n|r^i|n\rangle$, calculated in an arbitrary co-ordinate system, should be replaced by,

$$r_{nn}{}^i = \langle n|r^i|n\rangle - \langle g|r^i|g\rangle$$

The above quantity is essentially the difference between the dipole moments of the excited state and the ground state of the molecule. The non-diagonal matrix elements, $r_{nm}{}^i = \langle n \mid r^i \mid m\rangle$ are unaffected by the transformation to the charge centroid system. In the case of the third order tensor γ_{ijkl}, a more complicated rearrangement of terms is sometimes necessary to avoid problems associated with the implicit presence of the ground state in the summation[2].

2 THE SINGLE EXCITED STATE MODEL

In a typical intra-molecular donor-acceptor molecule (such as pNA) where it can be taken as a good approximation that only polarization along the molecular axis will contribute significantly to β and when it is also assumed that only one excited state, the charge transfer state, contributes to the summation, equation (2) reduces to,(See, for example, ref.4),

$$\beta_{xxx} = (3e^3/2\hbar^2) \cdot |r_{gn}{}^x|^2 r_{nn}{}^x \cdot \omega_{gn}{}^2 \cdot (\omega_{gn}{}^2 - 4\omega^2)^{-1} (\omega_{gn}{}^2 - \omega^2)^{-1}$$

(4)

Equation (4) has been the basis of much of the interpretative work on organic molecules. All the quantities in the formula have direct physical significance. $|r_{gn}{}^i|^2$ is essentially the oscillator strength of the transition $g \rightarrow n$ and r_{nn}, as we saw above, is the change in dipole moment on excitation. The quantum energy of the transition and the photon energy of the applied field appear in the denominator, which has

poles when either the input photon energy or the second
harmonic photon energy equals the transition energy.
Line broadening effects, leading to an absorption band of
finite width, are neglected in this treatment.
Equation (4) gives the criteria for the selection of
molecules likely to have high β values. There should be
an excited state, with high oscillator strength for the
transition from the ground state, which has a dipole
moment substantially different from that of the ground
state. The excited state should, for passive
applications, be as low lying as is compatible with the
requirement that there should be no actual absorption at
ω or 2ω, so that the pre-resonant enhancement, from the
energy denominators, is as large as possible. Clearly,
the optimum conditions can only be decided when a
knowledge of the shape of the absorption band is
available, but the calculations reported here do not take
account of this criterion.
       The above theory is central to the
understanding of the exceptionally high-valued β -tensors
which occur in polar aromatic molecules for $\lambda \sim 1$ micron.

3 QUANTITATIVE DEVELOPMENT OF THE SOS THEORY

       While the above theoretical interpretation has had a
crucial role in stimulating interest in organics [4], it by
no means follows that the SOS method is the most suitable
for the quantitative calculations required to distinguish
between the potential value of different molecules within
the general category identified by the qualitative
theory. Indeed, the method has been criticised on a number
of grounds [5].
       Firstly, it has been pointed out that the complete
set of excited states invoked in the perturbation theory
is infinitely large and that there is no guarantee of
rapid convergence [5]. Exact results are obtainable for
the linear polarizability of atoms and small molecules
and it is known that a large number of states must be
included to get good results from the SOS method. As an
example, 13% of the static α-value for atomic hydrogen
comes from virtual transitions to the ionization
continuum. It will be argued below that convergence for
β is , in fact, much faster than for α, but even for the
latter, it seems probable that the large effects produced
by delocalized π -electron systems can be accounted for
in terms of a limited number of transitions.
       It is also relevant to consider the degree of
accuracy which is required to obtain useful results. In
the case of β , where the aim is to screen large numbers

of molecules to pick out those which have unusually large β -values,the accuracy need not be high. The case of α is completely different, since this parameter will normally be needed for the purpose of calculating principal values of refractive indices of molecular solids at different frequencies, with a view to predicting, for example,phase matching properties. The accuracy necessary is much higher than for the case of β and cannot as yet be attained.

While the more fundamental criticisms of the method, at least for β-calculations, are unjustified, it would nonetheless be an extremely inconvenient approach if the intention was to make ab initio calculations. There are many better ways of proceeding if the results have to be calculated without reference to experimental data. The alternative methods-the finite field technique and various forms of coupled perturbation theory using polarizing functions rather than excited states-are themselves impracticable at present as ab initio methods for molecules of the size of those of interest, although some of them have been used successfully in a semi-empirical way [6,7]. The great advantage of the SOS method is that it can be closely calibrated against spectroscopic data and data relating to the molecular charge distribution, which is far easier to obtain than data from direct determinations of non-linear optical molecular parameters.

To provide the underlying quantum chemical framework, a semi-empirical spectroscopic method must be chosen. The parametrizations used for structural and spectroscopic calculations in semi-empirical molecular quantum theory differ in the choice of various local integrals. The difference is supposed to reflect, primarily, different electron correlation effects for an excited state containing two singly occupied orbitals in the molecular wave-function as compared with the closed shell ground state. The standard semi-empirical spectroscopic method is CNDO/S (as opposed to CNDO/2 which is used for structural work) [8]. The Pariser-Parr--Pople method which treats only the π-electron system has also been used for first hyperpolarizability [9]. There appears to be nothing available for general application to spectroscopic calculations equivalent to the elaborate MINDO/MOPAC system currently used extensively for structural/conformational calculations.

Albrecht and Morrell [10] first applied the CNDO/S method to the calculation of β for pNA. They included a number of excited states and reported good convergence. Lalama and Garito[3] applied a slightly different version of the theory [11] to nitrobenzene, aniline and pNA.

In this work and the subsequent studies by Teng and
Garito[12], the method was reparametrized to produce a very
close fit to the spectrum of the molecules, in the latter
case specifically for the solution spectrum appropriate
for the EFISH measurements reported. Very good agreement
for β and its frequency dependence was obtained.

4 THE CNDOVS/B METHOD[13,14]

The CNDO/S method has been reparametrized following
a method due to Francois, Carles and Rajzmann[15]. The
one-centre electron-electron repulsion integral and the
core parameters have been adjusted to produce the best
correlation with the ground state dipole moments, μ_g, and
the wavelength of the first intense absorption band for a
group of 6 molecules, having the general structure
associated with high β values. These molecules are
marked with an asterisk in table 1, which also includes a
list of all the molecules referred to in Figs. 1 and 2.
The agreement between measured and calculated ground
state dipoles and transition wavelengths is shown in the
two figures for the complete set of molecules. Since the
object is to be able to screen large numbers of
molecules, it is not practicable to look at each one in
sufficient detail to identify the most important
electronic transitions. We have therefore made a general

Table 1

(1) aniline   (2) nitrobenzene   (3) 4-cyanoaniline
(4) 3-nitroaniline   (5) 4-nitroaniline*
(6) 2,4-dinitroaniline   (7) 4-nitroanisole
(8) 3-methyl-4-nitropyridene-N-oxide
(9) 4-nitro-trans-stilbene
(10) 4-amino-4'-nitro-trans-stilbene*
(11) 4-[(1-methyl-4(1H)-pyridinylidene)ethylidene]-2,5-
                        -cyclohexadien-1-one*
(12) 4-dimethylamino-β-nitrostyrene*
(13) 1-(4-dimethylaminophenyl)-4-nitrobuta-1,3-diene*
(14) methyl-N-(2,4-dinitrophenyl)alaninate
(15) 1,3-diphenyl-2-pyrazoline
(16) 1-(4-nitrophenyl)-3-(4-methoxyphenyl)pyrazol-2-ine
(17) 1-(4-methoxyphenyl)-3-(4-nitrophenyl)-2-pyrazol-2-ine
(18) trans-4-styrylpyridine
(19) fluorobenzene
(20) indigo*   (21) pyridine
(22) 4-dimethylaminocinnamaldehyde
(23) 1-(4-cyanophenyl)-4-(4-dimethylaminophenyl)buta
                        -1,3-diene

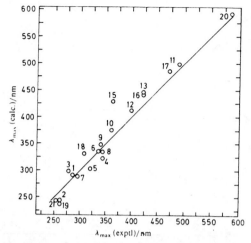

Figure 1 Calculated and experimental wavelengths of first major absorption bands for the molecules of table 1. The line is drawn at 45° to the axes.

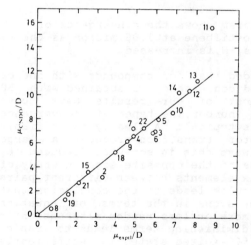

Figure 2 Calculated and experimental dipole moments μ_g for the molecules of table 1. The dipole moments are over-estimated by a nearly constant factor in the CNDO methods and the slope of the best straight line is here 1.3.

study of the convergence of the series of equation (2) as a function of the number of excited states included. Examples of the convergence behaviour are shown in fig 3.

In the following discussion, the quantity referred to is usually β_x, the component of the vector part of the tensor in the direction of the molecular ground state dipole moment. This is the quantity measured in electric field induced second harmonic generation (EFISH) measurements in solution.

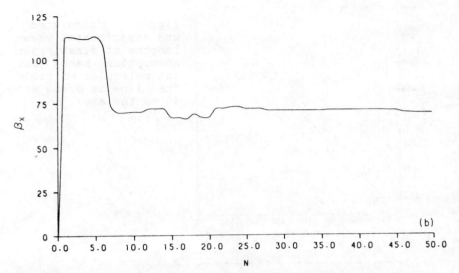

Figure 3   The diagram shows the convergence of
β_x for 4-amino-4'-nitrostilbene at 1.06 micron as the
number of excited states, N, is increased.

It has been concluded that for compounds with one or
two aromatic rings good convergence is obtained with 50
excited states and most of our results have been
calculated using this number. The type of convergence
illustrated in fig. 3 is typical. A clearly identifiable
low lying donor-acceptor transition produces a large
effect, but almost always this is somewhat reduced by
contributions from terms of the opposite sign. A study of
the signs of the matrix elements between different pairs
of m.o.'s in these molecules leads to the conclusion that
this alternation of the signs in the terms of equation
(2) is inevitable. A more or less random alternation in
the signs of gradually diminishing terms leads to rapid
convergence. The set of 50 excited states is sufficiently
large to ensure that the comparatively small number of
states that make a substantial contribution are included.
The behaviour found in the calculations of β is in
marked contrast to that found for α. Indeed it can be
seen from equation (3) that , if $\hbar\omega$ is less than the
excitation energy of the first excited state, then the
terms of the series all have the same sign and α
increases monotonically with n. Convergence is therefore
very slow. Nevertheless, it is interesting to note that
quite accurate results for the in-plane principal values
of α for aromatic compounds have been obtained [16] , but

only after the inclusion of a much larger number of
excited states.
     In the following sections we report the results of
two applications of the method: to the study of β for
long chain polyenes and polyphenyls and to β-calculations
on pyrazoline derivatives.

5 β-HYPERPOLARIZABILITIES OF POLYENES AND POLYPHENYLS

     Increasing the conjugation length in a molecule
affects the value of β, firstly, through the greater
change in dipole moment on excitation which results from
the transfer of charge over a longer range and, secondly,
through the shift of the absorption bands to longer
wavelengths. There are also changes in the oscillator
strengths of the transitions and variations in the
behaviour of β that depend on more detailed structural
features of the sequence, such as the extent of the
alternation in bond lengths which occurs in most linear
systems.
     The relevant structural parameter for assessing the
effectiveness of the molecule is the value of β_x divided
by the molecular volume. We have carried out a study of
β_x/V for the three classes of compounds I, II and III.

(I)   (II)

(III)

     The structures were built on a calligraphic screen
display from molecular fragments.The geometry of the
fragments was taken from crystallographic data from the
Cambridge Data Base [17]. This compilation contains many
examples of bi-,ter- and quater-phenyls. The structures
of the polyenes have been derived from crystallographic
data on related systems, such as dimethyl trans-1-(NN
dimethylamino) hexa -1, 3, 5 triene-6, 6 - dicarboxylate.
The polyenes built in this way exhibit their
characteristic bond alternation; the single bonds vary
between 1.40 and 1.44 Å and the double bonds from 1.35 to
1.38 Å. The bond length alternation leads ultimately to a
finite band gap in the infinite chain, so that the lowest
energy electronic transition should approach a limiting
value as the length of the chain is increased.
     The computations have been carried out
for n-mers up to n=10. For molecules of this length, the

convergence with number of excited states must be carefully monitored and up to 200 states have been included for the longer polymers. An example of the kind of convergence observed is shown in figure 4.

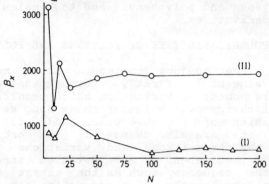

<u>Figure 4</u> (I) $\beta_x \times 10/10^{-30}$ cm^5 esu^{-1} for polyphenyls and (II) $\beta_x/10^{-30}$ cm^5 esu^{-1} for the polyenes. The values are plotted against the number of excited states included in the calculation.

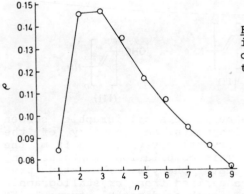

<u>Figure 5</u> Effect of increasing chain length on β–polarizability of the polyphenyls. $\rho = \beta_x/V$.

Results for $\rho = \beta_x/V$ for polyenes and polyphenyls are shown in figs (5) and (6). The plots demonstrate very clearly that the amino nitropolyenes are much more effective than the polyphenyls and that nothing is achieved by increasing n above 3 in the latter series. The internal field effect in the crystal tends to reduce $\chi^{(2)}$ for long chain compounds substantially[18] and this effect must also be taken into account in assessing the optimum length of the chain.

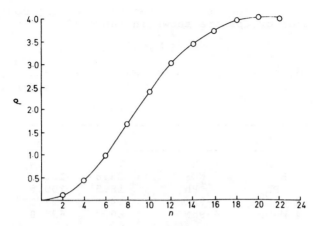

<u>Figure 6</u>　Effect of chain length on the density, ρ , of the polyenes.

6 β -HYPERPOLARIZABILITIES OF 2-PYRAZOLINE AND RELATED COMPOUNDS

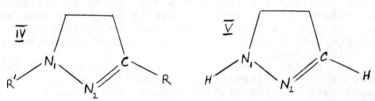

The structures of these compounds may be described by reference to the general structure (IV). In the parent molecule (V), 2-pyrazoline, R=R'=H; derivative compounds considered have donor or acceptor groups in the R and R' positions. Here we consider the cases where methoxy-, or nitro- groups are substituted on the benzene rings in 1,3-diphenyl-2-pyrazoline.
The structure of 1,3 diphenyl-2-pyrazoline has been determined crystallographically and has provided the template from which other derivatives have been constructed by molecular graphics. Because the actual ring structure of 2-pyrazoline could differ substantially from that found in the 1,3-diphenyl derivative, structure optimization studies were carried out using the STO-3G <u>ab</u> <u>initio</u> method.Derivative structures were built around the templates using molecular graphics techniques. Results for β_x for 1,3-

diphenyl-2-pyrazoline   substituted with methoxy donor and
nitro acceptor groups are shown in table (2).

Table 2 β_0 is the value of β_x calculated at ω = 0, to
exclude resonance effects. β_ω is the value for input
wavelength 1.06 micron.

R	R'	β_0	β_ω
H	H	1.7	2.9
H_2N	NO_2	28.6	133.6
Ph	Ph	18.5	100.5
$4-O_2NPh$	4-PhOMe	14.5	37.1
4-MeOPh	$4-PhNO_2$	60.3	624.0

There is a remarkable change in the β-values when
the donor and acceptor groups are interchanged between
the left and right hand rings. The table includes the
values obtained for the case ω = 0, which demonstrate that
the effect is not entirely due to near resonance at 2.
The origin of the effect can be traced to the
characteristics of the main contributing excitation in 2-
pyrazoline itself.
    In table 3 we show the net charges on the two N
atoms and the adjacent C, as labelled in (IV), in the
ground and excited states. There is a large transfer of
charge on excitation from the N_1 nitrogen to the N_2
nitrogen and the carbon bonded to it. The changes in β
for the substituted molecules can be accounted for in
terms of the enhancement or reduction of the charge

Table 3 Net atomic charges for 2-pyrazoline. g is the
          ground state and n the dominant excited state.

	N_1	N_2	C
g	-0.188	-0.172	0.066
n	0.365	-0.515	-0.221

transfer effect by the action of the donor and acceptor
groups. The effect is also enhanced (see table 2) by
the greater conjugation length introduced by the phenyl
groups.

ACKNOWLEDGEMENTS

The authors wish to to thank the E.E.C. for support under the ESPRIT scheme and the Science and Engineering Research Council and Department of Industry for support through the JOERS initiative.

REFERENCES

1. J. Ward ,Rev. Mod. Phys., 1965, 37, 1.
2. B. J. Orr and J. Ward, Mol. Phys., 1971, 20, 513.
3. S. J. Lalama and A. F. Garito, Phys. Rev., 1979, A20, 1179
4. J. Zyss, in: Nonlinear Optical Properties of Organic Molecules and Crystals, Vol. 1, Academic Press, 1987, 1.
5. H. F. Hameka and E.N Svendsen, Int. J. Quantum Chem: Quantum Chemistry Symposia, 1984, 18, 525
6. J.Zyss, J. Chem. Phys., 1979, 70, 3333
7. M. G. Papadopoulos, J. Waite and C. A. Nicolaides, J. Chem. Phys., 1982, 77, 2572
8. J. Del Bene and H. H. Jaffe, J. Chem. Phys., 1968, 48, 1807
9. C. W. Dirk, R. J. Tweig and G. Wagniere, J. Am. Chem. Soc.,1986, 108, 5387
10. J. A. Morrell and A. C. Albrecht, Chem. Phys. Letters, 1979, 64, 46
11. N. O. Lipari and C.B. Duke, J.Chem. Phys.,1975,63,1748
12. C. C. Teng and A. F. Garito, Phys. Rev. , 1983, B28, 6766
13. V. J. Docherty, J. O. Morley and D. Pugh, J.Chem. Soc., Faraday Trans. 2, 1985, 81, 1179
14. D.Pugh and J. O. Morley, in: Nonlinear Optical Properties of Organic Molecules and Crystal, Vol 1, 1987, 193
15. P. Francois, P. Carles and M. Rajzmann, J. Chim. Phys.,1977, 74, 606
16. J. O. Morley and D. Pugh, to be published
17. J. O. Morley, V. J. Docherty and D. Pugh, J. Chem. Soc. Perkin Trans. II, 1987, 1351
18. M. Hurst and R. W. Munn, J. Mol. Elec., 1986, 2, 101
19. J. O. Morley, V. J. Docherty and D. Pugh, to be published
20. Cambridge Structural Data Base, Cambridge Crystallographic Data Centre, Cambridge

Strong Field Theory of Degenerate Multiwave Mixing: Application to Polydiacetylenes

F. Charra and J.M. Nunzi

CENTRE D'ETUDES NUCLÉAIRES DE SACLAY, 91191 GIF-SUR-YVETTE CEDEX, FRANCE

INTRODUCTION

The usual treatment of multiwave mixing supposes the existence of a nth order susceptibility tensor $\chi^{(n)}_{\gamma, \mu_1 \ldots \mu_n}$ ($\Sigma\omega_i, \omega_1, \ldots \omega_n$). $\chi^{(n)}$ coefficients are generaly obtained by identification of a Volterra expansion to a time dependent perturbation expansion. In most cases, the first $\chi^{(n)}$ producing the studied effect is retained.

However, at high light intensities, the influence of higher order terms has been proved[1]. Moreover convergence of perturbative expansions is limited. In addition, in the degenerate case (where all the fields have the same frequency), classical third order perturbation expansions exhibit unphysical infinite terms. A development proposed by Orr and Ward[2] removes these so-called "secular terms" by considering steady states. These are semiclassical analogs of the dressed states. However such series expansion diverge at high light intensities.

In this paper we propose an exact solution in the framework of Floquet theories which is also a semiclassical dressed atom picture[3]. In this formalism we derive the intensity dependent polarizability of a N-level system, and the optical Stark shifts of the levels. We progressively apply the formalism to polydiacetylenes modeled by a three level system. Original experiments on optical Stark shift in a polydiacetylene (4-BCMU red gel) are discussed.

EXACT INTENSITY DEPENDENT POLARIZABILITY

The problem of a N-level system with hamiltonian periodic in time can be treated by Floquet theories: By applying Floquet theorem and developing the solutions in Fourier-series, it reduces to an eigenvalue-eigenvector problem. In this formalism the dressed state energies are the eigenvalues of a hamiltonian (Floquet hamiltonian) in the form of an infinite, time independent hermitian matrix.

We consider here a system with time dependent hamiltonian of the form:

$$H(t) = H_o - 2D.E.\cos(\omega t + \varphi) \tag{1}$$

Where H_o is the hamiltonian of the isolated system, D is the dipole moment operator and $E = |E\omega|$ is half the local electric field amplitude. Applying Hellmann-Feynman theorem to a dressed state $|u>>$ with energy e_u we find the ω-Fourier component of the dipole moment in the form $d_R \cos(\omega t + \varphi) + d_I \sin(\omega t + \varphi)$ where:

$$d_R = -\left(\frac{\partial e_u}{\partial E}\right) \text{ and } d_I = \frac{1}{E}\left(\frac{\partial e_u}{\partial \varphi}\right) \tag{2}$$

Since H_o is time independent, e_u does not depend on φ and $d_I = 0$. The complex ω-dipole moment can then be written in the form:

$$d_\omega = -\left(\frac{de_u}{dI}\right) \cdot E_\omega \quad ; \quad I = E_\omega.E_\omega^* \tag{3}$$

Thus variation of the intensity dependent polarizability of dressed state $|u>>$ is readily deduced from its energy plots.

DRESSED STATES ENERGIES OF POLYDIACETYLENES

Polydiacetylenes can be modeled by a λ-type three level system[4] (see fig. 1a). Matrix representation of unperturbed hamiltonian and dipole moment operators are respectively:

$$H_o = \begin{bmatrix} E_2 & & \\ & E_1 & \\ & & 0 \end{bmatrix} \quad ; \quad D = \begin{bmatrix} & & d_2 \\ & & d_1 \\ d_2 & d_1 & \end{bmatrix}$$

Floquet theorem states the existence of eigensolutions in the form:

$$s(t) = \exp(-ie_u t/\hbar) \cdot \sum_n s_n \cdot \exp(in\omega t)$$

Carrying it in the Schrödinger equation we find the "Floquet hamiltonian":

$$H_F = \begin{bmatrix} H_0 + \hbar\omega & D.E & & \\ DE & H_0 & DE & \\ & D.E & H_0 - \hbar\omega & DE \\ & & DE & H_0 - 2\hbar\omega \end{bmatrix} \qquad (4)$$

Diagonalization of this hermitian matrix is a well known problem. The eigenvalues for a particular frequency ω are given in fig. 1b (solid lines). Since eq. (3) is applicable only for a pure dressed state we must assume the field amplitude variations to be slow enough for the system to remain in the dressed state corresponding to the ground state (thick line on fig. 1b).

A strong nonlinearity (proportional to slope variations) is observed when light interaction brings two dressed levels close together. If they correspond to an allowed transition those two curves form an "anticrossing". Such an anticrossing corresponding to the 2-photon $0 \to 2$ transition appears in fig. 1b. Dressed energies plots contain enough information to predict Stark shifts and nonlinear degenerate effects via eq. (3) for all local field strengths.

LOCAL EFFECTS AND RELAXATIONS

Up to this point we can only relate dipole moment d_ω to the local field E_ω. Since we are interested in the measurable macroscopic quantities P (polarization) and E_m (macroscopic electric field), material polarization function $P(E_m)$ is computed from the implicit equations:

$$P = Nd(E) \quad \text{and} \quad E = E_m - rP \qquad (5)$$

where N is the number of species per unit volume and r is the local field factor ($1/3\varepsilon_0$ in Lorentz model).

Energy dissipation in the environment (e.g. solvent) can be formally accounted for by introducing delayed local effects. Adding an imaginary part to r

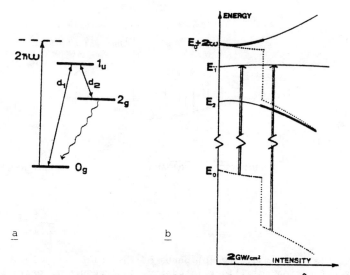

a: 3-level model for a red PDA. $E_1=2.28$eV; $E_2=2.19$eV; $d_1=6$Åe; $d_2=3.9d_1$
b: Dressed energies for $2\hbar\omega = 2.33$eV. Dotted line is the ground state dressed energy including a 100 fs relaxation time. Arrows show the exciton Stark shift.

Figure 1

($r = r'+ir''$) yields relaxations from state 1 to ground state at a rate $N.r''.d_1^2/\hbar$ and yields the corresponding lorentzian shaped absorption spectrum. In the particular case of fig. 1 the most relevant dissipation process is relaxation from state 2 to ground state which are resonantly coupled. Since this transition is not dipolar allowed we must introduce delayed quadrupolar local effects. In this picture energy dissipation of the 2-photon level results from friction with the environment of the 2ω-oscillating quadrupole drived by the 2-photon transition. In that case eq. (1) contains an additional quadrupolar term and both eq. (2) are to be used. The dotted curve in fig. 1b represents the dressed state energy of the system including relaxations. This original relaxation description provides continuity with low intensity classical effects and appears powerful at high intensities, when dressed states are necessary.

Fig. 2 depicts the variations of real and imaginary parts of the polarizability $\alpha(I)$ versus intensity at various field frequencies. All degenerate nonlinear

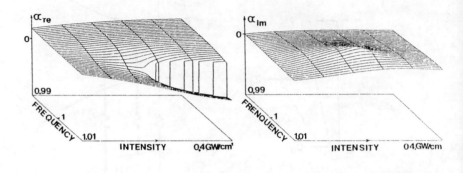

a   b

Real (a) and imaginary (b) parts of the molecular polarisability
as a function of light intensity and frequency near the
2-photon 0→2 resonance.

Figure 2

optical responses are deduced from these variations. As
an example, due to its spatial filtering properties,
phase conjugation reflectivity expresses simply in
terms of an effective intensity dependent $\chi^{(3)}$:

$$\chi^{(3)}_{P.C.}(I) = \frac{N}{\varepsilon_o} \cdot \frac{1}{2I} \cdot \frac{d}{dI} \left(I^2 \frac{d\alpha(I)}{dI} \right) \tag{6}$$

OPTICAL STARK EFFECT IN POLYDIACETYLENE

EXPERIMENT

Fig. 1b predicts also the variation of state 1 - ground
state energy difference with light intensity. In order
to check the validity of the formalism, we investigate
excitonic absorption changes of a 0,2% 4-BCMU red gel
prepared in chlorobenzene solvent[56]. The sample, contained in a 1mm thick cell is excited by the 33ps/1064nm
pulses of a Nd-YAG LASER. The absorption is probed
using weak probe pulses at five different wavelengths,
selected near the maximum absorption and obtained by
frequency doubling and Raman-shifting on part of the

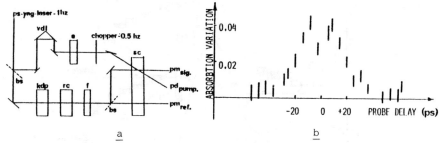

a
b

a: Experimental setup. bs-beam splitter; kdp-doubling crystal; rc-raman
  shifting cell; f-filters; vdl-variable delay line; a-attenuator;
  sc-sample cell; pm-photomultiplier tube; pd-photodiode.
b: Time dependence of absorption variations probed at 532nm with
  1.7 GW/cm^2 pump.The error bars represent 70% confidence limits.

Figure 3

LASER pulse. Transmission is measured alternatively
with and without the pump pulse (fig. 3a) and studied
as a function of pump-probe delay.

A typical time evolution of absorbance change is reported in fig. 3b. Up to our experimental time scale, the
response of the system is faster than about 10ps. The
pump intensity dependence of absorption changes is
depicted at each probe wavelength on the insets of
fig. 4.

Fit of the absorption changes using Gaussians or Loren-
tzians (it is insensitive to the test function) gives
the following results: there is very little bleaching
(< 0,03%). The absorption band is broadened and shifted
towards blue (fig. 5). Study of its pump/probe polarization angle dependence reveals a 1D effect which is
cubic bellow 1,5 GW.cm^{-2}. Those facts and the time response rule out the possibility of an effect due to
2-photon heating. However, above 1,5 GW.cm^{-2} the shift
is no more linear in pump intensity.

DISCUSSION

Isotropization produces a 5 fold decrease of the linear
effect. The linear Stark shift is thus $\Delta\lambda$ = a.I where
a = 0,75 nm/GW.cm^{-2} for a polarization parallel to the
polymer chain. For given E_2, E_1 and d_1, the model gives
a as a function of d_2. Thus, the measure allows a
determination of d_2 : d_2 = 3.9 d_1. Similarly the absorp-

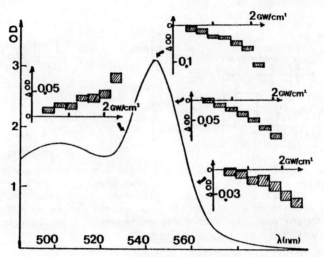

Sample absorption spectrum. Insets: absorption variations as a function of pump intensity probed at 532 nm, 545 nm, 551 nm and 562 nm. No variations are measured at 632 nm.

Figure 4

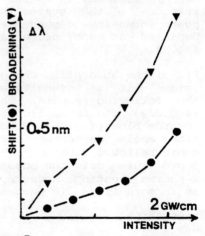

Stark shift and broadening as a function of pump intensity estimated using a Gaussian fit.

Figure 5

tion broadening can be explained by a 30nm inhomogeneous broadening of the 2-photon state 2. The non linear Stark-shift positive curvature is probably due to the anticrossings of the systems having the highest 2-photon levels.

CONCLUSION

Experimental results give evidence of the necessity of our model at light intensities > 1.GW cm^{-2} for the polydiacetylenes. The model predicts strong polarizability variations at anticrossings. This is a useful prediction for numerous degenerate effects such as phase conjugation and bistability. In addition, in view of the atomic levels and of the driving frequency a simple hand plot of the Stark shifted energies allows direct estimation of the sign, strength and power dependence of the nonlinearity. Experiments are now in progress in our laboratory with other polydiacetylenes.

REFERENCES

1. J.M. NUNZI and D. GREC, J. Appl. Phys. 62(6), p. 2198 (1987).
2. J. ORR and J.F. WARD, Mol. Phys. 20(3), p. 513 (1971).
3. S.I. CHU, in Advances in atomic and molecular physics, Vol. 21, p. 197 Academic (1985).
4. F. KAJZAR and J. MESSIER, Polymer J. 19(2), p. 275 (1987) and NATO advanced research workshop on "Non linear optical effects in organic polymers" Nice - June 20-24 (1988).
5. Ref. 1 and J.M. NUNZI and F. CHARRA, these proceedings.
6. F. CHARRA, J.M. NUNZI and F. KAJZAR, to be published.

Molecular Orbital Calculations of Third-order Polarizabilities for π-Electron Conjugated Molecules

B.M. Pierce

HUGHES AIRCRAFT COMPANY, BLDG. A-I, M/S 3C 924, PO BOX 9399, LONG BEACH, CA 90810, USA

1 INTRODUCTION

The theoretical understanding of the relation between the electronic structure of an organic molecule and the non-resonant, electronic component of its third-order polarizability is important to the development of organic molecules as nonlinear optical materials. Organic molecules and polymers with π-electron conjugated bonding networks (e.g., linear conjugated polyenes) are of particular interest because of their large third-order susceptibilities [1-3].

Our theoretical study concentrates on the calculation of the non-resonant, electronic component of the molecular third-order dynamic polarizability for third-harmonic generation [$\gamma(-3\omega; \omega, \omega, \omega)$] for ethylene; linear conjugated polyenes: all-trans-1,3-butadiene, all-trans-1,3,5-hexatriene, all-trans-1,3,5,7-octatetraene, cis-butadiene, 3-cis-hexatriene, and 3,5-dicis-octatetraene (hereafter designated as ATB, ATH, ATO, CB, CH, and CO); ATB and ATH with methyl group substituents at the ends of the polyene chain: 2,4-trans-hexadiene, 2,5-dimethyl-2,4-trans-hexadiene, 2,4,6-trans-octatriene, and 2,7-dimethyl-2,4,6-trans-octatriene (hereafter designated as 2M-ATB, 4M-ATB, 2M-ATH, and 4M-ATH); and benzene.

The above selection of molecules allows us to study the effect of chain length, conformation, and methyl substitution on the calculation of $\gamma(-3\omega; \omega, \omega, \omega)$ for π-electron conjugated molecules. Other important issues that are addressed are the effect of the treatment of electron-electron interactions, or electron correlation, on the calculation of $\gamma(-3\omega; \omega, \omega, \omega)$ [4], and the origin of the non-resonant third-order nonlinear optical response of the linear conjugated polyenes.

48

2 THEORETICAL

Molecular Third-order Polarizability

The $\gamma(-3\omega; \omega, \omega, \omega)$ was partitioned into a σ-electron component (γ_σ) and a π-electron component (γ_π). This separation was motivated by several reasons: 1) restricted π-electron molecular orbital procedures like the Pariser-Parr-Pople method have been successful in the treatment of excited π-electron states [5] and second-order nonlinear optical properties of π-electron conjugated molecules [6,7]; 2) the π-electrons are expected to dominate the polarization response of these molecules; and 3) measurements of γ for π-electron conjugated molecules have been successfully interpreted in terms of γ_σ and γ_π components [8,9].

The γ_σ for an unsaturated hydrocarbon was estimated by using the bond-additivity approximation [10] and measured values of γ for saturated hydrocarbons. Because the values of γ measured for saturated hydrocarbons are defined completely by σ-electrons, these γ's can be used to determine γ_σ's for π-electron conjugated hydrocarbons like the linear polyenes. The bond-additivity approximation makes it possible to express the γ for an alkane (C_nH_{2n+2}) in terms of third-order polarizabilities for C-C and C-H bonds:

$$\gamma = (n-1)\gamma(C\text{-}C) + (2n+2)\gamma(C\text{-}H). \tag{1}$$

A fit of this equation to values of γ measured for the alkanes, yields $\gamma(C\text{-}H) \cong +0.12 \times 10^{-36}$ esu and $\gamma(C\text{-}C) \cong +0.11 \times 10^{-36}$ esu [9]. Thus, as an example, the γ_σ estimated for ATB with 3 C-C bonds and 6 C-H bonds is $\gamma_\sigma \cong 3\gamma(C\text{-}C) + 6\gamma(C\text{-}H) \cong +1.05 \times 10^{-36}$ esu.

The γ_π for a π-electron conjugated molecule was explicitly calculated using the sum-over-states expression derived by Orr and Ward [11]. Comments regarding our use of this expression are given in Ref. 4. The only difference between the expression used in the present study and that given in Ref. 4 is that the present study treats both dipolar and non-dipolar molecules. This is possible by replacing the electric dipole moment operator (μ_α) in Eq.(2) of Ref. 4 by $\mu_\alpha{}^* = \mu_\alpha - <o \mid \mu_\alpha \mid o>$.

Molecular Electronic Structure

The electronic structure calculations were for isolated molecules unperturbed by solvent or crystalline environments. The $\gamma_\pi(-3\omega; \omega, \omega, \omega)$'s computed using the results of our electronic structure calculations are therefore best compared with measurements of the molecules in the gas phase.

Standard geometries were assumed for *ethylene* and the *linear conjugated polyenes*: $r(C_\pi=C_\pi)$ = 1.35 Å, $r(C_\pi-C_\pi)$ = 1.46 Å, $r(C_\pi-H)$ = 1.08 Å, all bond angles 120°, planar; *methyl-substituted linear conjugated polyenes*: $r(C_\pi=C_\pi)$ = 1.35 Å, $r(C_\pi-C_\pi)$ = 1.46 Å, $r(C_\pi-H)$ = 1.08 Å, $r(C_\pi-C_{meth})$ = 1.52 Å, $r(C_{meth}-H)$ =1.09 Å, bond angles of 120° with C_π as center atom, bond angles of 109.5° with C_{meth} as center atom, all carbon atoms lie in a plane; *benzene*: $r(C_\pi-C_\pi)$ = 1.397 Å, $r(C_\pi-H)$ = 1.08 Å, all bond angles 120°, planar. The molecule-fixed z-axis is defined to be along the chain backbone of the given molecule, with the x-axis in the molecular plane, and the y-axis perpendicular to this plane.

An all-valence-electron, semi-empirical INDO (Intermediate Neglect of Differential Overlap) self-consistent field (SCF) molecular orbital (MO) procedure combined with either single-excitation configuration interaction (SCI) or single- and double-excitation configuration interaction (SDCI) was used in the theoretical treatment of the electronic structures of the molecules of interest. Because only the properties of the excited π-electron singlet states are needed to calculate γ_π, we only included singlet π-electron configurations in our INDO-SCI ($^1\pi\pi^*$) and INDO-SDCI ($^1\pi\pi\pi^*\pi^*$) calculations. The *complete* number of π-electron configurations and states was always used in the calculation of $\gamma_\pi(-3\omega; \omega, \omega, \omega)$. For example, an INDO-SDCI calculation of ATB required that 4 singly-excited and 10 doubly-excited π-electron configurations were generated to define 14 excited π-electron states. More details concerning the parameterization of the INDO-SCI and INDO-SDCI formalisms are given in Ref. 4.

3 RESULTS AND DISCUSSION

In order to investigate the importance of the treatment of electron-electron interactions (EEI) on the calculation of γ_π for π-electron conjugated molecules, two sets of calculated values of γ_π for ethylene, linear polyenes, methyl-substituted linear polyenes, and benzene are presented in Table 1. The set under the heading of INDO-SCI only included the treatment of the interaction of a given electron with the averaged field of all other electrons, whereas the second set under the heading of INDO-SDCI included interactions between pairs of π-electrons as well as the averaged field interaction.

Several criteria can be used to determine whether the INDO-SCI formalism is sufficient to calculate γ_π for π-electron conjugated molecules: 1) experimental studies indicate that γ_π is *positive* for all molecules listed in Table 1; 2) the linear conjugated polyenes and related molecules are most polarizable along the chain backbone (z axis), and so the γ_{zzzz} component of γ_π should dominate all others; 3) the π-electron bonding network of the *trans* linear polyenes are more extended than those of the *cis* linear polyenes, and hence $|\gamma_{zzzz}|$ for a given *trans* linear

Table 1.   Calculated $\gamma_\pi(0;0,0,0)$ for ethylene, linear polyenes, methyl-substituted linear polyenes, and benzene

INDO-SCI (10^{-36} esu)

MOLECULE	γ_{zzzz}	γ_{zzxx}	γ_{xxxx}	γ_π
ETHYLENE	-1.042	0.000	0.000	-0.208
ATB	-1.143	-1.032	-0.129	-0.667
CB	3.417	-1.225	-1.432	-0.093
ATH	1.244	-3.040	-0.294	-1.026
CH	-22.12	3.661	0.556	-2.848
ATO	10.42	-9.192	-0.786	-1.750
CO	-57.54	5.659	0.722	-9.099
2M-ATB	-5.118	-0.316	-0.053	-1.161
4M-ATB	-8.586	-1.087	0.104	-2.131
2M-ATH	-3.812	-3.321	-0.299	-2.151
4M-ATH	-9.107	-5.747	-0.301	-4.180
BENZENE	-5.391	-1.797	-5.391	-2.875

INDO-SDCI (10^{-36} esu)

MOLECULE	γ_{zzzz}	γ_{zzxx}	γ_{xxxx}	γ_π
ETHYLENE	0.169	0.000	0.000	+0.034
ATB	3.805	0.331	0.039	+0.901
CB	2.155	0.070	0.467	+0.552
ATH	20.26	1.140	0.101	+4.528
CH	15.04	1.350	0.339	+3.616
ATO	59.05	2.161	0.049	+12.68
CO	40.85	2.466	0.393	+9.235
2M-ATB	6.486	0.437	0.010	+1.474
4M-ATB	6.845	0.600	0.166	+1.642
2M-ATH	28.67	1.232	0.027	+6.233
4M-ATH	32.89	1.451	0.085	+7.176
BENZENE	1.378	0.459	1.378	+0.735

polyene should be greater than $|\gamma_{zzzz}|$ for the corresponding *cis* conformer.

We see that all the INDO-SCI-calculated values of $\gamma_\pi(0; 0, 0, 0)$ are *negative*, while all the INDO-SDCI-calculated values are *positive*. The INDO-SCI-calculated values of $|\gamma_{zzxx}/\gamma_{zzzz}|$ range from 0.1 to 2.44, while those calculated using the INDO-SDCI formalism never exceed 0.09. The INDO-SCI method calculates $|\gamma_{zzzz}|$ of the *cis* linear polyenes to be greater than that of the corresponding *trans* conformer, while the opposite ordering is the case for the INDO-SDCI calculations. It is concluded that the treatment of EEI at the level of SCI is *not sufficient* to calculate γ_π for a π-electron conjugated molecule. The correct calculation of γ_π requires that EEI is treated at an SDCI, or higher, level of configuration interaction [4].

A comparison of INDO-SDCI-calculated values of non-resonant $\gamma(-3\omega; \omega, \omega, \omega)$ at ω = 0.656 eV (1.89 μm) for our set of molecules, and selected measured values of non-resonant $\gamma(-2\omega; \omega, \omega, 0)$ [12] and $\gamma_\pi(-3\omega; \omega, \omega, \omega)$ [8,13,14], is given in Table 2. It is possible to compare our calculated values of $\gamma(-3\omega; \omega, \omega, \omega)$ and the measured values of $\gamma(-2\omega; \omega, \omega, 0)$ because these polarizabilities are equivalent in the non-resonant, zero-frequency limit [11,12]. This statement is supported by the two measurements for benzene, where $\gamma(-2\omega; \omega, \omega, 0)$ at ω = 1.79 eV [+2.06 ± 0.05 x 10^{-36} esu] [12] is very close in value to $\gamma(-3\omega; \omega, \omega, \omega)$ at ω = 0.66 eV [+2.4 ± 0.72 x 10^{-36} esu] [14].

The agreement between theory and experiment is very good. As expected, the closest match exists between the calculated values of $\gamma(-3\omega; \omega, \omega, \omega)$ for ethylene, ATB, ATH, and benzene and the measured values of $\gamma(-2\omega; \omega, \omega, 0)$ for these molecules in the *gas phase* [12]. In addition to validating the importance of the treatment of EEI at the level of SDCI, the success of our theoretical formalism leads us to several conclusions regarding the calculation of $\gamma(-3\omega; \omega, \omega, \omega)$ for planar, π-electron conjugated hydrocarbons: 1) the separation of γ into σ- and π-electron components is a good approximation; 2) it is not necessary to employ extended atomic orbital basis sets (*e.g.*, d orbitals on carbon atoms and p orbitals on hydrogen atoms); 3) the same set of INDO semi-empirical parameters that have been used successfully in calculating one- and two-photon electronic absorption spectra for π-electron conjugated molecules [15] are adequate for calculating γ. The third point is reassuring because Orr and Ward's expression for γ [11] is given in terms of spectroscopic quantities - transition moments and transition energies.

Table 2 also indicates how chain length, conformation and methyl substitution affect the calculation of $\gamma(-3\omega; \omega, \omega, \omega)$. The γ increases in a nonlinear fashion with increasing chain length. The ratio γ_π/γ also increases with chain length. In the case of ATO, γ_π/γ = 0.89, and so γ can

Table 2.   Comparison of INDO-SDCI-calculated and measured values of the non-resonant γ for the molecules

MOLECULE	THEORY (10^{-36} esu)			EXPT (10^{-36} esu)
	γ_σ [a)	γ_π [b)	γ [c)	γ
ETHYLENE	0.59	0.04	+0.63	+0.758 ± 0.017 [d)
ATB	1.05	1.05	+2.10	+2.30 ± 0.13 [d)
CB	1.05	0.65	+1.70	
ATH	1.51	5.49	+7.00	+7.53 ± 0.7 [d)
				+9.1 ± 0.4 [e)
CH	1.51	4.37	+5.88	
ATO	1.97	15.9	+17.9	
CO	1.97	11.5	+13.5	
2M-ATB	1.75	1.74	+3.49	
4M-ATB	2.45	1.99	+4.44	+7.4 ± 0.1 [e)
2M-ATH	2.21	7.65	+9.86	
4M-ATH	2.91	9.00	+11.9	+9.7 ± 1.7 [f)
				+38 ± 0.5 [e)
BENZENE	1.38	0.81	+2.19	+2.06 ± 0.05 [d)
				+2.4 ± 0.72 [g)

a) σ-electron component of γ determined using the bond-additivity approximation

b) π-electron component of γ at $\omega = 0.656$ eV (1.89 μm) calculated using INDO-SCF-SDCI molecular orbital theory combined with the sum-over-states expression for $\gamma(-3\omega; \omega, \omega, \omega)$

c) $\gamma = \gamma(-3\omega; \omega, \omega, \omega) = \gamma_\sigma + \gamma_\pi(-3\omega; \omega, \omega, \omega)$, $\omega = 0.656$ eV

d) $\gamma(-2\omega; \omega, \omega, 0)$, *gas phase*, $\omega = 1.79$ eV (694.3 nm), Ref. 12

e) $\gamma(-3\omega; \omega, \omega, \omega)$, *liquid phase*, $\omega = 0.65$ eV (1.908 μm), Ref. 13

f) $\gamma(-3\omega; \omega, \omega, \omega)$, *liquid phase*, $\omega = 0.66$ eV (1.89 μm), Ref. 8

g) $\gamma(-3\omega; \omega, \omega, \omega)$, *liquid phase*, $\omega = 0.66$ eV (1.89 μm), Ref. 14

be approximated very well by γ_π for ATO and longer chain polyenes.   The presence of *cis* linkages in a given polyene diminish γ with respect to that for the all-*trans* conformer.   Methyl group substitution at the ends of the polyene chain increases γ_π, but not to the same extent that is achieved by increasing the conjugation chain length.   For example, 2M-ATB and

ATH have the same number of carbon atoms, but γ_π for ATH is approximately three times that of 2M-ATB. These findings are in agreement with a similar theoretical study [16], and support the consensus that π-electron conjugated molecules with all-*trans* bonding networks are among the strongest candidates as nonlinear optical materials [1-3].

The non-resonant, third-order nonlinear optical response of the linear conjugated polyenes is defined principally by two excited π-electron singlet states: 1) the 1^1B_u excited state, which has the strongest one-photon coupling (or greatest one-photon transition moment) with the ground state (o), and 2) an n^1A_g (n>2) excited state, which has the strongest one-photon coupling with the 1^1B_u state and is *higher* in energy than this state. *It is stressed that the lowest-lying* ^1A_g *state, which is important in the photophysics and photochemistry of linear polyenes* [17], *is not the most important* ^1A_g *state contributing to* γ *in the non-resonant case.* The two state approximation of $\gamma_\pi(0; 0, 0, 0)$ for the linear conjugated polyenes is therefore:

$$\gamma_\pi^* \cong a_1 \, (\, | \, \mu^z_{o,B} \, |^2 \, /\Delta E_B^2) \, [\, (\, | \, \mu^z_{B,A} \, |^2 \, /\Delta E_A) - (\, | \, \mu^z_{o,B} \, |^2 \, /\Delta E_B) \,] \qquad (2)$$

where a_1 is a constant, $\mu^z_{i,j}$ is the z-component of the transition moment between states i and j; ΔE_k is the transition energy between the ground state and the excited state k; o, B, and A represent the ground, 1^1B_u, and n^1A_g states, respectively. In terms of one-photon absorption cross sections $(\sigma^{(1)}_{i,j})$ for transitions between the various states, Eq.(2) becomes

$$\gamma_\pi^* \cong a_2 \, (\, \sigma^{(1)}_{o,B} \, /\Delta E_B^3) \, [\, (\, \sigma^{(1)}_{B,A} \, /\Delta E_{B,A} \, \Delta E_A) - (\, \sigma^{(1)}_{o,B} \, /\Delta E_B^2) \,] \qquad (3)$$

where a_2 is a constant and $\Delta E_{B,A}$ is the transition energy between the B and A states.

The two-state approximation of γ_π improves with increasing chain length: γ_π^*/γ_π = 0.70, 0.77, 0.83 for ATB, ATH, ATO, respectively. The two-state approximation also implies that γ_π^*, and hence γ_π, will reach a limiting value as the chain length increases. The reason is that $| \, \mu^z_{o,B} \, |^2$ and ΔE_B are known to reach a limiting value [18], and it is assumed that $| \, \mu^z_{B,A} \, |^2$ and ΔE_A will also reach a limiting value. Additional excited state absorption spectra and two-photon absorption spectra should be measured for the linear polyenes in order to quantify the dependence of $| \, \mu^z_{B,A} \, |^2$ and ΔE_A on chain length. A more extensive discussion of the origin of γ for the linear conjugated polyenes will be presented in a subsequent paper.

4 CONCLUSIONS

An accurate and computationally tractable theoretical procedure for the calculation of the non-resonant, electronic component of $\gamma(-3\omega; \omega, \omega, \omega)$ can be constructed that is based on the combined use of the semi-empirical INDO all-valence-electron molecular orbital method including full single- and double-excitation configuration interaction (SDCI) of excited π-electron singlet states and Orr and Ward's sum-over-states expression for $\gamma(-3\omega; \omega, \omega, \omega)$.

The successful application of this theoretical procedure to the calculation of $\gamma(-3\omega; \omega, \omega, \omega)$ for planar, π-electron conjugated hydrocarbons indicates that 1) the correct calculation of γ requires that the repulsive π-electron interaction is treated at an SDCI, or higher, level of configuration interaction; 2) the separation of γ into σ- and π-electron components is a good approximation; 3) extended atomic orbital basis sets are not needed; 4) a single set of INDO semi-empirical parameters can be used for the calculation of γ *and* one- and two-photon electronic absorption spectra.

Calculations of the effect of chain length, conformation, and methyl substitution on the values of γ for π-electron conjugated hydrocarbons conclude that γ is most strongly influenced by conjugation chain length. As γ increases with chain length, so does the ratio γ_π/γ. The presence of *cis* linkages in the chain does act to diminish $|\gamma|$, and should be avoided if one wishes to maximize $|\gamma|$. Methyl group substitution at the ends of a polyene chain increases $|\gamma|$ to a lesser extent than achieved by increasing conjugation chain length.

The non-resonant, third-order nonlinear optical response of the linear conjugated polyenes is defined principally by two excited π-electron singlet states: 1^1B_u, and n^1A_g (n>2). *This n^1A_g state is not the lowest-lying ^1A_g state that is important in the photochemistry and photophysics of linear polyenes.* A simple two-state approximation of the non-resonant γ can be derived in terms of one-photon transition moments and transition energies involving the 1^1B_u, and n^1A_g states. This approximation implies that γ will reach a limiting value as the polyene chain length increases.

ACKNOWLEDGEMENTS

B.M.P. thanks Mr. K. Bates, Dr. D. Chang, Dr. C. Dirk, Dr. G. Meredith, and Professor M. Ratner for interesting and helpful discussions.

5 REFERENCES

1. *Nonlinear Optical Properties of Polymers*, edited by A.J. Heeger, J. Orenstein, and D.R. Ulrich, Materials Research Society Symposium Proceedings Vol. 109, MRS, Pittsburgh, 1988.
2. *Nonlinear Optical Properties of Organic Molecules and Crystals*, edited by D.S. Chemla and J. Zyss, Academic, New York, 1987, Vols. 1 and 2.
3. *Molecular and Polymeric Optoelectronic Materials: Fundamentals and Applications*, edited by G. Khanarian, SPIE, Vol. 682, 1986.
4. B.M. Pierce, in *Nonlinear Optical Properties of Polymers*, edited by A.J. Heeger, J. Orenstein, and D.R. Ulrich, Materials Research Society Symposium Proceedings Vol. 109, MRS, Pittsburgh, 1988, pp. 109-114.
5. P. Tavan and K. Schulten, J. Chem. Phys. **85**, 6602 (1986).
6. C.W. Dirk, R.J. Twieg, and G. Wagniere, J. Am. Chem. Soc. **108**, 5387 (1986).
7. D. Li, T.J. Marks, and M.A. Ratner, Chem. Phys. Lett. **131**, 370 (1986).
8. J.P. Hermann and J. Ducuing, J. Appl. Phys. **45**, 5100 (1974).
9. B.F. Levine and C.G. Bethea, J. Chem. Phys. **63**, 2666 (1975).
10. A.D. Buckingham and B.J. Orr, Chem. Soc. Quart. Rev. (London) **21**, 195 (1967).
11. B.J. Orr and J.F. Ward, Molec. Phys. **20**, 513 (1971).
12. J.F. Ward and D.S. Elliott, J. Chem. Phys. **69**, 5438 (1978).
13. S.H. Stevenson, D.S. Donald, and G.R. Meredith in *Nonlinear Optical Properties of Polymers*, edited by A.J. Heeger, J. Orenstein, and D.R. Ulrich, Materials Research Society Symposium Proceedings Vol. 109, MRS, Pittsburgh, 1988, pp. 103-108.
14. J.P. Hermann, Opt. Commun. **9**, 74 (1973).
15. R.R. Birge, J.A. Bennett, L.A. Hubbard, H.L. Fang, B.M. Pierce, D.S. Kliger, and G.E. Leroi, J. Am. Chem. Soc. **104**, 2519 (1982).
16. A.F. Garito, J.R. Heflin, K.Y. Wong, and O. Zamani-Khamiri, in *Nonlinear Optical Properties of Polymers*, edited by A.J. Heeger, J. Orenstein, and D.R. Ulrich, Materials Research Society Symposium Proceedings Vol. 109, MRS, Pittsburgh, 1988, pp. 91-102.
17. B. Hudson, B.E. Kohler, and K. Schulten, in *Excited States*, edited by E.C. Lim, Academic, New York, 1982, Vol. 6, pg. 1.
18. H. Suzuki, *Electronic Absorption Spectra and Geometry of Organic Molecules*, Academic, New York, 1967.

Polaron and Bipolaron Formation in Model Extended π-Electron Systems: Potential Non-linear Optics Applications

C.W. Spangler, L.S. Sapochak, and B.D. Gates

DEPARTMENT OF CHEMISTRY, NORTHERN ILLINOIS UNIVERSITY, DEKALB, IL 60115, USA

INTRODUCTION

In the late 1970s, the discovery[1] that oxidative or reductive "doping" of either cis or trans polyacetylene (PA) films with iodine or AsF_5 produced dramatic increases in electrical conductivity led to an explosion of research in this field which continues to the present day. Such studies have ranged from various spectroscopic investigations of the structure of the doped conducting polymer and the chemical nature of dopant-polymer interaction,[2] to a wide-ranging and continuing discussion of the nature of the charge carriers (solitons, polarons, bipolarons) and the relative roles of interchain versus intrachain charge transport.[3,4] Recently, it has been recognized that electroactive polymers may also exhibit significant nonlinear optical (NLO) activity,[5] and that in particular the third order susceptibility $\chi^{(3)}$ is influenced by π-electron delocalization. Since this NLO activity is basically electronic in nature, there is the real possibility that electro-optic devices depending on fast switching times can be built using organic nonlinear materials.

Unfortunately, many electroactive polymers in pristine (undoped) or doped states are intractable and insoluble. Much recent work has been focused on solubilizing the polymer in a doped state, or manipulating the polymer into final desired form via soluble precursor polymers. Two such examples are Feast's[6] preparation of "Durham" polyacetylene and Karasz's and coworkers'[7] use of soluble polyelectrolyte intermediates in the preparation of oriented poly [p-phenylene vinylene] (PPPV) films. The use of such techniques has recently allowed other polymers

with formal copolymeric structures, such as poly [2,5-
thienylene vinylene], to be prepared.[8] The ability of
these techniques to produce free-standing, stable, and
often transparent films has allowed study by a variety of
techniques,[9] such as UV-visible absorption, infrared absorption, photoluminescence, and photoinduced absorption.
Such studies have allowed estimates of the polaron and
bipolaron energy levels, $\pi-\pi*$ transitions for the various
conjugated oligomeric segments within the polymer film,
and the number of conjugated segments giving rise to the
electron or charge delocalization upon doping to metallic
conductivity. In certain polymers, like polyacetylene,
it has been recognized that there is intrinsic (solitonic) delocalization as well as active or dynamic delocalization brought about by doping to a polaronic or
bipolaronic state.[5,6] Dalton has suggested that for NLO
activity considerations, short polymer chains or indeed
oligomeric materials may be quite effective due to the
relatively short intrinsic delocalization distances.[10]

Oxidative Doping of α,ω-Diphenylpolyenes

    In recent companion studies, we have studied the
oxidative doping of model α,ω-diphenylpolyenes by $FeCl_3$
and $SbCl_5$ in dilute solution by absorption spectroscopy.[11] In all cases a new intense band is rapidly
formed which decays into a second band structure characterized by a higher transition energy. Both new bands
are intensely red-shifted from the original $\pi-\pi*$ polyene
transition. We have interpreted these results as successive polaron and bipolaron formation, with the latter
being the more stable state. The bipolaron states for
these polyenes are stable in solution for more than 24
hours at 0° without any special precautions regarding the
presence of O_2 and H_2O vapor.

    The polaron band is distinguishable from the bipol-
aron by observing the rise and decay of ESR signals
associated with the paramagnetic species.[12] Progressive
red shifts for both new absorption bands (polaron and
bipolaron) are observed as the conjugation length in-
creases $(5 \rightarrow 6 \rightarrow 7)$, indicating increasing delocalization.

Stabilization and Destabilization of Bipolaron States

We would now like to report that the bipolaron states induced in diphenylpolyenes by oxidative doping can be significantly stabilized by substitution of polar electron-donating groups on the para positions of the terminal phenyl rings. These EDGs greatly stabilize the delocalized bipolaron, as evidenced both by the decreasing transition energies of the corresponding mid-gap states as well as the relative lifetimes in the presence of water vapor and air.

Table 1   Stabilized Bipolaron Absorption Spectra

$$EDG-C_6H_4 \; \text{-} (CH=CH)_n C_6H_4-EDG$$

EDG	n	Absorption Spectra (nm)			Approximate Lifetime[a]
H	5	564,	612		0.5h
Me	5	590,	650		> 2 h
MeO	5	640,	692,	760	>12 h
Me_2N	5	658,	714,	776	>24 h
H	6	615,	685		> 2 h
Me	6	640,	700		>12 h
MeO	6	676,	728,	816	>24 h
Me_2N	6	700,	752,	832	>72 h

[a]Approximate time for at least 90% decay of band.

The increased delocalization and stabilization can be rationalized on the basis of the mesomeric electron releasing abilities of the substituent groups:

In addition, it can be readily seen that the bipolaron bands are not only sensitive to the mesomeric electron-donating ability of the substituent groups (Me_2N>MeO> Me>H), but are also highly dependent upon the pi-electron conduit delocalization length.

One would predict, on the basis of the above observation that electron-withdrawing groups (EWG) would significantly reduce the stability of the positively charged bipolarons. This, in fact, is exactly what we observe, as shown in Table 2.

Table 2   Destabilized Bipolaron Absorption Spectra

$$EWG-C_6H_4 \text{+} CH=CH \text{+}_n C_6H_4-EWG$$

EWG	n	Absorption Spectra (nm)
H	5	564, 612
Cl[a]	5	574, 632
CF$_3$	5	not formed
NO$_2$	5	not formed
H	6	615, 685
Cl[a]	6	636, 690, 760
CF^3	6	602, 650
NO$_2$	6	605, 652

[a]Cl can function as an EWG by an inductive effect, but seems to function as an EDG (resonance) in oxidative doping.

These results hint at an intriguing possibility for the stabilization of delocalized charged states for third harmonic generation, $\chi^{(3)}$, as originally proposed by Dalton[10] and Heeger[13] for polyacetylene and polythiophene, both of which have large resonant third order nonlinear optical effects. Theoretical calculations[14] suggest that $\chi^{(3)}$ should be related to the sixth power of the electron delocalization length, and it can be argued that the large third order effects in conjugated polymers are related to the size of the solitonic, polaronic, or bipolaronic domains, and their concentration. The question remains, of course, as to whether the extent of delocalization necessary for measurable third order nonlinearity is independent of monomer, polymer or oligomer type.

Further studies of how substituent groups in aromatic rings can transmit bipolaronic charge down-chain and stabilize bipolaronic states in molecules of varying size will allow us to determine how the pi-electron network of a nonlinear oligomer or polymer system should be designed and assembled to maximize third order effects. DeMelo and Silbey[15] have most recently considered the relationship problem of conjugation in polyenes, solitons, polarons and bipolarons to nonlinear polarizabilities. While theoretical treatment and prediction of third order effects must be considered as embryonic, at the present time, these authors have recognized that there are virtually no available experimental data for large polyene chains. However, they do suggest that the presence of polar substituents, such as the polyenes described in this study, may yield materials capable of high nonlinear hyperpolarizabilities. Although much work remains to be done, we believe the results of this study give us a firm foundation for predicting whether extension of the pi-system to longer lengths, stabilization of delocalization by EDGs or EWGs, or ease of charge transmission and storage down the polymer or oligomer backbone are of similar or complimentary importance. Accumulation and understanding of such relationships will eventually allow us to tailor third order nonlinear properties in a true molecular engineering sense.

REFERENCES

1. H. Shirakawa, et al., J. Chem. Soc. Chem. Commun., 1977, 578; C. Chiang, et al., J. Am. Chem. Soc., 100, 1978, 1013; C. Chiang, et al., J. Chem. Phys., 69, 1978, 5098.

2. S. Hsu, et al., J. Chem. Phys., 69, 1978, 106; I. Harada, et al., Chem. Lett., 1978, 1411; S. Lefrant, et al., Solid State Commun., 29, 1979, 191; S. Etemad, Phys. Rev. B., 23, 1981, 5137; R. Baughman, et al., J. Chem. Phys., 68, 1978, 5405; J. Chien, et al., Macromol., 15, 1982, 1012; C. Rieskel, J. Chem. Phys., 77, 1982, 4254.

3. E. Mele and M. Rice, Phys. Rev. Lett., 45, 1980, 926; W. Su, J. Schrieffer and A. Heeger, Phys. Rev. Lett., 42, 1979, 1698; J. Ritsko, Phys. Rev. Lett., 46, 1981, 849; S. Kivelson, Phys. Rev. Lett., 46, 1981, 1344; M. Kertesz and P. Surjan, Solid State Commun., 39, 1981, 611; G. Crecelius, et al., Phys. Rev. Lett., 51, 1983, 1498; J. Bredas, et al., Phys. Rev. B. 26, 1982, 5843; J. Bredas, et al., Phys. Rev. B., 29, 1984, 6761; J. Bredas and G. Street, Acc. Chem. Res., 18, 1985, 309.

4. For a general overview of polyacetylene and conducting polymers in general, see G. Wegner, Angew. Chem. I. E., 20, 1981, 361 and A. MacDiarmid and A. Heeger, Syn. Met. 1, 1980, 101; J. Frommer and R. Chance, Encyc. Pol. Sci. Eng., 5, 2nd ed., John Wiley, 1986, pp 462-507; "Polyacetylene-Chemistry, Physics and Material Science," J. C. W. Chien, Academic Press, 1984, 634 pp.

5. (a) "Nonlinear Optical Properties of Organic and Polymeric Materials", ACS Symp. Ser. No. 233, D. Williams, ed., Washington, 1983.

(b) D. Williams, Angew. Chem. I. E., 23, 1984, 690.

(c) "Nonlinear Optical Properties of Organic Molecules and Crystals", Vol. 1 and 2, D. Chemla, and J. Zyss, Eds., Academic Press, Orlando, 1987.

6. J. Edwards, W. Feast and D. Bott, Polymer, 25, 1984, 395; W. Feast, M. Taylor and J. Winter, Polymer, 28, 1987, 593.

7. J. Capistran, D. Gagnon, S. Antoun, R. Lenz and F. Karasz, Polymer Preprints, 25(2), 1984, 282; D. Gagnon, J. Capistran, F. Karasz and R. Lenz, Polymer Preprints, 25(2), 1984, 284; F. Karasz, J. Capistran, D. Gagnon and R. Lenz, Mol. Cryst. Liq. Cryst., 118, 1985, 327; D. Gagnon, F. Karasz, E. Thomas and R. Lenz, Syn. Met., 20, 1987, 85; D. Gagnon, J. Capistran, F. Karasz, R. Lenz and S. Antoun, Polymer, 28, 1987, 567.

8. K.-W. Jen, M. Maxfield, L. Shacklette and R. Elsenbaumer, J. Chem. Soc. Chem. Commun., 1987, 309; S. Yamada, S. Tokoto, T. Tsutsui and S. Saito, J. Chem. Soc. Chem. Commun., 1987, 1448.

9. J. Obrzut and F. Karasz, J. Chem. Phys., 87, 1987, 2349; K. Wong, D. Bradley, W. Hayes, J. Ryan, R. Friend, H. Lindenberger and S. Roth, J. Phys., C20, 1987, L187, D. Bradley and R. Friend, Syn. Met., 17, 1987, 645; D. Bradley, G. Evans, and R. Friend, Syn. Met., 17, 1987, 651.

10. L. Dalton, J. Thompson and H. Nalwa, Polymer, 28, 1987, 543.

11. C. Spangler, E. Nickel and T. Hall, Polymer Preprints, 28(1), 1987, 219.

12. C. Spangler, E. Nickel and L. Dalton, to be published.

13. A. Heeger, D. Moses and M. Sinclair, Syn. Met., 17, 1987, 343 and references therein; M. Sinclair, et al., Solid State Commun., 61, 1987, 221.

14. K. Rustagi and J. Ducuing, J. Optics Commun., 10, 1974, 258; G. Agrawal, C. Cojan and C. Flytzanis, Phys. Rev. B., 17, 1978, 776.

15. C. DeMelo and R. Silbey, Chem. Phys. Lett., 140, 1987, 537.

The Effect of Close Ordered Packing upon the Transition Moments and Energies of Rigid-rod Dipoles and Pseudo-linear Donor–Acceptor Systems

C.L. Honeybourne

MOLECULAR ELECTRONICS AND SURFACES GROUP, BRISTOL POLYTECHNIC, BRISTOL BS16 1QY, UK

The non-linear optical (NLO) performance of organic films of wave-guiding thickness (0.5 - 2 micron) is inadequate when currently available materials (e.g. MNA,POM,DAN, PTS(1,2)) are used. However, the NLO performance in thick crystal slices, wherein the optical length is sufficiently large to obtain the required phase-shift from,say,MNA(3) is showing promise. Clearly, to obtain the same phase-shift from a thin film, particularly in the circumstance of the light propagating normal to the plane of the film, demands values of NLO coefficients 100 to 1000 times larger than those of MNA.The following is a non-exhaustive list of suggested means by which enhancement of NLO coefficients might be achieved.
(i)   Enhancement of transition dipole moments.
(ii)   Reduction of state-state transition energies.
(iii) Enhancement by chemically "tuning" the position of
        a major contributing transition.
(iv)   Increasing the active species per unit volume.
(v)   Enhancement of microscopic contributions to bulk
        NLO performance by enforcing favourable alignments
        for vector addition.
(vi)   Enhancement of the number of contributing transit-
        ions by generating extra,charge-transfer bands.

The foregoing list has been put forward on the basis that expressions for CHI-2 and CHI-3 contain numerators with the product of field terms by transtion moments, and denominators involving products of transition energies.

The purpose of this paper is to assess, in accord with
 (i) and (ii) above, how ordered arrays of pseudo-linear donor-acceptor molecules interact with special reference

to the perturbation of transition moments and transition
energies.

A Pedagogic Two-level Model.

A simple heteronuclear diatomic dipole, C=X, can be used
as a model for the consequences of close, ordered packing
in an application of the crystal orbital method(4). The
mode of packing may be as follows:
A. A pseudolinear array (..C=X C=X..) that is energetically favourable and non-centrosymmetric.
B. A parallel,regular array with like facing like. This
   is energetically unfavourable, but is non-centrosymmetric.
C. A parallel regular array with like and unlike in
   alternation. Although energetically favourable, this
   is centrosymmetric, and is therefore unsuitable as
   a model for second-order NLO activity.

In earlier work(5),all repulsion terms were ignored, and
all(attractive) hopping integrals set to zero except
β_1 for C..X adjacent and β_2 for C=X. Agrawal et al.(5)
and Flytzanis(6) respectively give expressions for NLO
coefficients and transition moments.In the case of a
small β_1 , which corresponds to a highly alternating
chain, CHI-2 has a maximum value, for a fixed spacing,
when$(a_1- a_2)/(2\beta_2)$ = 0.5, in which a_1 and a_2 are Hückel
Coulomb Integrals.

In the simplest version of the crystal orbital method,
the above treatment is modified by the introduction of
three further hopping integrals spanning three adjacent
units in the infinite Born-von-Karman chain.These are
β_3(remote CC), β_4(remote XX) and β_5(remote CX).

The values of the valence band width(dV),conduction
band width(dC), band gap(dE) and transition moment(M_z)
for a bond of length R(Å) along the z-axis are:

$$dV = (2(\beta_3+\beta_4)+(a+b+c)^{\frac{1}{2}} + a^{\frac{1}{2}})/2$$

$$dC = (2(\beta_3+\beta_4)-(a+b+c)^{\frac{1}{2}} + a^{\frac{1}{2}})/2$$

$$dE = \frac{1}{2\pi}\int_{-\pi}^{\pi} (a + b\cos k + c\cos^2 k)^{\frac{1}{2}} dk$$

$$M_z = 4.803R(GE-N)/((G^2+N)^{\frac{1}{2}}(E^2+N)^{\frac{1}{2}})$$

in which the wave vector,k, is defined as $2\pi p/M$ where

M is the very large number of interacting units in the chain and p is an integer. The values of a, b and c are given by the following expressions in which t is the amount by which a_2 is larger than a_1:

$$a = t^2 + 4(\beta_1^2 + (\beta_2-\beta_5)^2)$$

$$b = 4(2t(\beta_3+\beta_4)-2t\beta_3 + 2\beta_1(\beta_2+\beta_5))$$

$$c = 4((\beta_3+\beta_4)^2 + (\beta_2+\beta_5)^2 - (\beta_2-\beta_5)^2 - 4\beta_3\beta_4)$$

$$N = \beta_1^2 + (\beta_2+\beta_5)^2\cos^2 k + (\beta_2-\beta_5)^2 + 2\beta_1(\beta_2+\beta_5)\cos k$$

$$G = a_1 - e_v + 2\beta_3\cos k$$

$$E = a_1 - e_c + 2\beta_3\cos k$$

where e_v and e_c are the valence and conduction level eigenvalues for a specific value of k. In the case of stacking mode (ii), similar expressions are obtained by setting A and β_5 to zero in the above equations. The meanings of $\beta_1 \ldots \beta_4$ are explained in the Figure.

mode(i)   mode(ii)

In his treatment of weakly interacting chromophores, using the delocalisation of excitation(which is not overlap dependent) Murrell(7) **found** small energy changes with both of the above modes producing a x1.4 increase in transition moment. Mode (i) gives a lowered transition energy whereas mode(ii) gives an increased transition energy. For a spacing of 0.35 nm we find, in our foregoing treatment, increases in transition moments of 11%. If repulsion terms are included in an SCF calculation, using the Fock matrix operator as defined and used by us elsewhere(**4**), we obtain a transition moment of 2.723 Debye for an isolated unit. For mode (i) we obtain 2.814D, and for mode (ii) 2.820D. The energy levels are **perturbed** in the following way: the transition energies of both modes are virtually identical and circa 0.4 eV smaller than in the isolated unit. Therefore, in accord with the findings of others (5-7), we find that increasing the proximity of dipoles will tend to enhance non-linear optical coefficients unless too close a spacing is enforced.

Cyanoacetylene:a model four level system.

The crystal orbital Hartree-Fock operator matrix,F, has
been defined in complex form by Del Re(9) as

$$(\text{Real})F^R = F^O + 2F^+\cos k$$

$$(\text{Imag})F^I = \quad - 2F^-\sin k$$

In the most general case, the off-diagonal terms of F^I
are non-zero, whereas, in perfectly columnar stacking,
these terms are zero. In all cases the diagonal elements
of F^I are zero.

$$F^R_{ii} = H^O_{ii} - \quad (P^O_{jj} - Q_j)G^O_{jj} + 0.5P^O_{ii}G^O_{ii}$$

$$+ 2 \quad (P_{jj} - Q_j)G^+_{ij} + \cos k(H^+_{ii} + H^-_{ii} - P^{+R}_{ii}G^+_{ii})$$

$$F^R_{ij} = H_{ij} - 0.5P^O_{ij}G^O_{ij}$$

$$+ \cos k(H^+_{ij} + H^-_{ij}) - 0.5(P^{+R}_{ij}G^+_{ij} - P^{-R}_{ij}G^-_{ij})$$

$$+ 0.5\sin k(P^{+I}_{ij}G^+_{ij} - P^{-I}_{ij}G^-_{ij})$$

$$F^I_{ij} = \sin k(H^+_{ij} - H^-_{ij}) - 0.5(P^{+R}_{ij}G^+_{ij} - P^{-R}_{ij}G^-_{ij})$$

In the foregoing the symbols H,P and G refer to matrices
of attractive terms, densities and repulsion terms respectively. The superscripts "o","+" and "-" refer to the
central,one forwards and one backwards units in the
chain. All other terms, and the parameterisation, are
given in detail elsewhere(4).

The results of applying the foregoing method to a
model four-level system are as follows(see Table 1).
The strongest transition has an enhanced intensity, as
does the weakest. The two transitions of intermediate
intensity are weakened. However, inspection of the wave
functions for the isolated molecule show that only the
transition with an M_z of 0.498D is associated with a
large change in molecular dipole moment as the one-
electron excitation takes place. Refering again to the
single molecule,only one excited state contributes

significantly to β_{zzz} (10), and this state is composed of over 98% of the one-electron transition with M_z = 0.498D.

TRANSITION MOMENTS OF 4 π-π^* MODEL SYSTEM OF C≡C.C≡N(D).

Single Molecule	MODE(i) End-to-End	MODE(ii) Ladder
0.498	0.494	0.378
-0.959	-0.790	-0.804
-4.627	-4.850	-4.807
-0.057	0.107	0.210

Table 1. SCF Crystal Orbital results for Cyanoacetylene using gas-phase geometry, and integrating the transition moments over the wave vector.

On the basis of our results, packing mode(i) is the more favourable.

Push-Pull Acetylenes and Di-acetylenes.

The archetypal rigid-rod push-pull molecule of this general type is 4-nitrophenyl-4-aminophenyldiacetylene (11). However, we have found that rather unconventional parameters are required for the nitro-group if our solid-state iterative procedures are to be forced to converge. We therefore choose to report our results for the similar species 4-cyanophenyl-4-aminophenylacetylene and the related diacetylene(CYAMACET and CYAMDIAC respectively). In our calculations, we used a standard set of parameters apart from a hopping integral for the CN triple bond of -2.86 eV. All bond lengths were set to 0.14nm with the following exceptions: CC(triple),0.119 nm; CN(triple), 0.1158 nm;CC(single,acet.),0.1382 nm. The molecules are projected in the xy plane, with the long molecular axis being the y-axis.

We must now consider three packing modes: namely, mode (i), in-line; mode(ii), face-to-face columnar stacking; mode(iii), side-by side in the xy plane. In all cases the distance of closest approach of heavy atoms is set to 0.35 nm. State compositions were calculated from the interactions between 48 one-electron excitations between molecular orbitals of π-symmetry with respect to the xy plane of an isolated molecule. To render comparisons valid(12), we compare energy gaps with single excitation energies(see Table 2).

| | C Y A M A C E T | | | | C Y A M D I A C | | |
Single	M(i)	M(ii)	M(iii)	Single	M(i)	M(ii)	M(iii)
3.31	3.12	2.35	2.61	3.56	3.02	2.45	2.39
1.88	1.74	2.38	2.25	2.08	2.07	2.74	2.64
6.81	7.40	7.65	7.61	7.73	8.77	9.01	9.00
2.82	2.57	1.87	2.03	2.98	2.50	1.72	1.86

Table 2.Values of large M_y and M_y integrated over $k(D)$.

In CYAMACET, only one state contributes significantly to β_{yyy}; this state has 88% of the transition with $M_y=6.81$. The transition energy is lowered by 0.6 eV in mode(ii), and by 0.2 eV otherwise. Therefore, face-to-face packing is the most favourable at the distance chosen, and will enhance the value of β_{yyy}. All the M_y cited above are associated with dipole changes with excitation.
The results for CYAMDIAC are similar, because only one state contributes significantly to β_{yyy}; this state has 88% of the transition with $M_y=7.73$. The transition energy is lowered by 0.8 eV in mode(ii) and by 0.3 eV otherwise.Face-to-face packing is the most favourable for the enhancement of β_{yyy}.

REFERENCES.
1. Non-linear Optical Properties of Organic Molecules & Crystals(Edt.Chemla and Zyss),Academic Press,NY, 1987.
2.J.T.Lin and C.Chen,Lasers and Optr.,1987, 59.
3.G.F.Lipscomb,A.F.Garito and R.S.Narang,J.Chem.Phys. 75, 1509 (1981).
4.C.L.Honeybourne, Molec.Phys., 50, 1045 (1983).
5.G.P.Agrawal,C.Cojan & C.Flytzanis,P.Rev.17B,776(1978).
6.C.Flytzanis,Treatise on Quantum Electronics,1A,1(1970).
7.J.N.Murrell,The Theory of the Electronic Spectra of Organic Molecules,Chapman & Hall,London,1963.
8.S.J.Lalama and A.F.Garito, P.Rev.20A, 1177 (1979).
9.G.Del Re,J.Ladik and G.Biczo,P.Rev.,115, 997(1967)
10.D.Pugh and J.O.Morley,Vol I of Ref I,p.193.
11.C.Fouquey,J.-M.Lehn and J.Malthête, J.C.S.Chem. Commun., 1987, 1424.
12.J.Ladik, S.Suhai, P.Otto and T.C.Collins, Internat. J.Quantum Chem., QBS4, 55 (1977).

Crystals and Crystal Structure

The Growth, Perfection, and Properties of Crystals of Organic Non-linear Optical Materials

J.N. Sherwood

DEPARTMENT OF PURE AND APPLIED CHEMISTRY, UNIVERSITY OF STRATHCLYDE, GLASGOW G1 1XL, UK

1 INTRODUCTION

The search for high performance organic non-linear optical materials has followed two principal avenues; the development of theoretical search programmes using quantum mechanical calculations directed towards the evaluation of the second order molecular hyperpolarisability,[1] and the use of chemical intuition based on empirical criteria defined by the performance of existing materials which show good characteristics[2]. Both approaches have identified series of suitable materials the potential of which has been confirmed by assessment of the relative efficiencies of second harmonic generation (SHG) of the powdered material. The ultimate proof of the utility of these materials can only come from the more detailed optical assessment of high quality single crystal specimens. Additionally, only from such examinations will come the definition of molecular structure-optical property relationships with which to define and advance the understanding of these systems.

One series of materials which has been identified by empirical criteria and powder assessments as having good potential is the range of structures based on the substitution of 5-nitropyridine and nitrobenzene moieties.[3,4] In two cases (-)2-(α -methyl benzyl amino)-5-nitropyridine (MBA-NP)[5] and 2-N,N-dimethyl amino-5-nitro acetanilide (DAN)[6,7] this potential has been confirmed by optical assessment of small crystals. As part of a general programme directed towards the better understanding of the non-linear optical properties of organic materials and their eventual commercial use we have developed existing methods for the purification and growth of organic crystals, applied them to these materials and made initial examinations of the relationship between structural perfection and ultimate performance.

2 PURIFICATION[8]

An essential prerequisite for crystal growth is the availability of material of the highest purity attainable. In the present cases zone refining and gradient sublimation were precluded by the thermal instability of the solids at high temperatures (MBA-NP) or in the melt (MBA-NP, DAN). Fractional recrystallization was effective but inefficient.

The most effective procedure proved to be column chromatography, using dichloromethane on a silica gel column. Repetition of the process twice plus a final recrystallization from the same solvent gave material of purity 99.98% compared with 99.7% (MBA-NP) and 96.2% (DAN) for the initial synthetic materials.[3,4]

3 CRYSTAL GROWTH[9]

The principal commercial requirement of the crystals; that they should be of large size ($1-100cm^3$), free from strain and crystal imperfections and produced over a short time scale, limits the method of production to low temperature solution growth and melt growth. Vapour growth is a slow process and the method, which involves growth on a substrate, inevitably causes a gradient of strain in the crystal.

Of the previous two methods low temperature solution growth at near ambient temperatures offers the best prospect of producing crystals under near equilibrium conditions and free from thermal strain. Whether or not this can be achieved is defined by the limitations of the growth system. A typical system[10] of 2l capacity usually operates with a temperature stability of $\pm 0.001 - 0.005K$ over long periods and can be subjected to cooling rates (to achieve supersaturation) of $0.1 - 1K$ per hour. For such a system we have shown empirically that satisfactory crystals can be produced if the "solubility ratio"[9]

$$SR = (dS/dT)/S_a \approx 0.01 - 0.03K^{-1}.$$

Where S is the solubility at temperature T and S_a is the average solubility in the temperature range used. Also the metastable zone width should be < 5K. Where these conditions do not hold then poor uncontrolled growth occurs; the crystal develops strain and solvent inclusions . In this event the control in the system must be improved or an alternative method sought. One alternative, melt growth by the Bridgman technique[9] involves preparation in closed vessels under extreme temperature gradients both of which can generate strain in the grown crystal. There is also the problem of melt stability. In contrast the extreme temperature gradients employed give high supersaturation and hence quite large samples can be produced in short growth periods. Where the solid has suitable mechanical properties the crystals produced can be strain free.

MBA-NP[11]

MBA-NP is unstable in the melt and in the solid-state at high temperatures. Solution growth offers the most suitable means to crystal production.

The primary step in solution growth is the identification of a suitable solvent. This should yield crystals of a prismatic habit and satisfy the criteria noted above. Experiments with solvents of a range of polarities yielded the results shown in Table 1. It was also confirmed that all solvents produced the desired structure.

The best results were achieved with methanol which yielded seeds of generally excellent quality but with some imperfection in the (010), polar, sectors. Further experiments showed that this combination gave a solubility ratio of SR < $0.01K^{-1}$ and a metastable zone width of ≈15K. Hence the prismatic seeds should

Table 1 Solvent effects on the Morphology of MBA-NP

Solvent	Dipole Moment $/10^{30}$ Cm	Solubility	Morphology	Quality
Acetone	10.01	high	prismatic	fair
EtOH	5.7	medium	prismatic	excellent
MeOH	5.5	medium	prismatic	excellent
Toluene	1.3	low	triangular	good
O-xylene	2.0	low	plate	good
Hexane	0.0	insoluble	-	-

be capable of development to large sizes in a standard growth system at temperature lowering rates of ~ 0.05K per hour.

Under these conditions and using selected seeds crystals could be produced in sizes up to $7 \times 7 \times 10 cm^3$ of high visual quality (Figure 1).

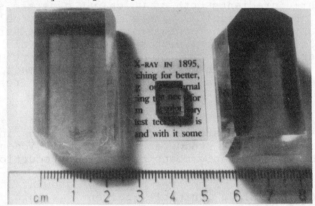

Figure 1.   Crystals of MBA-NP prepared by the slow cooling of saturated methanol solutions.   The optical clarity can be seen from the cut and polished blank.

DAN[12]

This compound suffers from slow melt decomposition and is only slightly soluble in many solvents. Although slow evaporation of saturated methanol solutions gives satisfactory seeds, the low solubility coupled with a shallow solubility curve yields a solubility ratio SR ≪ $0.01K^{-1}$. Thus growth occurs in an unstable region, very slowly and to small overall sizes; a seed $1 \times 1 \times 4 mm^3$ develops to $3 \times 3 \times 8 mm^3$ in 3 months.   Although suitable for optical assessment the low production rate is inhibitive to the eventual use of the material.   Supersaturation by slow solvent evaporation yields an improvement in production rate and crystal size (Figure 2).

The decomposition of DAN at high temperatures in the melt is more of a problem with the impure than with purified material.   Using ultra-pure material and with careful control of the period for which it is held in the molten state satisfactory crystals in sizes up to

2-3cm diameter x 2cm long can be prepared by the Bridgman technique.

 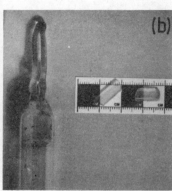

Figure 2. Crystals of DAN. (a) Crystal prepared by slow evaporation of a saturated solution. (b) A melt-grown boule and cleaved specimen.

4 CRYSTAL PERFECTION

In order that such materials achieve optimal optical performance the perfection of the crystals must be defined▪ and its role assessed. Equally, in the preparation programme the nature and distribution of defects must be assessed in order that improvements in processing can be directed towards the preparation of structurally more perfect samples. X-ray topography[13] using monochromatic or synchrotron radiation offers the best method for the evaluation of bulk samples.

The role of the technique is well exemplified by the series of synchrotron X-ray topographs shown in Figures 3 and 4.

Figure 3b shows a topograph of a typical seed crystal grown by rapid solvent evaporation of methanol solution as described above. The original specimen is shown in Figure 3a. The volumes A and B are missing from the image because they are of a different crystallographic orientation to the rest. Region A is a region of general low perfection which is obvious anyway from Figure 3a. Region B is a twinned volume.

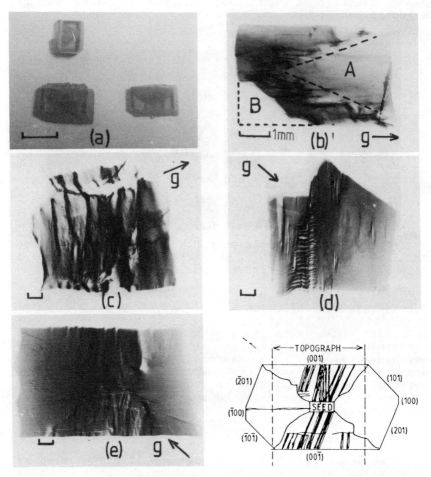

Figure 3.   (a) Photograph of typical MBA-NP seed
crystals,   (scale   mark   1cm)   (b),   (c)   (d)   and   (e)
Synchrotron   "white   radiation"   topographs   of   MBA-NP
crystals (scale marks 5mm):   (b) one of the crystals
depicted in (a) showing two out of contrast portions A
and B due to imperfection, (c) a crystal grown from a
good seed under non-ideal conditions showing evidence of
strain (d) and (e) better defined images obtained from a
specimen prepared under more ideal conditions; (d) (101)
Section (e) (010) section, (f) schematic diagram of the
(010) section of the crystal showing the general defect
structure of the specimen.

The direct use of such a seed could yield a multiply-twinned larger crystal which would be of unsatisfactory optical quality. The definition of better quality seeds gives improved results.

When the better seeds are developed too rapidly or too slowly, i.e. under extreme conditions of supersaturation, the crystal may include a high degree of strain even though it appears visually perfect. This can be detected immediately by topographic examination. Figure 3c shows the result. The variations in strain (mosaic spread ~ 200-400 sec arc) yield a "folded" image due to the irregular diffraction of X-rays from regions of differing strain.

Correction of the growth conditions to give more ideal growth yields excellent samples which give good topographs containing well-defined dislocation images (mosaic spread <100 sec arc) (Figures 3d and 3e). These allow the defect structure of the crystal to be "mapped" against growth history and seed geometry and allows further improvements to be made.

A parallel situation can be seen in Figure 4 which compares the strained crystal of DAN developed under non-equilibrium growth conditions from solution with the much better specimen obtained by melt growth.

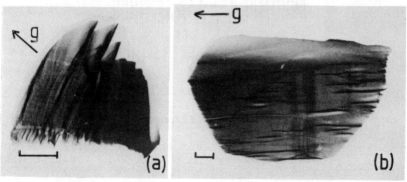

Figure 4. Synchrotron "white radiation" topographs of DAN crystals (scale mark 1mm): (a) crystal grown from methanol solution showing considerable strain arising from growth fluctuations, solvent inclusion etc. (b) A more perfect specimen of high perfection grown from the melt showing only a few dislocations aligned along the core of the boule and which have developed from the seed point. The nature of the additional linear imperfections, which lie in the cleavage plane has yet to be identified.

5 OPTICAL PERFECTION

The high visual perfection of the crystals is
well-exemplified by Figures 1 and 2. Samples of this
quality are readily oriented, cut and polished to yield
samples of good quality parallelism (30 sec. arc,
flatness λ /10, scratch width <0.5µm) without the
introduction of further imperfections into the bulk.
Maker fringe examinations of these yield patterns with
well-defined symmetrical envelopes and near zero minima
(Figure 5). The departure of the fringe minima from
zero is a consequence of the short coherence length
(2µm) for these materials and the degree of flatness
finish achieved. To improve on this would require an
increased (but probably unattainable) flatness of
λ/200.

Both solids give well-defined phase-matched SHG,
(DAN types 1 and 2, MBA-NP type 1). The excellent fit
of the angular dependence of the intensity again with
the predicted shape (Figure 6) testifies to the high
optical quality of the material.

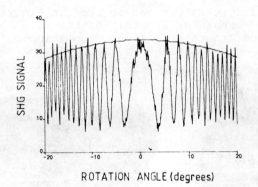

ROTATION ANGLE (degrees)

Figure 5. Maker fringes and theoretical envelope
for d_{22} from an X cut crystal of MBA-NP; rotation axis
Y, E (1064nm) and E (532nm) both parallel to Y.

The intensity of SHG depends markedly on the degree
of structural perfection of the solid. Using a cut and
polished section of the crystal, the topographs of which
are shown in Figures 3d and 3e, it was possible to map
the relative efficiency of SHG in regions of differing
perfection from highly perfect to highly dislocated.

All regions of the sample other than in the close vicinity of the seed were optically transparent and allowed transmission of the laser beam with no observable distortion. Figure 6 shows the comparison between the phase-matched peaks obtained from the best and worst regions. A factor of 3 difference in intensity was found. This confirms the need for high quality materials in order to achieve ultimate performance. The degree of coincidence between the experimental and observed variation is similar in both cases; however the more defective sample shows a greater departure in the wings of the peak. In the better regions of the phase-matched cut crystal the SHG efficiency has been quantified as 50% (0.4cm thick, input flux 117 MW.cm^{-2}). In comparison a non-phase-matched cut of DAN gives an efficiency of 2% (0.15cm thick, input flux 27 MW.cm^{-2}). At low power densities the SHG increased linearly with power. Thus both materials offer the prospect of excellent performance at higher power rates provided that the damage threshold is as high as is anticipated.

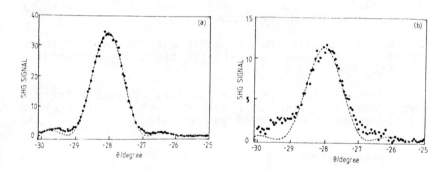

Figure 6. The coincidence between the observed and calculated angular variation in SHG intensity for MBA-NP. a) perfect region, lateral sectors Figure 3e, b) imperfect region, (001) sectors Figure 3e.

A parallel variation has also been noted for the dependence of optical damage threshold on structural quality. In less well-defined specimens the optical damage threshold increases with increasing perfection.

We have no quantitative comparison as yet and this relationship may be more complicated than for SHG intensity since the perfection of surface finish also seems to play a large part in this process. Detailed studies are in progress on this phenomenon and its relationship to the thermal and optical properties of the material. At present we would quote minimum values of 120 $MWcm^{-2}$ and 70 $MW\ cm^{-2}$ for a phase-matched cut of MBA-NP and non phase-matched cut of DAN respectively. Full publications on the optical assessment of these materials are in preparation.[14]

6 SUMMARY

Studies the growth of MBA-NP and DAN have led to the production of well-defined and well-characterized large crystalline samples for optical assessment. Indications have been obtained as to how further improvements in quality might be achieved. Optical assessment demonstrates that the optical perfection of the crystals is high but that residual imperfections in lower quality samples can have a significantly deleterious effect on performance.

The best quality material has attractive properties compared with those of existing organic and inorganic non-linear optic crystals.

In order to test the utility of such materials simple devices such as frequency doublers and Pockels cells have been constructed and operated.

7 ACKNOWLEDGEMENTS

We gratefully acknowledge the financial support of British Telecom and the SERC through its JOERS initiative in the performance of this work. The work described has been drawn from the cumulative efforts of a number of people whose names are recorded in references 5, 8, 10, 11 and 14.

REFERENCES

1. J. Morley and D. Pugh in "Non-linear Optical Properties of Organic Molecules and Crystals", Editors, D.S. Chemla and J. Zyss, Academic Press, New York 1987, Volume 1 p193.
2. J.F. Nicoud and R.J. Twieg, ibid, P.227.

3. R.J. Twieg, A. Azema, K. Jain and Y.Y. Cheng, Chem. Phys. Letts., 1982, 92, 208.
4. A. Azema, R.J. Twieg and K. Jain in 'New Optical Materials', Editors S. Musikant and J. Du Puy, Proc.Soc. Photo-opt. Instrum. Eng. 1983, 456, 183 (1983).
5. R.T. Bailey, F.R. Cruickshank, S.M.G. Guthrie, B.J. McArdle, H. Morrison, D. Pugh, E.E.A. Shepherd, J.N. Sherwood, C.S. Yoon, R. Kashyap, B.K. Nayar and K.I. White, Optics.Comms., 1988, 65, 229.
6. P.A. Norman, D. Bloor, J.S. Obhi, S.A. Karaulov, M.B. Hursthouse, P.V. Kolinsky,, R.J. Jones and S.R. Hall, J.Opt.Soc.Am., 1987, B4 1013.
7. J.C. Baumert, R.J. Twieg, G. Bjorklund, J.A. Logan and C.W. Dirk, Appl. Phys.Letters., 1987, 51 1484.
8. B.J. McArdle and J.N. Sherwood, Chemistry and Industry, 1985, P268.
9. B.J. McArdle and J.N. Sherwood in 'Advanced Crystal Growth', Editors P.M. Dryburgh, B. Cockayne and K.G. Barraclough, Prentice Hall, New York, 1987 p179.
10. R.M. Hooper, B.J. McArdle, R.S. Narang and J.N. Sherwood, in 'Crystal Growth', Editor B. Pamplin, Pergamon, Oxford 1985, 2nd Edition 395.
11. H. Morrison, E.E.A. Shepherd, J.N. Sherwood and C.S. Yoon, in preparation.
12. E.E.A. Shepherd, G.S. Simpson, J.N. Sherwood and C.S. Yoon in preparation.
13. B.K. Tanner, 'X-ray Diffraction Topography', Pergamon, Oxford 1976.
14. R.T. Bailey, F.R. Cruickshank, S.G. Guthrie, G. McGillivray, D. Pugh, J.N. Sherwood, G.S. Simpson, in preparation.

Molecular Shape and Crystal Packing Modes for Organic Molecules: a Computational Approach

A. Gavezzotti

DIPARTIMENTO DI CHIMICA FISICA ED ELETTROCHIMICA E CENTRO CNR, UNIVERSITÀ
DI MILANO, MILANO, ITALY

1 PURPOSE

In a vivid account of the physical and chemical consequences of the existence of polar axes in organic crystals,
D.Y.Curtin and I.C.Paul, with perhaps a mild understatement,
express their opinion that "in spite of the discernible
tendency for certain types of compounds to prefer particular
symmetries, we seem to be some time away from being able
to control or even to predict with real assurance the
packing a compound will adopt when it crystallizes or the
relationship of the crystal symmetry to molecular structure"[1] . Some years later, not much has really changed;
not even the relatively modest task of predicting whether
a given molecule will form a non-centrosymmetric crystal
structure has been accomplished. A number of leading
concepts are however beginning to emerge out of an
aggregate of scattered observations and sparse theoretical
models, from both molecular and crystal studies. The purpose
of this communication is to illustrate these concepts,
with an emphasis on computational models, especially in
view of their application to classify and predict the
mechanical, optical, electrical and magnetic properties
of molecular crystals, to the (still very limited) extent
to which this is possible.

2 POTENTIALS

Looking for equilibrium in a chemical system means finding
the extreme of an appropriate thermodynamic   function.

In practice, the minimum is sought of potentials which
have been empirically fitted to experimental crystal properties, mostly sublimation enthalpies. The best fit is
usually obtained by employing functions of the type:

$$E_{ij} = A \exp(-BR_{ij}) - CR_{ij}^{-6} \tag{1}$$

where E is an intermolecular atom-atom energy, R is the
corresponding interatomic distance, and A, B, C are the
parameters of the fit. The total lattice energy is obtained
by summation over the entire network of intermolecular
contacts. A vast literature deals with details and relative
merits of various parameter schemes, whereby various
viewpoints are, sometimes hotly, debated[2-6].

One frequently overlooked, but crucial, property of
eq. (1) is that the potential is isotropic in space: hence,
the forces it is fitting must also be largely isotropic.
One striking example of the extent to which this is true
comes from a computational experiment on naphthalene.
Figure 1 shows the real crystal structure, an orthorhombic,
hypothetical structure[7], and a planar-sheet structure
which is in complete violation of all known principles
on the packing of aromatic molecules, but still can count
on 84% of the total lattice energy of the true structure.
Nevertheless, solids form in strict obedience to precise
directional requirements. Thus, small but effective forces
superimpose on the bulk of the isotropic potential, and
dictate the observed crystal packing; an appropriate name
for them would be "tug-boat" forces. These can conveniently
be identified as:
i) molecular shape: any perturbation from an isotropic
distribution of mass, volume or surface in the molecule
is ipso facto a directional pointer. Thus, even using
isotropic potentials, the observed crystal structure of
naphthalene is calculated to be the most stable one;
ii) formal charge distributions: these correspond to no
quantum mechanical observable, and must be obtained by
such schemes as population analysis or fitting, by site
charges, of calculated electrostatic potentials. The
perverse effects of changing the quality of the wavefunct
ion in population analysis are seen in Table 1. A safe
attitude is therefore to consider point charges as useful,
but conventional, parameters;

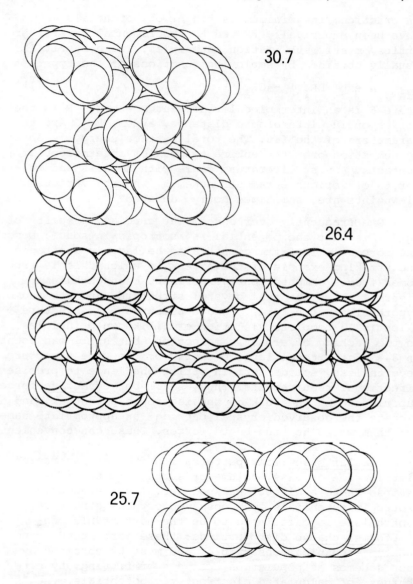

Figure 1. Packing diagrams for the true monoclinic structure of
naphthalene (top) and hypothetical orthorhombic (middle) or planar
sheet (bottom) structures, with calculated packing energies (kcal
mol^{-1}).

Table 1. Carbon atom charges from population analysis in different
MO approximations. From a compilation in S.Fliszar,'Charge
Distributions and Chemical Effects', Springer, New York,
1983, pp. 17-19. 10^{-3} electron units.

Molecule	INDO	PCILO	EHMO	STO-3G	7s3p/3s
Methane	+43	-43	-503	-72	-790
Ethane	+77	-10	-326	-26	-570
Propane C'	+67	-14	-343	-31	-580
C''	+94	+20	-164	+19	-380

iii) weak through-space orbital interactions (HOMO-LUMO
type interactions). Directionality results from symmetry
requirements on the interacting orbitals; strength increases
with orbital overlap, which, in the range of non-bonded
distances, is one order of magnitude smaller than for bonding distances. Also, effective mixing implies some electron
migration, so that these forces are more commonly invoked
in charge-transfer complexes (see Figure 2).
iv) hydrogen bonding: such a vast subject requires separate
treatment, and will not be considered here.

What makes a priori crystal structure prediction so
difficult is, that there is no unique way of disentangling
and of quantifying separately these forces. Their balance,
upon which crystal equilibrium rests, is thus not easy to
strike. Besides, it is very difficult to introduce a sound
geometrical basis for the model, unless appropriate key
structural parameters are identified.

3 STRATEGY

Crystal packing analysis and prediction may go by two
routes. The first involves the following steps: i) the
accurate optimization of force parameters, using generalized interaction formulas; ii) the exact calculation of
the packing energy and its derivatives; iii) a least-squares search for zero-gradient states, to obtain equilibrium
structures, and iv) calculation of the properties (e.g.,
vibrational spectra) which depend on the second derivatives

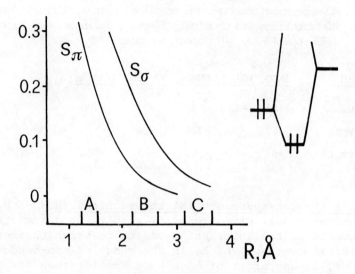

Figure 2. p-orbital overlap, S, at distance R; A, bonding region, B, intramolecular and C, intermolecular non-bonding regions. Slater-type orbitals, exponent 1.625 (as for carbon).

of the potential. The analytical power of this approach is large, but its predictive power is limited to a trial and error comparison of crystal energies for any given compound. Input to such calculations consists in fact of molecular coordinates and of a postulated crystal symmetry, requiring, mostly, previous diffraction experiments.

We advocate the construction of a crystal packing model dispensing with accurate calculation of potentials, representing packing forces, whatever they be, by simple molecular or crystal descriptors. This approach has broader scope as far as general principles of packing are concerned, since it is more readily adaptable to statistical work on a large body of crystal data, as is by now available in the Cambridge Crystallographic Data Files. This opens the way to the exploitation of the considerable potential of these Databases in the field of materials science.

A provocative example of how some properties do not

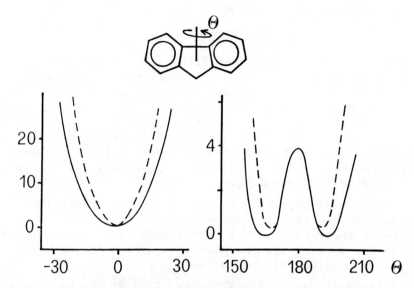

Figure 3. Packing energy profiles for in-plane molecular rotation in fluorene. Full lines, correct non-bonded potentials; broken lines, "random" parametrization (see text; energy scale to be multiplied by 10 in this last case). kcal mol^{-1} units.

depend on accurate potentials is given in Figure 3, showing that the qualitative features of the potential energy curve for molecular rotation in the fluorene crystal are reproduced even using randomly chosen numbers as potential energy parameters (the number of students at the University of Milano in various courses, and the average number of exams taken per student per year).

4 SIZE AND SHAPE

The first rough recognition of molecular shape goes through the structural formula of the compound. Naphthalene is more evenly shaped than biphenyl, and biphenyl more than diphenylmethane. An extended molecular geometry, useful in packing analysis, is built starting from intramolecular coordinates, to include molecular volume[8] and surface[9], and the moments of the distribution of mass, M_x, M_y, M_z, in decreasing order. One can then define:

$$F_c = (M_x + M_y)/2M_z \quad \text{cylindric index}$$
$$F_s = M_x/M_z \quad \text{spherical index,} \quad (2)$$

a molecule being cylindrical if $F_c > 2.5$, or spherical if $F_s < 1.5$ (although there are some subtleties in the use of these indices[10]). At the same time, one has to recognize some structure-defining crystal parameters, like the angle formed by the direction of molecular moments in neighbour molecules, or interplanar distances (stacking distances), which sometimes are reflected in short cell edge values; for example, a 4 Å cell edge is compellingly indicative of aromatic ring stacking.

The total packing energy is in itself a descriptor of packing, in that it depends on molecular size and shape, although these two factors cannot usually be deconvoluted; other descriptors are the packing coefficient, C_K, and the bulk modulus, K_o, evaluated as[11]:

$$K_o = 4.27 \times 10^{-7} \Delta H_s / V_m (31.63 D_c V_m / W_m - 14.2) \text{ dyn cm}^{-2} \quad (3)$$

where the sublimation enthalpy is approximated by the computed packing energy and D_c and W_m are density and molecular mass, respectively.

The total energy can be apportioned among atoms in the molecule, and the contribution of each atomic species to the packing energy is constant in the limit of the close-packing regime (homomeric principle[12]). Atoms which contribute less or more than the average share of their species represent breakdowns of the close-packing principle at each site, helping to visualize directional effects.

Two comments are appropriate at this point: i) one thing is the crystal structure, and another the space group in which it is described: small molecular tilts or asymmetries may change the space group, but not the main structural framework; ii) none of the analyses and energy dissections we propose requires sharply optimized potential functions. Any reasonable set of empirical parameters available in the literature would do for these purposes. The reverse, namely that any potential would do for any purpose, does not follow; for example, the vibrational analysis requires carefully chosen potential functions.

5 DIPOLES

The rule is simple in the case of permanent molecular dipoles, namely, these will tend to a head-to-tail arrangement to optimize coulombic interactions. In a typical example, calculations show that dipolar forces may split the degeneracy of size-equivalent molecular orientations in a molecule like 1,2,3-trichloro-trimethylbenzene[13]. This requirement usually produces a drive towards centrosymmetry in the crystal which is very difficult to overcome. It has been pointed out however[14] that molecular polarizability does not necessarily result in a permanent molecular dipole; then, the predictive value of shape arguments is enhanced, as, conceivably, shape effects may dominate again the orientation of polarizable molecules in the limit of vanishing permanent dipole, if the charge distribution is smeared over the entire molecule and hence is more or less isotropic. All this awaits for confirmation from a statistical analysis on available crystal structures.

6 STATISTICS

A database containing nearly all the available, well characterized, disorder-free hydrocarbon crystal structures has been extracted[10] from the 1986 release of the Cambridge data files. It contains 18249 atoms in 391 crystal structures. All the indices and descriptors mentioned in the foregoing sections of this paper, plus some derived ones, have been calculated. Averages, estimated standard deviations and skewness have been calculated for the overall distributions, as well as for distributions partitioned over classes of isostructural compounds. Least-squares analysis and factor analysis have been fruitfully applied. The following is a presentation of the main results as an illustration of the relative methodologies.

Center of symmetry

The center of symmetry in a crystal is a powerful shape effect optimizer, and therefore appears more frequently in shape-dominated crystals. In Table 2, one can see a noticeable and significant increase of the occurrence of centrosymmetric space group in hydrocarbon crystals. The effect of the molecular center of symmetry can be seen from the data in Figure 4: there, the bulk modulus is

used as an indicator of tight packing, and is always
higher for the more symmetric molecules.

Table 2. Space group frequencies (percent).

Space group	Hydrocarbons (ref. 10)	General organic compounds (ref.15)
$P2_1/c$	42.4	36.0
$P\bar{1}$	14.9	13.7
C2/c	10.3	6.6
Pbca	7.7	4.3
$P2_12_12_1$	4.6	11.6

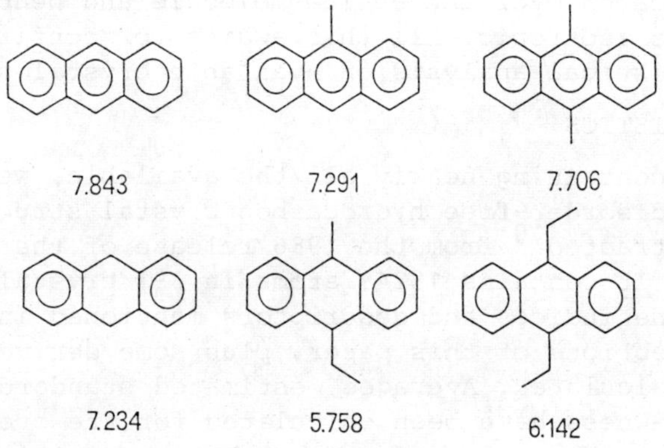

7.843   7.291   7.706

7.234   5.758   6.142

Figure 4. Packing efficiency, as judged by the value of the bulk
modulus, equation (3).

Homomeric principle

    The extent to which the principle of a constant
contribution to the packing energy for each atomic species,
irrespective of the crystal environment, is observed in
hydrocarbon crystals can be judged from Table 3. The standard deviation of the sample for each type of carbon atom
is rather small, but some spread remains. Those atoms
which are most exposed to intermolecular contacts provide
contributions higher than the average. This analysis
generalizes the concept of local packing density at atoms

Table 3. Average atomic contributions to the packing energy (esd's in parentheses; from ref. 10).

Atom	Number of entries	$\overline{E}$,kcal mol^{-1}
Hydrogen	8666	0.67(16)
Tertiary carbon	7202	1.60(40)
Quaternary carbon	2303	1.54(41)

in a molecule[16], providing a sound basis on which this important factor in crystal chemistry can be gaugeed.

Size, shape and packing

Figure 5 is an illustration of the effects of molecular structure on packing efficiency. Flat, evenly-shaped aromatic hydrocarbons pack much more efficiently than their geminal phenyl-substituted analogues. The difference is dramatic, and provides a striking illustration of the increasing difficulties in obeying the close-packing principle for flexible and bumpy molecules. For any class of compounds, the higher the slope of the energy-surface curve, the higher the packing efficiency. If trends in sublimation and melting enthalpies are similar, and if melting entropies are constant, a higher slope also means higher melting points.

Many bivariate correlations can be established among molecular and crystal descriptors. A more general treatment of the variance of the sample is provided by factor analysis, which shows that 13 indices of molecular and crystal structure can be incorporated in only 3 factors with only a 7% loss of information[10]. These factors can be roughly identified as size, packing efficiency and "packing underscore" (see ref. 10 for further details).

Orientation parameters

Statistics on molecular orientation can be best done when the molecular shapes are such that main molecular planes or preferential axes can be clearly identified. This is the case for flat condensed aromatics, or for cylindrical molecules. Since the packing energy correlates with molecular surface, it is possible to subdivide the

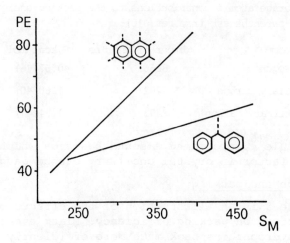

Figure 5. Packing energy, PE (kcal mol^{-1}) as a function of total molecular surface, S_m (Å^2) for flat aromatic compounds and for geminal polyphenyl substituted compounds (from ref. 10).

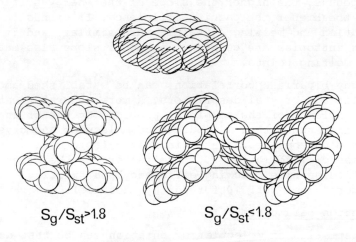

Figure 6. Rim surface, S_g (dashed) and core surface, S_{st}, for a flat molecule. Below, different packing modes according to their ratio: if it is smaller than 1.8, stacking of molecular planes occurs, eventually leading to complete layering (4 Å short cell axis).

total surface into a stack-promoting (core) part, S_{st}, and a glide-promoting (rim) part, S_g (see Figure 6). If $S_g/S_{st} > 1.6$, the formation of molecular stacks (4 Å inter planar distance) is impossible[17]. Besides, for cylindric molecules, $F_c > 2.5$, the packing must be such that the elongation axes of neighbour molecules are parallel[10]. Together, these two rules explain the herringbone packing mode (see Figure 1, top) of a very large class of molecular objects, whose prototype is anthracene, and there is evidence that they apply also to heteroatom-containing molecules of moderate polarity. This is at least one example of a clear link between molecular properties and crystal packing mode. Work along these lines, and using these coordinates, is in progress, and appears very promising in the field of organic materials science.

REFERENCES

1. D.Y.Curtin and I.C.Paul,Chem.Revs.,1981,81,525.
2. A.J.Pertsin and A.I.Kitaigorodski,'The Atom-Atom Potential Method',Springer-Verlag,Berlin,1987.
3. S.Ramdas and N.W.Thomas, in 'Organic Solid State Chemistry', G.R.Desiraju,Ed.,Elsevier,Amsterdam,1987,p.433.
4. D.E.Williams and D.J.Houpt,Acta Cryst.,1986,B42,286.
5. T.V.Timofeeva,N.Y.Chernikova and P.M.Zorkii,Russ.Chem.Revs., 1980,6,49.
6. K.V.Mirksy,Acta Cryst.,1976,A32,199.
7. J.Bernstein,A.Gavezzotti and J.A.R.P.Sarma, work in progress.
8. A.Gavezzotti,J.Amer.Chem.Soc.,1983,105,5220.
9. A.Gavezzotti,J.Amer.Chem.Soc.,1985,107,962.
10. A.Gavezzotti, submitted for publication.
11. A.Bondi, 'Physical Properties of Molecular Crystals, Liquids and Glasses',Wiley and Sons,New York,1968,chapters 4 and 5.
12. A.Gavezzotti,Nouveau J.Chim.,1982,6,443.
13. A.Gavezzotti and M.Simonetta,Acta Cryst.,1976,A32,997.
14. J.Zyss,D.S.Chemla and J.F.Nicoud,J.Chem.Phys.,1981,74,4800.
15. A.D.Mighell,V.L.Himes and J.R.Rodgers,Acta Cryst.,1983,A39,737.
16. N.W.Thomas,S.Ramdas and J.M.Thomas,Proc.Roy.Soc.London, 1985,A400,219.
17. A.Gavezzotti and G.R.Desiraju,Acta Cryst., in the press.

Characterization of Non-linear Optical Properties

Development, Comparison, and Limitations of Non-linear Optical Characterization Methods

G.R. Meredith*, H. Hsiung, S.H. Stevenson, H. Vanherzeele, and F.C. Zumsteg

CENTRAL RESEARCH AND DEVELOPMENT DEPARTMENT, E. I. DU PONT DE NEMOURS AND CO., EXPERIMENTAL STATION-356, WILMINGTON, DE 19898, USA

1 INTRODUCTION

Nonlinear optics (NLO) has proven to be a richly diverse field due not only to the variety of optical phenomena which can occur on breakdown of the linearity approximation, but also due to the fascinating ways in which materials can be probed and due to the interesting ways in which various materials become important. Appreciation of details which cause specific materials to be well suited for certain NLO effects lies in 1) understanding the types of responses which accentuate the relevant nonlinear polarization P^{NLS} or 2) knowing properties which can be exploited to construct a high finesse structure for which the impact per unit P^{NLS} is maximized. In generating this appreciation it is conceptually convenient to think of NLO to be two separate areas: optics and wave propagation, and materials. Unfortunately, these cannot really be separate and distinct areas. Nonlinear optical physicists require knowledge of materials and their processing in order to avoid describing unrealistic situations and to allow design and construction of feasible devices. Materials workers must have a knowledge of optics to guide them through research choices, to insure recognition of the significance of new materials, and to assure that characterization methods are developed, used and interpreted in intelligent and valid ways. As will be shown below, NLO provides both 1) nonlinear extensions of well known linear optical processes, elements and devices and 2) optical phenomena which are fundamentally new and different. Characterization methods can therefore be based on nonlinear adaptations of linear situations or based on these totally new processes. The range of choices is extremely broad. Our goal is to present a tutorial on characterization methods, plus some examples of our own implementations.

Polarization Density Expansion

In the electric dipole approximation deviations from linearity in optics is phenomenologically expressed in the constitutive relationships

$$\mathbf{P} = \mathbf{P}_0 + \mathbf{P}^{(1)} + \mathbf{P}^{(2)} + \mathbf{P}^{(3)} + \quad \cdot \quad \ldots$$

$$= \mathbf{P}_0 + \chi^{(1)} \cdot \mathbf{E} + \chi^{(2)} \cdot\cdot \mathbf{EE} + \chi^{(3)} \cdot\cdot\cdot \mathbf{EEE} + \quad \ldots \qquad (1)$$

As chemists we expect, based on success with linear dielectric behavior, that the susceptibility tensors can be understood in terms of molecular analogs

$$p = \mu + \alpha \cdot \mathbf{F} + \beta \cdot\cdot \mathbf{FF} + \gamma \cdot\cdot\cdot \mathbf{FFF} + \quad \ldots \qquad (2)$$

Besides quantifying $\chi^{(n)}$ our goals include pursuit of techniques to probe the nature and structure-property relations of polarizability in molecules.

General principles. Consequences of nonlinearity are twofold. First, a material will begin to oscillate coherently at harmonic, combination and difference tones. Consequences depend on the spatial distributions, being significant if oscillations match with optical modes. Second, intensity- or dc field-dependent terms arise in the wave equations. $\mathbf{P}^{NLS}$ will thus modify both amplitudes and retardation of E waves. The problem must be reviewed to ensure that all coherence and interference effects are considered. Many simplifications make problems tractable, but it is wise to ask if any physics is missing. Examples of such narrowed vision appear in the literature. For example, to focus only on $\mathbf{P}$ which is induced by $\chi^{(1)}$ and $\chi^{(3)}$ would be a mistake if $\chi^{(2)} \neq 0$; cascading in NLO can be significant.[1-3] Also, third-order experiments, though similar to second-order, are more complex.[4]

Intensity Dependent Refraction

A second picture in NLO has become important as emphasis on "all-optical" processes has increased. Intensity dependence of the optical indicatrix is the central issue. Largest effects are seen when a material is substantially altered by optically induced perturbations having characteristic times very long compared to optical oscillation times (e.g. excited state population, heating, redistribution of molecular orientations, electrostriction, etc.). In such "evolutionary" processes perturbations from some original condition often evolve on time scales comparable to or longer than optical-pulse rise and fall times. Thus the $\mathbf{P}$-expansion is inappropriate for two reasons: 1) expression of $\mathbf{P}$ in modified media requires higher-order contributions (often-quoted $\chi^{(3)}$ and the related n_2 describe behavior in the low-intensity limit), and 2) $\chi^{(n)}$ (which are functions of frequencies, being Fourier transforms of temporal response) are appropriate to steady-state or quasi-cw situations, where material response times are much shorter than optical-pulse envelope times. Study of evolutionary NLO requires specialized knowledge about a specific systems' dynamics. For example, the Kramers-Kronig transformation relates refractive index $n(\omega)$ and absorption spectrum $\alpha(\omega)$, thus transient absorption spectroscopy and nonlinear refractive index changes are interrelated. However, in a phenomenological manner the nonlinear responses can be characterized by the methods categorized below.

2 CATEGORIZATION OF OPTICAL EFFECTS

In Table 1 various induced **P** are listed. We adopt three classifications: Types A, B and C. In Type A **P** and one of the inducing **E** components have equal frequency, and for analogous reason are spatially distributed similarly. Consequently propagation of that wave is affected, as seen in the generalized optical dielectric function

$$\varepsilon(\omega_r) = 1 + 4\pi \{ \chi^{(1)} + 2\chi^{(2)} \cdot E_0 + 3\chi^{(3)} \cdot\cdot E_0 E_0$$
$$+ 3\chi^{(3)} \cdot\cdot E_r E_r{}^* + 6\chi^{(3)} \cdot\cdot\cdot E_s E_s{}^* + ... \} . \quad (3)$$

Propagation is thus dependent on dc fields and light intensities. In scalarized version the change in n is

$$\delta n \approx \{ (E_r \cdot [\varepsilon - 4\pi \chi^{(1)}] \cdot E_r / |E_r|^2) - 1 \} / 2n . \quad (4)$$

Table 1 Induced Polarizations and Related Optical Processes. The field convention is $E = \Sigma_r [E_r \exp\{i\omega_r t\} + E_{-r} \exp\{-i\omega_r t\}]$; $E_{-r} = E_r{}^*$. The table key is **A**: Nonlinear variations of refraction (n, Snell's law), absorption ($\kappa = \text{Im}\{\varepsilon\}/2n$), reflection, scattering, interference, diffraction, polarization ellipse evolution, focussing, waveguiding, cavity modes, etc. **B**: Frequency conversion, propagating phase mismatch, and coherent interferences. **C**: Rectification.

E Fields	Induced Polarization		Optical Consequences or Enabled Processes
	First-Order		
E_r	P_r	$= \chi^{(1)} \cdot E_r \quad \exp\{i\,\omega_r t\}$	A
	Second-Order		
E_r	P_{rr}	$= \chi^{(2)} \cdot\cdot E_r E_r \quad \exp\{i\,2\omega_r t\}$	B
E_r	P_{r-r}	$= 2\,\chi^{(2)} \cdot\cdot E_r E_r{}^* $	C
E_r, E_s	P_{rs}	$= 2\,\chi^{(2)} \cdot\cdot E_r E_s \quad \exp\{i\,(\omega_r + \omega_s) t\}$	B
E_r, E_s	P_{r-s}	$= 2\,\chi^{(2)} \cdot\cdot E_r E_s{}^* \quad \exp\{i\,(\omega_r - \omega_s) t\}$	B
E_r, E_0	P_{rs}	$= 2\,\chi^{(2)} \cdot\cdot E_0 E_r \quad \exp\{i\,\omega_r t\}$	A
	Third-Order		
E_r	P_{rrr}	$= \chi^{(3)} \cdot\cdot\cdot E_r E_r E_r \quad \exp\{i\,3\omega_r t\}$	B
E_r	P_{r-rr}	$= 3\,\chi^{(3)} \cdot\cdot\cdot E_r E_r{}^* E_r \quad \exp\{i\,\omega_r t\}$	A
E_r, E_s	P_{rrs}	$= 3\,\chi^{(3)} \cdot\cdot\cdot E_r E_r E_s \quad \exp\{i\,(2\omega_r + \omega_s) t\}$	B
E_r, E_s	P_{s-sr}	$= 6\,\chi^{(3)} \cdot\cdot\cdot E_s E_s{}^* E_r \quad \exp\{i\,\omega_r t\}$	A + gain
E_r, E_s, E_u	P_{rsu}	$= 6\,\chi^{(3)} \cdot\cdot\cdot E_r E_s E_u \quad \exp\{i\,(\omega_r + \omega_s + \omega_u) t\}$	B
E_r, E_0	P_{rr0}	$= 3\,\chi^{(3)} \cdot\cdot\cdot E_r E_r E_0 \quad \exp\{i\,2\omega_r t\}$	B
E_r, E_0	P_{r00}	$= 3\,\chi^{(3)} \cdot\cdot\cdot E_r E_0 E_0 \quad \exp\{i\,\omega_r t\}$	A

In the full tensor form,   knowing that with this field convention

$$I_r = (nc/2\pi) |E_r|^2 , \tag{5}$$

electro-optic r and s or B tensors and n_2 and n_2' are obvious.   Recall

$$n = n_0 + n_2 I_r + n_2' I_s . \tag{6}$$

With general tensors, the dielectric vibration directions are redetermined by mixed index terms.   When χ's are complex, phase of the polarization oscillation is altered.   The component oscillating in-phase with the E-wave determines phase velocity, while the out-of-phase component causes gain or loss.   Thus multiphoton absorption and gain are described naturally.

Analysis of Type A processes can be straightforward adaptation of the linear optical treatments.   Be they modifications of propagation through birefingent media, retardation in interferometry, spatial modulation of n and α in diffraction, propagation constants in guided waves, refraction across interfaces, focussing, etc., descriptions are available in standard texts on optics, where ε is taken as a constant.   Modification for variable ε is straightforward.   Use of these effects in characterization requires familiarity with linear optics and insightfulness in selecting processes where the induced changes are easily and reliably seen and quanitified.

3 TYPE B NONLINEAR OPTICAL PROCESSES

Type B polarization causes novel effects because new frequencies may appear and the spatial dependence of $\mathbf{P}^{NLS}$ does not necessarily coincide with that of propagating light.   Generic effects are seen in the situation of an isolated interface between linear and nonlinear media.[5]   Assuming normal plane waves ($E = \mathcal{E} e^{ikz}$, etc) transmitted and reflected waves due to nonvanishing $\mathbf{P}^{NLS}$ are

$$E(z) = E_b + E_f = (1 - A e^{i\Delta kz}) E_b \tag{7}$$

$$\mathcal{E}_f = - \mathcal{E}_b (\varepsilon_b^{1/2} + \varepsilon_f^{1/2}) / (\varepsilon_f^{1/2} + \varepsilon_f^{1/2}) = - A \mathcal{E}_b \tag{8}$$

$$\mathcal{E}_r = (1 - A) \mathcal{E}_b = - 4\pi [(\varepsilon_f^{1/2} + \varepsilon_f^{1/2})(\varepsilon_b^{1/2} + \varepsilon_f^{1/2})]^{-1} \mathcal{P}^{NLS}$$
$$\sim - \pi \mathcal{P}^{NLS}/\varepsilon \tag{9}$$

$$\mathcal{E}_b = 4\pi \mathcal{P}^{NLS} / (\varepsilon_b - \varepsilon_f) = - 8\pi \mathcal{P}^{NLS} \mathcal{L}_c / \lambda (\varepsilon_b^{1/2} + \varepsilon_f^{1/2}) \tag{10}$$

$$\Delta k = k_f - k_b = \pi / \mathcal{L}_c = \pi (\ell_c^{-1} + i \ell_a^{-1}) ;$$
$$\ell_c = \lambda / 2 (n_f - n_b) , \qquad \ell_a = 2\pi / (\alpha_f - \alpha_b) \tag{11}$$

Interference between "bound" and "free" waves in Eq. 7 results in amplitude oscillations in the transmitted E-wave as function of z. Corresponding intensity fringes can be seen by rotating plane parallels or translating wedges of nonlinear material across the optical beam. The reflected wave is small, but less critically dependent on dispersion of ε. When attenuation of p^{NLS} or light at this frequency occurs, fringe contrast is diminished. Clearly if E_b is totally attentuated only E_f is observed at constant intensity and vice versa. Scaling of E(z) with dispersion and attentuation is described by $\mathscr{L}_c$. Interpretation of E(z), and thus the underlying $\chi^{(n)}$, requires determining this "coherence length" and establishing the exact pattern of fringing. Very high accuracy can be achieved in some cases with good quality samples. On the other hand, reflection is attractive for less good samples, when absorption negates fringe observation and when $\mathscr{L}_c$ falls near or below λ.

When multiple interfaces exist the treatment can be made recursive,[6] but care must be taken to consider whether reflection interferences (both of the nonlinear and linear variety) as well as forward travelling interferences are significant. When non-plane-wave situations arise, as with focussed laser beams, we have found it convenient to use vibration diagrams to predict the behavior of Type B processes.[1,4,7] A vibration diagram is a two-dimensional integration device which accounts for variable $|p^{NLS}|$ and phase accumulation along an optical path. Portions from different optical situations may be spliced together according to simple rules.

4 MICROSCOPIC-MACROSCOPIC RELATIONSHIPS

To set a framework for understanding NLO response in molecular ensembles, a straightforward derivation of $\chi^{(n)}$ is summarized.[8] The key point is that microscopic fields which polarize molecules κ are the sums of the Maxwellian electric field plus contributions from dipole moments on all other molecules in the ensemble,

$$E_m{}^\kappa = E + \Sigma_\lambda L^{\kappa\lambda} \cdot p^\lambda \quad . \tag{12}$$

With some algebra one obtains a "local field tensor" N^κ such that

$$\Sigma_\kappa p^\kappa = \Sigma_\kappa \{ (N^\kappa)^T \cdot \mu^\kappa + \alpha^\kappa \cdot N^\kappa \cdot E$$
$$+ (N^\kappa)^T \cdot [\beta^\kappa \cdots E_m{}^\kappa E_m{}^\kappa + \dots]\} \quad . \tag{13}$$

A "dressed dipole transfer tensor" $Q^{\kappa\mu}$ can also be identified. Two relationships demonstrate these functionalities,

$$E_m^\kappa = N^\kappa \cdot E + \Sigma_\lambda Q^{\kappa\lambda} \cdot (\mu^\lambda + p^{(nl)\lambda}) \quad , \tag{14}$$

$$(N^\kappa)^T \cdot \mu^\kappa = \Sigma_\lambda (\delta_{\kappa\lambda} U - \alpha^\lambda \cdot Q^{\lambda\kappa}) \cdot \mu^\kappa \quad . \tag{15}$$

$$D^\kappa = \Sigma_\lambda Q^{\kappa\lambda} \cdot \mu^\lambda \quad , \tag{16}$$

another naturally occuring tensor, is seen to represent induced field at molecule κ due to permanent dipoles in the ensemble.

Frozen Ensemble Nonlinear Susceptibilities

By collecting terms in Eq. 13 by powers in E one identifies

$$P_0 = V^{-1} \Sigma_\kappa (N^\kappa)^T \cdot \{ \mu^\kappa + \beta^\kappa \cdot\cdot D^\kappa D^\kappa + \gamma^\kappa \cdots D^\kappa D^\kappa D^\kappa$$
$$+ 2\beta^\kappa \cdot\cdot D^\kappa (\Sigma_\lambda Q^{\kappa\lambda} \cdot \beta^\lambda \cdot\cdot D^\lambda D^\lambda) + \ldots \} \tag{17}$$

$$\chi^{(1)} = V^{-1} \Sigma_\kappa \{ (\alpha^\kappa)^T \cdot N^\kappa$$
$$+ (N^\kappa)^T \cdot [2\beta^\kappa \cdot\cdot D^\kappa N^\kappa + 3\gamma^\kappa \cdots D^\kappa D^\kappa N^\kappa$$
$$+ 2\beta^\kappa \cdot\cdot (\Sigma_\lambda Q^{\kappa\lambda} \cdot \beta^\lambda \cdot\cdot D^\lambda D^\lambda) N^\kappa + \ldots] \} \tag{18}$$

$$\chi^{(2)} = V^{-1} \Sigma_\kappa (N^\kappa)^T \cdot \{ \beta^\kappa \cdot\cdot N^\kappa N^\kappa + 3\gamma^\kappa \cdots D^\kappa N^\kappa N^\kappa$$
$$+ 2\beta^\kappa \cdot\cdot D^\kappa (\Sigma_\lambda Q^{\kappa\lambda} \cdot \beta^\lambda \cdot\cdot N^\lambda N^\lambda)$$
$$+ 4\beta^\kappa \cdot\cdot N^\kappa (\Sigma_\lambda Q^{\kappa\lambda} \cdot \beta^\lambda \cdot\cdot D^\lambda N^\lambda)$$
$$+ 6\delta^\kappa \cdots D^\kappa D^\kappa N^\kappa N^\kappa + \ldots \} \tag{19}$$

$$\chi^{(3)} = V^{-1} \Sigma_\kappa (N^\kappa)^T \cdot \{ \gamma^\kappa \cdots N^\kappa N^\kappa N^\kappa$$
$$+ 2\beta^{\kappa} \cdot\cdot N^\kappa (\Sigma_\lambda Q^{\kappa\lambda} \cdot \beta^\lambda \cdot\cdot N^\lambda N^\lambda)$$
$$+ 4\delta^\kappa \cdots D^\kappa N^\kappa N^\kappa N^\kappa + \ldots \} \tag{20}$$

Complexities. Terms arise due to the influence of 1) E and N^κ, 2) μ^λ, and 3) $p^{(nl)\lambda}$ in determining $E_m^{\ \kappa}$. If only the first mechanism were active, there would be a one-to-one relationship by order of nonlinearity between polarizabilities and susceptibilities. However, the other two sources of $E_m^{\ \kappa}$ respectively "pull down" higher-order molecular nonlinearities into lower-order susceptibilities and "cascade" lower-order molecular nonlinearities into higher-order susceptibilities. These two mechanisms are almost never included in descriptions of molecular behavior, only the first terms of Eqs. 8-11 usually being assumed. Considering the discovery that highly nonlinear molecular species exist, one can see the necessity of their inclusion, since mutual polarization phenomena are so important in condensed-phase media.

Limitations. These expressions are not very useful in their infinite summation form. Given the complexity of their calculation, it is realistic to acknowledge the form and adopt ensemble averaging processes, perhaps using them in a semiempirical manner. Several points need to be made. 1) Linear polarizability enhancement is as large as or larger than the hypothetical sum of nonlinear polarization assuming no mutual polarization. Thus uncertainty in this effect severely limits our ability to accurately and realistically a) predict macroscopic properties and b) extract molecular properties from macroscopic susceptibilities. 2) In crystals, since only one or a few crystallographically unique molecular sites occur, the averaging is trivial. However, the formalism requires that N^K be determined at sites, not at the unit cell level. It cannot be factored from inside the summation. Thus the formalism often used to discuss effects of molecular orientation[9] is not correct, though probably supportable. 3) In fluids there is a distribution of fluctuating environments. The process of averaging over environments and orientations requires giant leaps to generate the often used expressions of e.g. $\chi^{(3)}$ in which all five factors of the first term in Eq. 20 appear as independently averaged quantities. Are correlations and correlated fluctuations insignificant in these processes?

Oriented Gas Model Generalization

Eqs. 17-20 are useful only if no zero-frequency perturbations of the ensemble occur. Should this not be the case, "ensemble distribution perturbations" of the zero-perturbation expressions can be added by a straightforward Taylor expansion,

$$P_0 = \{ P_0 \}_{E=0} \tag{21}$$

$$\chi^{(1)} = \{ \chi^{(1)} + \partial P_0/\partial E \}_{E=0} \tag{22}$$

$$\chi^{(2)} = \{ \chi^{(2)} + \partial \chi^{(1)}/\partial E + \partial^2 P_0/\partial E^2 \}_{E=0} \tag{23}$$

$$\chi^{(3)} = \{ \chi^{(3)} + \partial \chi^{(2)}/\partial E + \partial^2 \chi^{(1)}/\partial E^2 + \partial^3 P_0/\partial E^3 \}_{E=0} . \tag{24}$$

Basic physical mechanisms can be seen within this picture. Strong dispersion is thus expected in NLO responses. This is also true of intramolecular responses.

5 INTERFERENCES, COHERENCE, CASCADING, DISPERSION, AND ANISOTROPY

Interferences. Interferences are a general effect in NLO. Within molecules different motions interfere to determine total polarizability.[10] Within ensembles different degrees of freedom interfere, as do contributions from different molecules, to determine total $\chi^{(n)}$.[11] In propagation, fields generated along the optical path interfere to determine the total E, both amplitude and phase, to be measured.[12] Thus it is possible to determine relative magnitudes and complex phases of components of hyperpolarizability and $\chi^{(n)}$ with these interferences. They can be observed by NLO spectroscopies, variation of solution composition, and varying details of interference between fields arising in different media.

Coherence. Coherence is important in NLO. Real light sources have noise characteristics related to departure from idealized space-time sinusoidal functions. Consequently, amplification of "bright" periods in *nonlinear* responses will overcompensate "dim" periods. Thus, SHG is twice as efficient with "incoherent" light.[5] Absolute measurements are thus always suspect. In Type B nonphase-matched processes one also worries whether, over a sample length, the various retardations may cause interfering parts to be shifted from different "brightness" regions, thus diminishing accuracy even of relative measurements. Finally, use of interferences to set up intensity, and thus n and α, gratings is error prone if the response time of the medium is long enough that it cannot track the changing grating pattern over the course of an optical pulse.

Cascading. Uncertainty in details of cascading in the molecular description of $\chi^{(n)}$, as outlined above, is a major limitation to determination of molecular properties from condensed-phase $\chi^{(n)}$. To date only the impact in THG has been observed,[13] though it is probably also important in EFISH (see below). Further, characterization of higher-order $\chi^{(n)}$ is complicated by interfering sequential lower-order NLO processes in macroscopic fields. This subject is straightforward, but detailed,[3] and should be considered whenever cascading routes are possible in an experiment. Fortunately, in contrast to microscopic cascading, macroscopic symmetry can sometimes prevent its occurrence. With definitions of $\chi^{(n)}$ given above, and their trivialization to the usually quoted versions, there is an over-counting problem for molecular polarization in macroscopic cascading. A brief description and a crude method to handle it has been outlined earlier.[1]

Dispersion and anisotropy. Dispersion and anisotropy of nonlinear polarizability are unavoidable complications. Contributions to polarizability by various motions fall-off only linearly at large detunings from their resonance frequencies. Since they are induced by forces at the optical and sum and difference frequencies, it is crucial to specify experimental conditions and consider dispersion. Anisotropy at the macroscopic level makes the nonlinear wave equation cumbersome since non-transverse polarization can result. Also, since experiments usually only monitor a few of the $3^{(n+1)}$ possible $\chi^{(n)}$ tensor elements and since contributions from the hyperpolarizability tensors are highly averaged, it is not possible to empirically obtain a very complete picture of molecular polarizability. Thus comparisons between results for different molecules is not necessarily simple. High quality molecular models and calculations, tested against empirical data, are valued windows into microscopic polarization behavior.

6 SOLUTION TECHNIQUES FOR MOLECULAR CHARACTERIZATION

One of our aims is to generate a database which will aid in creating physical generalities on which to base broad, intuitive pictures of structure-NLO properties.[14-16] To achieve this a methodology is required which in a simple, rapid, accurate and reproducible fashion can characterize $\chi^{(1)}$ and $\chi^{(3)}$ of solutions.[4,17] The former are obtained from capacitance, n, and SHG and THG (second- and third-harmonic generation) coherence-length measurements. The latter are determined by THG and EFISH (dc-electric-field-induced SHG). The apparatus used is described in detail elsewhere.[18] What follows is a summary of design and analysis considerations.

Wavelength and dispersion. A valid comparison of properties among a series of molecules which differ slightly in structure may be made only if the frequencies occur in the weakly dispersive region above vibrational fundamentals but well below electronic transitions. Since most organic molecules have substantial near-ultraviolet absorption, and since as will be described, determination of γ is essential to proper interpretation of EFISH results, for this type of study the fundamental wavelength must be significantly greater than 1 µm. The 1.908 µm Stokes shifted output from a Nd:YAG-pumped hydrogen-filled Raman cell is used. This is satisfactory for most compounds, but has the disadvantage that the EFISH signal is effectively weakened by the inefficiency of most photodetectors at 954 nm.

Cell design. In designing Type B $\chi^{(3)}$ measurements such as THG and EFISH, a major problem is interference of fields from all media in the path. Fields arise fairly independently of density (Eq. 10) thus not being limited to condensed-phases. If the dimension of the active region is several coherence lengths, contributions to the NLO signal effectively arise from each interface between materials. Complexities are also introduced by the phase behavior of focused optical beams. The key then is to design an apparatus which effectively limits nonlinear activity to the near field, and to include a minimum number of material boundaries in that region. A standard configuration for liquids is a wedge-shaped glass cell which is translated across the beam. For THG the number of interfaces is reduced to two by a large window thickness to confocal parameter ratio. For EFISH the windows need only extend past the region of applied electric field. But needing uniform, well-defined field strength at each inner interface dictates the extension of electrodes past the liquid region. If there are effectively two interfaces in the beam path, the resultant harmonic intensity displays the simple wedge fringing of Type B processes (Eq. 7), but only precisely if the confocal parameter is significantly longer than the liquid length and the focus is in the exact center (front-to-back), and if the liquid is perfectly transparent at fundamental and harmonic. Correction for absorption results in complex expressions which can, if required, be fit with aid of a computer. Accurate coherence lengths may still be obtained under imperfect transparency and focusing conditions, but relatively large uncertainties may be introduced into the extracted bound wave magnitudes.

Calibration.   For high-precision, it is important to incorporate a
reference standard .   Observation conditions should be as similar as possible.
Our wedge-shaped cells therefore are divided into two liquid chambers, one
containing a liquid of known properties and the other containing a sample.
The signal amplitude ratio is a function of $\mathscr{E}_b$ in liquid, reference and glass,

$$S = I_L / I_R = \{ [a_L \mathscr{E}_b{}^G - b_L \mathscr{E}_b{}^L] / [a_R \mathscr{E}_b{}^G - b_R \mathscr{E}_b{}^R] \}^2 . \quad (25)$$

a and b are products of Fresnel-like terms. $\mathscr{E}_b$, ℓ_c and n(3ω) yield $\chi^{(3)}$.

Simplified approach.   THG and EFISH measurements are simplified by a
single-interface configuration.   The beam is tightly focussed through a
thick front window onto the glass-liquid interface, defocussing through a
long liquid region.   Our cell is again two-chambered, each being stirred to
reduce heating effects.   Parallel gold electrodes are deposited on the inner
face of the front window to create a field at and near the interface.   This
window is sealed to the stainless steel cell body by a mask fabricated from an
insulating elastomer which is impervious to solvents and unfilled to avoid
leaching.   Under the adiabatic approximation described above, S obeys Eq.
25.   Coherence lengths are measured separately in a single-chambered
wedged cell.   Crystalline quartz windows provide large SHG intensity to
improve EFISH ℓ_c determinations.   Sum frequency THG in the quartz
windows does not substantially influence the appearance of THG fringes.

There are three major advantages to using these cells.   Foremost,
absolute location of the focal point is not critical to amplitude measurements
as long as the beam is normal to and is focused *near* the glass-liquid
interface.   Since precise determination of ℓ_c can also be accomplished under
noncritical focusing conditions, much experimental difficulty is eliminated.
Measurements are faster and inherently more precise than the
two-interface technique.   Weak absorption of 1.908 μm light, common to
organic compounds, no longer contributes to analysis uncertainty.   Error
analysis shows, for example, that precision for THG $\chi^{(3)}$ is better than 2%.
Second, precision in EFISH ℓ_c is increased, since the large quartz SHG
virtually eliminates noise in measurements of what would otherwise often
be extremely weak signals.   Decoupling amplitude determination also allows
superior data averaging.   Finally, use of our single-interface cell allows
simultaneous, and rapid, EFISH and THG (SET) measurement.[15,16]

Molecular properties.   Field factors correcting for average local
environment are computed with Lorenz-Lorentz and Onsager models and
solute polarizabilities extracted from the concentration-dependence of $\chi^{(3)}$.
SET and capacitance measurements on several nonpolar compounds provide
crude scaling factors for estimating molecular deformation contributions in
dielectric and EFISH susceptibilities, α and $\gamma_{e,v}$, from nonresonant values
for α_e (from n dispersion) and γ_e (from THG).   Short of more complete

spectroscopic work, these values are the best available. Derived values of μ are uncertain by a few percent even in highly polar compounds if α is not known. Similarly $\gamma_{e,v}$ can contribute several percent of the total EFISH $\chi^{(3)}$ even for compounds with a high value of $\mu\beta$. One should, for example, consider that in molecules where $\mu\beta$ is enhanced by intramolecular charge transfer, γ is generally also enhanced. If a value for β is required which is accurate to better than 10-15%, contributions from α and γ cannot be ignored.

7 CRYSTAL CHARACTERIZATION TECHNIQUES

To find potential new nonlinear crystals we employ a powder screening method. It has some semiquantitative utility, but is best used as a prescreen for single crystallite characterizations as developed by Velsko.[19] The complexity of nonlinear frequency conversion in biaxial crystals necessitates a study at this level of sophistication. Very accurate $\chi^{(2)}$ characterizations can be performed via the nonlinear fringe techniques if sufficient quality samples can be obtained. The utility of new crystals is determined by details of the linear optical properties. The Velsko method provides a direct probe of these features. Electro-optic (EO) r tensors are determined with standard analysis using the technique described below.

8 CHARACTERIZATION OF ORIENTATIONAL ELECTRETS

The idea to generate second-order NLO media by electric-field induced orientation of molecular units exhibiting large molecular hyperpolarizability in polymeric media was first proposed and demonstrated by Meredith, et al.[20] In their demonstration of the concept liquid-crystalline cooperativity was also utilized to enhance the alignment at low poling field strengths. A statistical mechanical model was proposed with which to describe the behavior and the anisotropy of the resultant $\chi^{(2)}$ was thereby derived. Poling dynamics, dielectric relaxation and alignment relaxation, variation of alignment with temperature (and the polymer microstructure), and attempts to improve polymer properties by structural and compositional variations were studied. We, and others, continue to invent similar new materials and study their behavior. SHG is the primary probe of orientation with attention given to poling, current-flow and temperature histories in order to better understand orientation processes. EO measurements are made using a modified Mach-Zehnder interferometer similar to that described by Sigelle and Nierle.[21] An electroded sample and a calibrated EO phase modulator are placed in separate arms. Optical output is adjusted to obtain a fringe pattern which is sampled through a slit. Amplified sinusoidal voltage waves from a dual-channel waveform synthesizer cause EO modulation of optical retardation by sample and reference. A null in the phase-synchronous interferometer output, determined with high sensitivity by a lock-in amplifier, is obtained by balancing this modulation. The nulling fields and thicknesses determine r coefficients. Quadratic EO information is obtained with voltage offsets and synchronous harmonic detection. The technique is particularly appealing since it is insensitive to optical losses and drift in the interferometer and to laser noise, and has a high sensitivity appropriate to thin film work.

9 CHARACTERIZATION OF MOLECULAR THIN FILMS

In the Langmuir-Blodgett (L-B) method organic thin films of thickness from one to hundreds of molecular layers are fabricated. Each molecular layer is first prepared at an air-water interface in a "Langmuir trough" and transferred onto a solid substrate layer by layer. Due to its controllability, such a technique is potentially useful for making organic thin films exhibiting good linear and nonlinear optical properties. We use two relatively simple optical techniques, SHG and ellipsometry, to characterize monolayers at air-water interfaces as well as multilayers transferred onto substrates. Both optical systems are installed next to a trough in a clean-room environment so that *in situ* characterizations can be made.

SHG. Our SHG system employs a simultaneously mode-locked (100 MHz) and Q-switched (500Hz) Nd:YLF laser operated at $\lambda = 1.053$ μm, with each Q-switched envelope consisting of about 30 mode-locked laser pulses. The indiviudal laser pulses have a duration ~ 60 ps and an average pulse energy of ~ 90 μJ at the sample. The laser beam, having a Gaussian spatial profile, is focused by a lens to an e^{-2} diameter of ~300 μm with an incident angle $\theta = 70°$ onto the surface. Glass filters, monochromator, polarizers and polarization rotator allow careful checks of wavelength and polarization. A quartz plate, mounted on a rail, is translated along the reflected beam path for determination of phase via interference. An optical reference channel monitors SHG from a quartz plate. Detector signals are fed to integrators gated open during each Q-switched pulse train. Processing on a personal computer with custom software provides both analog and photon-counting outputs. Using a room-temperature, 9-stage photomultiplier tube, we have achieved a noise level corresponding to about 0.05 2ω-photons generated at the sample surface per pulse train. To calibrate the absolute values of $\chi^{(2)}$ and its phase, the sample surface is replaced by a clean quartz surface.

Consider an L-B film to be a uniform dielectric sheet characterized by ε_m and $\chi^{(2)}$ and exhibiting rotational symmetry about a polar z-axis. For thickness $d << \lambda$, amplitudes of the s- and p-polarization components of the reflected harmonic are

$$E_s(2\omega) = \kappa \exp\{i \; \alpha_s \; d\} \; \sin\{2\phi\} \; L_{yyz} \; \chi^{(2)}{}_{yyz} \; \varepsilon_m{}^{-1} \; d \; E(\omega)^2 \tag{26}$$

$$E_p(2\omega) = \kappa \exp\{i \; \alpha_p \; d\} \; [\; \sin^2\phi \; L_{zyy} \; \chi^{(2)}{}_{zyy} + \cos^2\phi \; (L_{zxx} \; \chi^{(2)}{}_{zxx} \\ - L_{xxz} \; \chi^{(2)}{}_{xxz} + L_{zzz} \; \chi^{(2)}{}_{zzz} \; \varepsilon_m{}^{-2}) \;] \; \varepsilon_m{}^{-1} \; d \; E(\omega)^2 \tag{27}$$

where $\kappa = 8\pi i(\omega/c)^2$. L_{ijk} are functions of θ, e_1(air) and ε_2(substrate). α_s and α_p are functions of θ, ε_1, ε_2 and ε_m. ϕ, the polarization angle relative to the incident plane, can be varied to facilitate determination of individual $\chi^{(2)}_{ijk}$'s. However, accuracy is limited by uncertainty in ε_m. In principle molecular β is related to and deducible from $\chi^{(2)}$. In practice, however, such an approach is inaccurate since both orientational distribution functions and local-field factors are not known. On the other hand, if β is known, values of the $\chi^{(2)}_{ijk}$'s can provide a quantitative measurement of molecular ordering. In multilayer L-B films, d is proportional to the number of layers (N). If optical properties are preserved in each layer, as in the case of ideal Z-type deposition, intensity of SHG should scale as N^2. This property has been widely used in the characterization of optically nonlinear L-B films.[22]

Ellipsometry. Our ellipsometry system uses an amplitude-stabilized cw He-Ne laser at 632.8 nm, directed onto the horizontal sample surface at $\theta = 69.5°$. The polarizer is orientated 45° with respect to the incident plane, and the analyzer is set for minimium transmission. A photoelastic modulator (PEM), operated at 50 kHz with a peak-to-peak modulation amplitude of ~$\lambda/10$, and a Soleil-Babinet compensator (SBC) follow the polarizer, both with one axis aligned horizontal. A fused-silica window, selected and mounted to minimize residual birefringence, reflects some light into an optical reference channel which monitors drift due to thermal drift in the PEM and SBC. With a lock-in amplifier, the SBC is so adjusted that a bare substrate surface yields a null output. The measured signal from a surface coated by an L-B film is then proportional to the phase retardation $\Delta\phi$ it causes. The calibration of the ellipsometry system is also accomplished by the SBC. We have obtained a sensitivity of ~1 x 10^{-4} rad in the measurement of $\Delta\phi$.

Similar to surface SHG, theory of linear optical reflection from molecular thin films is not yet fully developed. However, straightforward treatment of films as macroscopic dielectric sheets yields reasonable results. This is particularly true when molecules are closely packed and hence the film's optical constant is close to the bulk constant. The phase retardation thus obtained for film thickness $d \ll \lambda$ is

$$\Delta\phi = (d/\lambda)\ f(\varepsilon_1, \varepsilon_2, \theta)\ g(\varepsilon_1, \varepsilon_2, \varepsilon_m) \tag{28}$$

$$f = 4\pi (\varepsilon_1)^{1/2} (\varepsilon_2 - \varepsilon_1)^{-1} [(1 + \varepsilon_1/\varepsilon_2) \sin^2\theta - 1]^{-1} \sin^2\theta \cos\theta \tag{29}$$

$$g = (\varepsilon_m)^{-1} (\varepsilon_m - \varepsilon_1)(\varepsilon_m - \varepsilon_2) \tag{30}$$

In deducing film thickness uncertainty arises mainly from the value of ε_m.

Characterization of an amphiphilic monolayer. We illustrate SHG and ellipsometry measurements of a monolayer with the amphiphilic compound

O_2N-ϕ-$NH(CH_2)_{11}CONH(CH_2)_{11}CONH(CH_2)_3CH(NH_3^+)COO^-$ at the air-water interface. The amino acid group is strongly polar and hence should attach to the water surface. β arises mainly from the p-nitroanilino (NA) group. Therefore, SHG reflects orientational ordering of NA-groups. The ellipsometric signal, $\Delta\phi$, provides information about uniformity and average thickness of the monolayer. Fig. 1 displays surface pressure Π, SHG intensity $I(2\omega)$, and $\Delta\phi$ as functions of the average surface area per molecule A_m. Only s-to-p polarization results are shown. In that case, only one component, $\chi^{(2)}_{zyy}$, contributes to the SHG. In the $I(2\omega)$-A_m curve development of ordering of NA-groups is seen starting at the onset of the surface-pressure increase, at $A_m \sim 40$ Å^2, but behavior of $\Delta\phi$ shows that the average thickness increases more gradually. Due to the inhomogeneity of the monolayer in the plateau region of Π, the optical data in this region appear more scatttered. Details of this study will appear elsewhere.

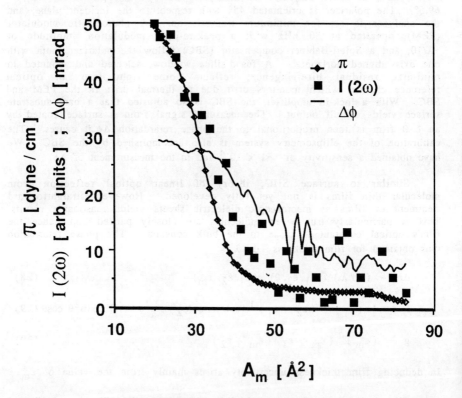

Figure 1   Surface pressure (Π), SHG ($I(2\omega)$) and phase retardation ($\Delta\phi$) for floating monolayer as function of average area per molecule (A_m) (see text).

10 NONLINEAR REFRACTIVE INDEX

For study of NLO properties, we have designed, constructed and characterized a versatile picosecond apparatus. Relevant "n_2" techniques are degenerate four wave mixing (DFWM) and power limiting. Both require good stable beam quality, moderate pulse energies and broad tunability. Short pulses are also required for delayed probing in DFWM to verify and probe mechanisms. A stable, cw-pumped, mode-locked Nd:YLF laser serves as master oscillator which synchronously pumps a dye laser and provides pulses to Nd:YLF regenerative amplifiers which also pump a four stage dye amplifier. With these various light sources and nonlinear conversions in β-barium borate and KTP crystals, the spectral range from 400 to greater than 4500 nm is covered. Details of this system are available elsewhere.[23]

REFERENCES

1.   G. R. Meredith, Phys. Rev. B, 1981, 24, 5522.
2.   G. R. Meredith, J. Chem. Phys., 1981, 75, 4317.
3.   G. R. Meredith, J. Chem. Phys., 1982, 77, 5863.
4.   G. R. Meredith, 'Nonlinear Optical Properties of Organic and Polymeric Materials', ed. D. J. Williams, Am. Chem. Soc., Washington, D.C., 1983, p. 27.
5.   N. Bloembergen, 'Nonlinear Optics', Benjamin, Reading, 1965.
6.   G. R. Meredith, B. Buchalter and C. Hanzlik, J. Chem. Phys., 1983, 78, 1533.
7.   G. R. Meredith, B. Buchalter and C. Hanzlik, J. Chem. Phys., 1983, 78, 1543.
8.   G. R. Meredith, Proc. SPIE, 1987, 824, 129.
9.   J. Zyss and J. L. Oudar, Phys. Rev. A, 1982, 26, 2028.
10.   R. M. Hochstrasser, G. R. Meredith and H. P. Trommsdorff, J. Chem. Phys., 1980, 73, 1009.
11.   R. M. Hochstrasser, G. R. Meredith and H. P. Trommsdorff, Chem. Phys. Lett., 1978, 53, 423.
12.   B. Buchalter and G. R. Meredith, Appl. Opt., 1982, 21, 3221.
13.   G. R. Meredith and B. Buchalter, J. Chem. Phys., 1983, 78, 1938.
14.   S. H. Stevenson, D. S. Donald and G. R. Meredith, 'Nonlinear Optical Properties of Polymers', ed. A. J. Heeger, J. Orenstein and D. R. Ulrich, Materials Research Society, Pittsburgh, 1988, p. 103.
15.   G. R. Meredith, H. Hsiung, S. H. Stevenson and H. Vanherzeele, Mat. Res. Soc. Symp. Proc., in press.
16.   G. R. Meredith and S. H. Stevenson, submitted for publication.
17.   D. S. Chemla and J. Zyss, 'Nonlinear Optical Properties of Organic Molecules and Crystals ', Academic, New York, 1987, Vols. 1 and 2.
18.   S. H. Stevenson and G. R. Meredith, Proc. SPIE, 1986, 682, 147.
19.   S. Velsko, Proc. SPIE, 1986, 681, 25.
20.   G. R. Meredith, J. G. Vandusen and D.J. Williams, 'Nonlinear Optical Properties of Polymeric Materials', ed. D. J. Williams, American Chemical Society, Washington, D.C., 1983, p. 109.
21.   M. Sigelle and R. Hierle, J. Appl. Phys., 1981, 52, 4199.
22.   R. Popovitz-Biro, K. Hill, E. M. Landau, M. Lahav, L. Leiserowitz, J. Sagiv, H. Hsiung, G. R. Meredith and H. Vanherzeele, J. Am. Chem. Soc., 1988, 110, 2672.
23.   H. Vanherzeele, to be published.

Efficiency and Time-Response of Three-wave Mixing Processes in Organic Crystals

R. Hierle, J. Badan, D. Josse, I. Ledoux, J. Zyss*

CNET, 196 AVENUE H. RAVÉRA, 92220 BAGNEUX, FRANCE

A. Migus, D. Hulin, and A. Antonetti

LABORATOIRE D'OPTIQUE APPLIQUÉE, ECOLE POLYTECHNIQUE-ENSTA, 91120 PALAISEAU, FRANCE

1 INTRODUCTION

Among the properties advocating the utilization of optically non-linear organic crystals in devices based on parametric three-wave mixing processes, their higher efficiency and fast response time are most often quoted[1-3]. Experiments reported thereafter probe, for the first time, the influence of the interacting pulse durations on the efficiency of non-resonant second-harmonic generation and parametric amplification or emission. In particular, the assumption that, owing to the optical power dependence of the parametric gain, one should expect a significant efficiency enhancement by reduction of the pump pulse duration required closer scrutiny : obviously the bandwidth limitation resulting from the angular phase-matching constraints should time expand the processed pulses and subsequently lower the nonlinear efficiency. It was therefore of interest to investigate three-wave mixing processes at pico- and subpicosecond regimes in various phase-matched configurations ranging from angularly to spectrally non-critical. The advantage of this latter configuration in terms of longitudinal group-velocity conservation has become clear in the course of this work. Another benefit to be derived from operating at ultrafast optical regimes results from the cumulative nature of the processes underlying the optical damage in materials : the high peak power (of the order of the gigawatt) and low average energy (of the order of the microjoule) pulses generated by mode-locked laser sources achieve an appealing combination in order to increase the nonlinear gain as well as the optical damage threshold. Two organic crystals of particular interest in view of their high figures of merit have been investigated throughout this work. Highly efficient second harmonic generation has been observed in 3-methyl-4-nitropyridine-1-oxide (POM)[4-6] at 1.32 μm in the picosecond regime. Second-harmonic generation, parametric amplification and emission experiments in N-(4-nitrophenyl)-L-prolinol (NPP)[7-11] in the subpicosecond regime will further be reported. In particular, high gain values of the order of 10^7 were achieved at 1.106 μm by pumping with a colliding pulse mode (CPM) laser emitting at 620 nm a sequence of two NPP crystals, each of them thinner than 2 mm[11]. Finally, a new spectroscopic technique termed PASS[10] after Parametric Amplification and Sampling Spectroscopy and based on the high nonlinear efficiency of such organic crystals as NPP will be discussed in terms of its unique potential to time-resolve, well below the picosecond, weak near infrared luminescence signals.

2 HIGHLY EFFICIENT SECOND HARMONIC GENERATION IN POM

The efficiency of a number of new organic nonlinear crystals has been shown to surpass that of Lithium Niobate in various three-wave mixing configurations : most of them are Paranitroaniline or even more transparent Nitropyridine derivatives which do not meet the stringent damage threshold requirements linked to high power, high gain second-harmonic generation (SHG) of the 1.064 μm emission line of the Nd^{3+}:YAG laser operating at nanosecond pulse duration regime. Better operating conditions are found when the fundamental wavelength is red-shifted and the harmonic frequency subsequently removed from the UV absorption region. The 1.32 and 1.34 μm emission lines of the Nd^{3+}:YAG laser are well located for this purpose as the 660 nm harmonic wavelength is safely removed by 170 nm from the 490 nm solid state absorption edge of POM. The combination of an excellent crystalline quality to be further ascertained by the determination of the angular width of the phase-matched harmonic intensity and of its high nonlinear coefficients ($d_{14}=d_{25}=d_{36}=23.10^{-9}$ esu)[4] have qualified POM for the experiment. POM[25] samples are grown by slow cooling close to room temperature of slightly supersaturated stirred acetonitrile solutions[12].

The source used in this work is a CNET home-made Q-switched mode-locked Nd^{3+}:YAG laser emitting both at 1.32 and 1.34 μm but preferentially at 1.32 μm. The Q-switch is an electrooptic Lithium-Niobate cell and the mode-locking device a silica acousto-optic cell switching at a frequency of 47.3 MHz. In the first series of experiments, the effective interaction length for SHG at 1.32 and 1.34 μm is probed in the Q-switched regime without additionnal mode-locking. The energy of the emerging pulses is a fraction of a mJ and their duration 80 ns. As seen from the type I phase-matching curve of POM from Ref. 4 an appealing feature is the favourable internal propagation angle $\theta_p=45.2°$ at $\lambda=1.32\mu$m.

This value brings the effective coefficient d_{eff} very close to its maximum value d_{14} as $d_{eff} = 2\sin\theta_p \cos\theta_p \, d_{14} \simeq d_{14}$.

Furthermore, the steeply ascending outlook of the phase-matching curve (i.e. $d\theta_p/d\lambda$ small) enlarges the spectral acceptance as required for broadband or short pulses nonlinear interactions. The actual transmittance of the sample, a 7.1 mm thick POM crystal cut perpendicular to the 1.0642 μm phase-matched direction is of the order of 50 % at λ=633nm. This coefficient comes close to the expected 56 % value resulting exclusively from boundary reflection losses. POM is an orthorhombic $P2_1 2_1 2_1$ crystal and phase-matching is obtained by propagating in the (100) plane with a fundamental (resp. harmonic) beam polarized in (resp. out-of) plane. The angular dependence of the SHG intensity (see Fig. 1) is related to the effective interaction length ℓ_{EFF} following the expression :

$$\delta = \frac{0.445}{\ell_{EFF}} \quad \frac{\cos\theta'}{\cos\theta} \left(\sin\theta' + \frac{\cos\theta'}{tg\rho}\right) \qquad (1)$$

wher ρ is the walk-off angle and θ (resp θ') is the external (resp. internal) propagation angle. At 1.32 μm (resp. 1.34 μm) a ℓ_{eff} value of 6.6 mm (resp. 6.9 mm) is found, coming close to the physical length and thus evidencing the good crystalline quality of POM.

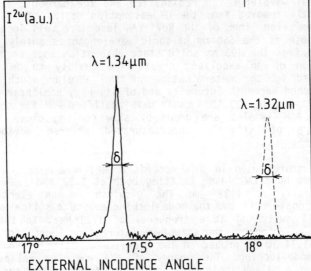

Figure 1 : Phase matched second-harmonic intensity as a function of the external incidence angle.

In a second series of experiments the laser is mode-locked, the emitted I.R. beam consisting of a periodic sequence of 120 ns overall duration composed of short pulses of 160 ps duration, separated by 10 ns intervals. The outgoing I.R. intensities in both phase-matched and unphase-matched orientations are compared. The I.R. beam emerging from the sample is detected by a photodiode and subsequently displayed on the oscilloscope, while the visible harmonic emission is removed by adequate filtering. In a non focused configuration, an effective conversion efficiency of 20% is obtained. However, a significant improvement can be reached by cylindrical focusing so as to benefit from power confinement and avoid the detriment of walk-off. A conversion yield of 50% is achieved, to be compared to that of 18% previously reported[13] for a 2mm long 4-(N,N-dimethylamino)-3-acetamidonitrobenzene (DAN) in an experimental configuration making use of a more complicated multistage 1.32 μm emitting source. The available intensity at 660 nm is of the order of 100 MWcm^{-2}, allowing for the efficient pumping, in the next stage, of a nonlinear crystal towards further parametric amplification or emission in the near infrared. Furthermore, this yield could be improved by depositing anti-reflective coatings on the faces of the crystal and by using a sample cut perpendicular to the phase matching direction while the crystal used here had been cut for 1.06 μm phase matching.

3 SUBPICOSECOND FREQUENCY DISPERSED THREE-WAVE MIXING IN NPP

The experimental set-up used for the subpicosecond probing of the nonlinear properties of NPP is depicted in Fig. 2.

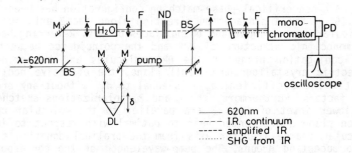

Figure 2 Experimental set-up for the probing of the tunability and time-response of molecular crystals in subpicosecond three-wave mixing parametric processes : M's mirrors ; BS's, beam splitters ; L's lenses ; F's, filters ; ND, neutral density ; DL, delay line ; PD, photodiode ; C, non-linear crystal.

The high power femtosecond source, set-up at ENSTA, compre-
hends an Argon pumped triangular colliding pulse mode-locked (CPM)
dye laser cavity and four successive frequency doubled YAG lasers
pumping four dye amplification stages. The outgoing pulses at 620
nm have a duration of 100 fsec and energy above 1 mJ for a pumping
energy of 300 mJ at 532 nm and 10 Hz repetition rate. The output
beam is split in two parts. One is vertically polarized and used
as the pump in the next amplification and emission experiments.
The other one is horizontally polarized and focused on a 20 mm
long water cell to generate, via self phase modulation, a white
light continuum with a conversion efficiency close to 100 %. A
reproducible and uniform continuum with a spectral range extending
from 0.2 to 1.6 μm and duration of the order of 180 fs is thus
obtained. In order to obtain second-harmonic generation, the pump
beam is removed and the sole continuum is allowed to shine the
crystal while the reverse is performed towards emission. Both
visible pump and infrared signals shine the crystal to allow for
the observation of parametric amplification. The time delay
between the visible pump beam and the infrared signal can be
adjusted by a translational stepping motor. The energy of the pump
is of the order of 1 μJ corresponding to an incident intensity of
1 GWcm^{-2}. The I.R. signal has to be considerably weaker in order
to avoid saturation phenomena which are prompt to occur in high
gain media such as NPP crystals. The emerging beams are collected
after the nonlinear crystal onto a monochromator and further
detected by a Silicon or Germanium diode. The second-harmonic
phase-matching curve of NPP has been plotted for a fundamental
beam ranging from 1.018 to 1.32 μm as shown in Fig. 3.

A θ-non critical phase-matching configuration has been found
at 1.15 μm for a propagation direction along the Z axis and perpendicular to the input and output faces of NPP. Referring back to
the monoclinic structure of NPP and the coincidence between the
mean conjugation plane of the NPP molecules and the principal
dielectric crystallographic (101) plane, the effective nonlinear
crystalline coefficient d_{eff} is equal to d_{21} without any projection factors. Furthermore, the ω and 2ω polarizations sketched in
the lower inset of Fig. 3 are parallel to the molecular conjugation plane and oriented near to optimal with respect to intramolecular charge transfer axis from the prolinol donating to the
nitro accepting groups. The pump-wavelength of the corresponding
amplification process is 575 nm falling within the emission domain
of the currently used Rhodamine 6G laser. This configuration
should thus be exploited in the future for nanosecond and if
possible CW operation of a parametric oscillator. When the phase-
matching direction departs from the Z axis within the ZX plane
while the wavelength is increasing, the slope of the phase-matching curve steepens-up leading to what has been called λ-non

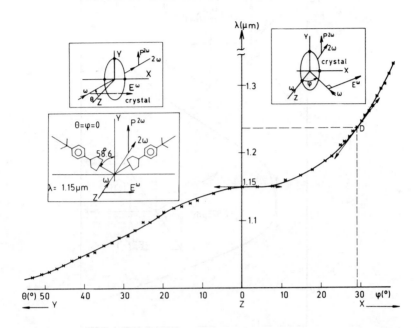

Figure 3 SHG phase-matching curve of NPP in the ZX and ZY planes exhibiting room temperature θ non-critical phase-matching at 1.15 μm. θ and φ are the external incidence angles of the IR beam. Point D corresponds to the 0.62 μm harmonic generation which is the inverse process to degenerated parametric amplification at 0.62 μm pump wavelength (see Fig. 4).

critical configuration. SHG as well as other three-wave mixing processes are then possible without drastic bandwidth reduction and subsequent increase of the interacting pulse durations. The parametric amplification phase-matching curve shown on Fig. 4 is obtained by angular tracking of phase-matched portions of the water emitted I.R. continuum. The experiment is initiated by locating point D (see Fig. 3), which corresponds to degenerated parametric amplification, by second-harmonic generation of a portion of the continuum centered at twice the pump wavelength i.e. 1.24 μm.

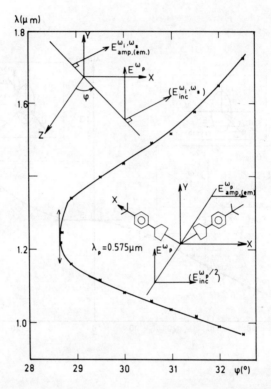

Figure 4 Phase-matching curve of parametric interaction in NPP relating the idler and signal wavelengths (λ_i, λ_s) to the external angle for a pump wavelength λ_p=620nm. The upper inset indicates the polarizations of the interacting beams while the lower inset depicts the θ noncritical parametric configuration at λ_p=575nm.

The temporal delay δ (see Fig. 2) is adjusted so as to allow for time coincidence of pump and signal in the NPP crystal. A relatively small variation over a 3° range of the external ϕ angle allows for a very large band tunability of the parametric process extending from 0.95 to 1.7 μm as shown on Fig. 4. The maximum gain value, obtained at degeneracy for a 1.5mm average quality NPP crystalline sample, is of the order of 10^4 cm^{-2} above which saturation occurs, the pump power being of the order of 1GWcm^{-2}. An estimation of d_{21} can be deduced from gain measurements and is found (after wavelength dispersion correction) to be of the order of 140 10^{-9} e.s.u. at 1.06 μm in fairly good agreement with previous nanosecond time scale measurements performed by SHG on thin single crystalline films[8] or from an oriented gas description of the crystal based on molecular β values from EFISH solution measurements[14-15]. This value is presently, to the best of our knowledge, the highest one reported for a phase-matchable crystalline coefficient and these experiments confirm the superior performances of NPP, over other nonlinear materials of similar transparency. Parametric emission was also obtained by pumping a NPP crystalline sample above a pump power density threshold of 3GWcm[-2], the emission bandwidth extending from 1.0 to 1.4 μm. Below 0.9 μm a strong Raman emission predominates over the parametric process.

Cross correlation experiments of pump and signal were performed so as to evaluate the influence of group velocity mismatch on the extension of the temporal profile of the I.R. pulse when undergoing amplification. As shown in Fig. 5, the amplified signal is plotted for various values of the dealy τ between pump (100 fsec duration) and signal (180 fsec duration) at degeneracy (λ = 1.24 μm).

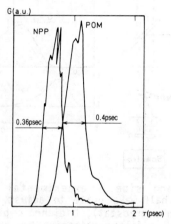

Figure 5 Cross-correlation plot of the pump and signal by amplification in 1.4 mm thick NPP sample and a 4 mm thick POM sample at degeneracy (λ=1.24μm).

The cross-correlation broadening due to group velocity mismatch is quite reduced (less than 100 fs) and slightly smaller for NPP than for POM as the interaction length is larger in the latter case. These experiments confirm the aptitude of NPP towards ultra-short signal processing over a large spectral range[16].

4 PASS SPECTROSCOPY FOR SUBPICOSECOND RESOLUTION OF INFRARED LUMINESCENCE

A new technique for subpicosecond time-resolution of near I.R. pulses such as originating from luminescence processes or from parametric interaction processes was inspired by the previous experiments. It is in principle feasible to substitute for the I.R. continuum generating water cell in Fig. 2 any sample which will emit luminescence after excitation by the 620 nm CPM laser emission. Frequency mixing in NPP with a variable delay of such I.R. signals with the pump pulse allows for temporal sampling and amplification of the incident luminescence signal. The principle of the experiment is sketched in Fig. 6.

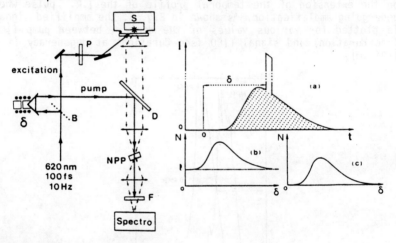

Figure 6 Left hand side : experimental scheme ; B is a beamsplitter, P a half-wave plate to adjust the polarization, S the sample (here in a cryostat), D dichroic plate reflecting only the pump beam and F a filter rejecting the pump beam. Right hand side : principle of the PASS scheme. The curve in (a) represents the luminescence signal after passing through the parametric crystal. The only part which has been amplified, by a factor of 100 or more, is that which is time-coincident inside the crystal with the pump pulse, which is variably delayed by δ (see Fig. 2). Curve (b) represents the signal collected by the photodiode (N photons) at the chosen luminescence wavelength as a function of delay δ. The baseline value ℓ corresponds to the dashed area in curve (a). Curve (c) is similar to the signal but at the more easily detected idler wavelength.

This technique has been termed PASS[10] after Parametric Amplification and Sampling Spectroscopy. By varying the time delay between the pump pulse and the onset of the signal to be analyzed, one can follow the time evolution of the signal at the selected wavelength with a time resolution limited only by the pump pulse duration. The contrast ratio is of the order of $G \, \tau_p/\tau_\ell$ reaching a magnitude of 10 with typical values of 100 fsec (resp. 100 ps) for the pump (resp. luminescence) duration τ_p(resp. τ_ℓ) and owing to the previously reported giant value of 10^4 for the parametric gain G over 1.5 mm of NPP.

The subpicosecond temporal features of various samples were studied by this method such as that of I.R. dyes and of Multiple Quantum Well Structures (MQWS) of alternating AlAs barriers and InAlAs wells. The latter is shown in Fig. 7 : the sample is cooled down to 15K and the luminescence is amplified and sampled at 1.44 μm, a spectral zone where no other technique would allow for such detailed time-resolution of a weak luminescence signal.

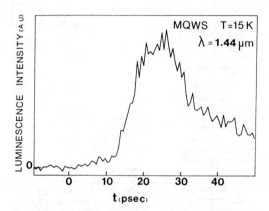

Figure 7 Time-resolved luminescence at 1.44 μm of an AlAs/InAlAs Multiple Quantum Well Structure at 15K excited by a 100 fsec duration 620 nm pump beam.

Previously inaccessible information, such as the delay between the arrival time of the photoexcitation and the actual onset of the luminescence, is now readily available from PASS experiments, and should lead to a better understanding of the migration of electron-hole photocarriers from the surface of the sample, through the buffer zone, onto their radiative recombination sites in the quantum wells.

Finally an experimental set-up for two-stage parametric interactions[11] in NPP crystals is sketched in Fig. 8. The output beam of the previously mentionned CPM source is split in three

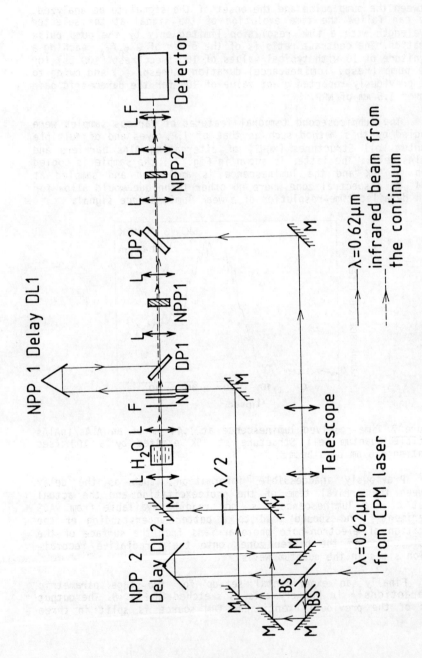

Figure 8 Experimental set-up for parametric amplification in two NPP crystals. M : mirror. DP1, DP2 : dichroic plates. L : converging lens. F : filter. ND : neutral density filter.

parts : two of them are used to pump, after being independently delayed through the variable optical path DL1 and DL2, crystals NPP1 and NPP2 while the third one generates an infrared continuum in the water cell. The energy of the pump is of the order of the μJ in the first crystal and about 10 μJ in the second one.

NPP2 can be utilized in this set-up to perform PASS temporal analysis of the parametric processes occurring in NPP1. A giant overall gain of the order of 2.10^7 is achieved with two samples of average quality and respective thicknesses 2.2 and 1.5 mm. Although individual gains of 10^4 and 10^5 have been independently measured, the theoretical limit of 10^9 cannot be reached owing to saturation phenomena. In this configuration, an incident I.R. beam of a few tens of Wcm^{-2} can be detected which corresponds to very low value of a few thousands of photons only per 100 fsec. This configuration was also used to demonstrate the ability of the PASS technique to solve the previously open question of detecting and time resolving very weak infrared pulses beyond 1.5 μm where the absence of sensitive photodetectors is resented. The principle is to observe the idler beam rather than the signal beam with no loss of information as it perfectly duplicates the signal beam (idler and signal photon are generated in pairs in parametric amplification processes). However the idler is easier to detect as it will fall within the domain of sensitive photo-detectors. The principle of the demonstration is shown in Fig. 9, where the idler (resp. signal) wavelength is 1.01 μm (resp. 1.62 μm) allowing for the use of a silicon photodiode.

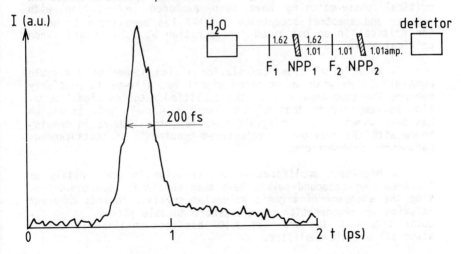

Figure 9 PASS profile of the amplified long wavelength pulse (λ_s=1.62 μm) observed at the idler wavelength λ_i=1.01μm.

Furthermore the signal to noise ratio of the PASS profile is increased by observing the idler rather than the signal as there is no baseline in the absence of incoming idler photons on the PASS performing crystal.

5 CONCLUSION

The high efficiency and fast response-time of POM and NPP in second-harmonic generation and parametric amplification processes have been confirmed through the various experiments depicted here namely :

- high quality POM has been shown to be a worthy candidate towards the emission of picosecond red pulses via the frequency doubling of a 1.32 μm emitting Nd^{3+}:YAG laser source. A 50 % yield has been demonstrated leading to the generation of up to 100 $MWcm^{-2}$ harmonic power in the red domain.

- the infrared continuum generated by a CPM laser source extends over the whole transparency range of paranitroaniline-like crystalline compounds and has been taken advantage of to fully explore the quadratic nonlinear properties of N-(4-nitrophenyl)-L-prolinol (NPP) confirming its superior figure of merit over a phase-matching range from 0.8 to 1.6 μm. The methodology developped for NPP can be applied to other materials.

- the concepts of angularly (θ) and spectrally (λ) non critical phase-matching have been explored in relation with angular and spectral acceptance and NPP has been shown to be θ-non critical in an optimized configuration at 1.15 μm and λ-non critical above 1.3 μm.

- an original cross-correlation mixing scheme of a visible pump at 620 nm with an infrared signal has allowed to precisely measure the time expansion of the amplified outgoing signal duration as compared to that of the incoming signal. This expansion has been shown to be negligible i.e. less than 100 fs in consistence with the previously conjectured hypothesis of instantaneous parametric interactions.

- high-gain amplification and emission in NPP crystals of infrared subpicosecond pulses have been obtained, thus demonstrating the adequacy of organic molecular crystals towards coherent emission or regeneration of ultrashort tunable pulses. In particular a gain of the order of 2.10^7 has been obtained in a two-stage all organic amplifier.

- a new configuration termed PASS (after Parametric Amplification and Sampling Spectroscopy) has been conceived and shown to be capable of ultrafast subpicosecond sampling of weak I.R. signals such as the luminescence from semiconductor structures. The PASS configuration is not presently challenged, in view of its speed and long wavelength (beyond 1.2 μm) sensitivity.

In conclusion, picosecond and subpicosecond experiments reported here have proved extremely instrumental in demonstrating the unique features of organics for ultrafast signal progressing applications, to be further exploited in a more compact format via integrated optics configurations.

REFERENCES

1. "Nonlinear Optical Properties of Organic and Polymeric Materials", D.J. Williams (ed.), ACS Symposium Series 233, Washington D.C., 1983.
2. "Nonlinear Optical Properties of Organic Molecules and Crystals", D.S. Chemla and J. Zyss (eds.), Academic Press, Orlando, 1987 (2 vols.)
3. "Nonlinear Optical Processes in Organic Materials", G. Carter and J. Zyss (eds.), feature issue of J. Opt. Soc. Am. B, 1987, 4(6).
4. J. Zyss, D.S. Chemla and J. Zyss, J. Chem. Phys., 1981, 4800, 74.
5. J. Zyss, I. Ledoux, R. Hierle, R.K. Raj and J.L. Oudar, IEEE J. Quantum Electron., 1985, 1286, QE-21.
6. D. Josse, R. Hierle, I. Ledoux and J. Zyss, "Highly Efficient Second-Harmonic Generation of Picosecond Pulses at 1.32 μm in POM", submitted to Applied Physics Letters.
7. J. Zyss, J.F. Nicoud and M. Coquillay, J. Chem. Phys., 1984, 4160, 81.
8. I. Ledoux, D. Josse, P. Vidakovic and J. Zyss, Opt. Eng., 1986, 202, 25.
9. I. Ledoux, J. Zyss, A. Migus, J. Etchepare, G. Grillon and A. Antonetti, Appl. Phys. Lett., 1986, 1564, 48.
10. D. Hulin, A. Migus, A. Antonetti, I. Ledoux, J. Badan, J.L. Oudar and J. Zyss, Appl. Phys. Lett., 1986, 761, 49.
11. I. Ledoux, J. Zyss, A. Migus, D. Hulin and A. Antonetti, "Two-Stage Parametric Amplification in NPP Crystals : Application to Subpicosecond Time Resolution of Parametric Emission", to be published in J. Appl. Phys.
12. R. Hierle, J. Badan and J. Zyss, J. Cryst. Growth, 1984, 545, 69.
13. P.V. Kolinsky, R.J. Chad, R.J. Jones, S.R. Hall, P.A. Norman, D. Bloor and J.S. Obhi, Electron. Lett., 1987, 791, 23.

14. M. Barzoukas, D. Josse, P. Frémaux, J. Zyss, J.F. Nicoud and
     J. Morley, J. Opt. Soc. Am. B, 1987, 977, 4(6).
15. M. Barzoukas, P. Frémaux, D. Josse, F. Kajzar and J. Zyss in
     "Nonlinear Optical Properties of Polymers" A. Heeger,
     T. Orenstein and D. Ulrich (eds.), Proceedings of the MRS
     fall Meeting 1987.
16. I. Ledoux, J. Badan, J. Zyss, A. Migus, D. Hulin,
     J. Etchepare, G. Grillon and A. Antonetti, J. Opt. Soc. Am.B,
     1987, 987, 4(6).

Small Organic Molecules

A Full Optical Characterization of the Organic Non-linear Optical Material (−)-2-(α-Methyl-benzylamino)-5-nitropyridine (MBA-NP)

R.T. Bailey, F.R. Cruickshank, S.M.G. Guthrie, B.J. McArdle,
G.W. McGillivray, H. Morrison, D. Pugh, E.A. Shepherd,
J.N. Sherwood, and C.S. Yoon

DEPARTMENT OF PURE AND APPLIED CHEMISTRY, THE UNIVERSITY OF STRATHCLYDE,
CATHEDRAL STREET, GLASGOW GI IXL, UK

R. Kashyap, B.K. Nayar, and K.I. White

BRITISH TELECOM RESEARCH LABORATORIES, MARTLESHAM HEATH, IPSWICH IP5 7RE,
UK

1. INTRODUCTION

Crystals of MBA-NP belong to the monoclinic system with
the non-centrosymmetric space group P2₁ , point group 2,
with unit cell parameters a=5.392Å, b=6.354Å,
c=17.924Å, β=94.6° and Z=2.[1]

The crystals were oriented using both the natural
facetting of the specimens and X-ray Laue techniques.
They were cut to give blanks ~1.5x1.5x0.5 cm³ with the
major faces parallel to the principal crystallographic
planes, (100), (010), and (001). An assessment[2] may be
made from the FWHM value of the phase-matched peak
signal, of an (010) cut, which is of the form $\sin^2 \psi/\psi^2$.
Fig. 1a i-ii shows typical curves taken from each of
four growth sectors opposite the sides of the
rectangular seed. X-Ray topographic analysis, Fig. 1b,
shows that sectors opposite the long edges of the seed
are badly defected. The corresponding curves (i) are of
generally lower intensity, poorer shape and noisier.
For sectors opposite the short edges of the seed of low
defect concentration (Fig. 1b), the curves (ii) are of
high intensity, near theoretical shape and low noise.
Calculation of the FWHM of these signal curves, using
the appropriate refractive indices and sample thickness
predicts a value of 0.94° , in very close agreement
with the observed data. The blanks used for subsequent
optical assessment were polished on a standard
"Logitech" machine. With care, it proved possible to

achieve a finish with parallelism to 30'' arc and
flatness in the range $\lambda/2$ to $\lambda/10$ and scratch width
below 0.5 μm. Studies on similar materials have shown
that polishing in this way can be achieved without
structural damage to the specimen. Orthoscopic
examination of these blanks in monochromatic light
shows complete extinction, as expected, for a-, b- and
c-cut samples.

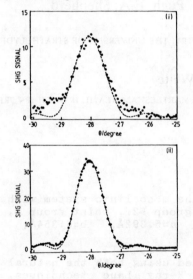

Figure 1b
a) Phase-matched signals as a
function of incidence angle;
(i) and (ii) from sectors
opposite long and short seed
edges respectively,
b) X-ray topograph showing
defects in these sectors
with the long seed edge
horizontal.

Figure 1a

2. OPTICAL ASSESSMENT

A feature of the optics of this crystal which
considerably complicates the analysis is the large
variation of orientation of the dielectric axes with
respect to the crystal lattice vectors as a function of
wavelength. The usual convention in crystal optics is
to label the principal values of the refractive index
n_x, n_y, and n_z in order of increasing magnitude at
each frequency. This definition is inconvenient and
confusing, however, in view of the above variation of
dielectric axes with wavelength, and the fact that it
transpires that, in this material, the largest
refractive index at both 1064 nm and 532 nm is, for all
frequencies, in the direction of the uniquely defined

monoclinic axis, b̃, which is also a crystal axis as
dictated by symmetry. It is acceptable to label the
piezoelectric axes along the conventional directions
(ã* , b̃, c̃) with coordinates (X,Y,Z) and to retain
the labels Y or y for all quantities pertaining to the
b̃ axis. The remaining two dielectric axes, which
rotate about b̃ as above with changing frequency, are
labelled x(λ) and z(λ), x(λ) being taken at each
frequency as the dielectric axis nearest in direction
to the piezoelectric X axis. The notation is summarised
in Table 1 below.

Table 1 Nomenclature.

ã,b̃,c̃	primitive lattice vectors
ã*,b̃,c̃	vectors defining the piezoelectric axes
X,Y,Z	piezoelectric coordinates
x(λ),y,z(λ)	dielectric axes at wavelength λ
ψ(λ)	rotation angle of dielectric axes at λ relative to piezoelectric axes
$n_x(λ)$, $n_y(λ)$, $n_z(λ)$	principal values of the refractive indices at λ

3. LINEAR OPTICS

    At present only one prism has been used to determine
the refractive indices as a function of frequency. This
has the prism edge along the b̃ direction, the light
ray entry face accurately aligned with the c̃ lattice
vector. The prism angle is accurately 15.43° . The n_y
values are easily and accurately obtained by means of
monochromatic light polarised with the E vector
parallel to the prism edge and application of the
minimum deviation technique. It is possible to obtain
the remaining n_x and n_z values by using monochromatic
light of the orthogonal polarisation, two different
angles of incidence and measurements of the
corresponding angles of deviation. Solving the geometry
yields n_{eff} values whence, by solving the simultaneous
equations;

$$n^2 = n_x{}^2 \cos^2 \theta + n_y{}^2 \sin^2 \theta,$$

we obtain the two refractive indices. In fact three
measurements were made and the extra solution used to
assess the error.

    Of these, the n_x value is least well measured,
since n_z is virtually orthogonal to the light
propagation directions at all frequencies. Further

prisms are currently being produced to yield accurate
n_x values. The value of ψ at 532 nm, required later in
the calculation of the phase-match locus, is 12.18° .
Values for refractive indices at wavelengths outside
the visible region could not be obtained from this
technique and were deduced from Maker fringe data
described below together with phase-match locus
measurements. The value of n_y (1064) is immediately
obtained from n_y (532) and the minima spacing of the
very clear set of Maker fringes for (100) and (001)
cuts with input and output polarisations parallel to
$\breve{b}$. The directional dependence of the intense
phase-matched second harmonic signal from the (010) cut
provides clearly interpretable information which helps
to fix the remaining linear parameters.

The plane of incidence for radiation entering the
(010) face was varied by rotating the (010) cut about
its normal ($\breve{b}$) axis by an angle $\emptyset$. This angle is
defined as positive about the $\breve{b}$ direction taken as the
inward normal to the entry face, from the position
where the $\breve{b}$, $\breve{c}$ plane was the plane of incidence. For
each angle $\emptyset$, the angle of incidence, θ_{PM} , at the
phase-matching intensity maximum was measured. $\emptyset$ and
θ_{PM} are plotted in Fig. 2. The phase-matching locus is
completely determined if full refractive index data are
available for both frequencies. The resulting function
is very insensitive to n_x(1064), but clearly the
dispersion in the x-direction is far less than that in
the y and z directions. Since no significant
pre-resonant behaviour is expected beyond the last long
wavelength data point used to calculate the Lorentz
equation for this orientation, we estimate n_x = 1.65
± 0.01.

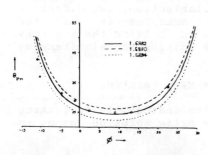

Figure 2 Angle of
incidence, θ for optimal
phase-matching in the (010)
cut as a function of
in-plane rotation angle $\emptyset$.

The phase-matching angles depend critically on the
difference between n_x (532) and n_z (1064), so that only
the single parameter n_z (1064) can be varied to fit

experimental data to the curve in Fig. 2. The angular variation of the data can be explained only if n_z (1064) is slightly less than n_x (532). Calculated curves for three neighbouring values of n_z (1064) are shown in Fig. 2 and n_z (1064) = 1.6882 seems to fit the data best. Complete refractive index data are summarised in Table 2.

Table 2 Refractive indices.

	1064nm	532nm
n_x	1.65 ± 0.01	1.6895 ± 0.02
n_y	1.7144 ± 0.005	1.8584 ± 0.0002
n_z	1.6882 ± 0.01	1.7632 ± 0.0005
ψ	$-14 \pm 1°$	$12 \pm 1°$

4. NON LINEAR OPTICS

Assuming Kleinman symmetry, we find only four independent tensor elements; $d_{21} = d_{16}$, $d_{23} = d_{34}$, $d_{25} = d_{36} = d_{14}$, and d_{22} . Maker fringes for d_{22} of MBA-NP, were collected with $0.2°$ resolution. The rotation axis was the $\bar{b}$-axis, with polarisations of both 1.06 μm laser and analyser parallel to this direction. The fringes were analysed by the equations developed for uniaxial crystals[3]. These are valid for the biaxial MBA-NP in cases where i=j in d_{ij}. $I(2\omega)$, found from the centre of the fringe envelope, is $33 \times I(2\omega)$ for quartz (d_{11}) . To convert this into a d_{ij} requires the value for n_y . The value of d_{22} obtained, relative to quartz d_{11} is 83 with a coherence length of 1.8 μm.(n(ω)= 1.7144, n(2ω) = 1.8584) The short coherence length for this material means that the flatness finish would have to be around $\lambda/200$ to achieve the same fringe signal to noise ratio as is obtained for example with a material such as quartz. This is impossible, even for the harder inorganic materials.

Phase-matched second harmonic generation was studied using the (100), (010) and (001) cut blanks polished flat and parallel to 30'' arc. Angular traces were made around all dielectric axes (aligned at 1064nm) in each face using 1064nm light, polarised either in or orthogonal to the plane of incidence. The transmitted 532nm light was also examined in these two polarisations for each incident polarisation. Incident light polarisation vectors at $45°$ to the rotation axis were also included in a search for type II phase matching, but this was not observed. Maximum SHG was

obtained by using an (010) sample with the ã* -axis as
rotation axis. The angle of incidence was 28.8° with
input polarisation in the b̃,c̃ incidence plane and the
analyser orthogonal to this.

All the phase-matching directions can be calculated
from the data in Table 2. These were found to give a
type I locus forming a closed loop around the c-axis.
The pattern approximates to a type XIII in Hobden's
classification[4], but, because of the angular movement
of the dielectric axes with wavelength, an asymmetry is
introduced, so that the locus is different in even and
odd octants.

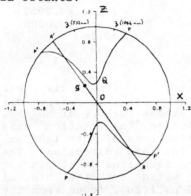

Y-AXIS INTO PAPER

Figure 3 Stereographic
projection   of   the
phase-matched locus in
MBANP.

In Fig. 3 the stereographic projection of the
phase-matched locus is plotted. In this projection, the
unique b̃ axis is orthogonal to the projection plane
and positive into this plane as drawn. Each direction
of propagation is represented as a point on a unit
sphere. The upper hemisphere is projected parallel to
b̃ on to the X-Z plane. Waves propagating in the X-Z
plane are thus represented by points on the
circumference of the unit circle. Planes of incidence
for the (010) cut are represented by diameters such as
ROR'. If r is the angle of refraction for a wave
incident on the 010 face in such a plane in the
direction S, then;
$$OS = \sin r$$
The curves PQP' represent the phase-matching locus. An
approximate interpretation of the phase-matching can be
given if the effects of the angular movement of the
dielectric axes can be neglected. The error introduced
by this approximation will be minimised if it is
assumed that there is a common set of dielectric axes

directed along the bisectors of the acute angles
between the 1064nm and 532m axes. A further fortunate
simplification then occurs because these bisectors lie
approximately along the piezoelectric axes! The
intensity of phase-matching maximises when the plane of
incidence for the 010 cut coincides with the Y-Z
piezoelectric plane. For propagation this is
represented by Q in Fig. 3. Within this scheme of
approximation, the phase-matching at Q for the 010 cut
is of the type $e + e = o$, the d_{14} coefficient
(referred to the piezoelectric axes) being entirely
responsible for the effect. At the points P and P',
where propagation is in the X-Z plane, and can only be
achieved on the (100) and (001) cuts, the
phase-matching is exactly of the type $o + o = e$, and
can only be generated by the coefficients d_{12} and d_{32},
which are zero by symmetry! This result is confirmed
by the failure to observe phase-matching in these (100)
and (001) cuts.

5. THE $\chi^{(2)}$ TENSOR

From the $\mathfrak{b}$ -axis Maker fringes d_{22} = 30×10^{-12}
m.V^{-1} and, from phase-matching at Q, d_{14} = $1\text{--}10 \times 10^{-12}$
m.V^{-1} , depending upon the interpretation of the data.
For this calculation, n_{eff} at phase matching is
calculated to be 1.6899. An approximate calculation of
the $\chi^{(2)}$ tensor is possible based on β-tensors from
CNDOVS/B calculations and an average Lorentz cavity
internal field correction, confirming that d_{22} and d_{14}
are at least of the correct order.

The efficiency of phase-matching for a 4.59 mm
thick, phase-matched cut varies as the square of the
laser flux, a maximum value of ~50% being predicted for
117MW.cm^{-2} with a spot diameter of 22.5 μm at
essentially single-shot repetition rate and 20ns pulse
width. This prediction, for a high quality cut, is
based on the improvement factor measured for poor
quality cuts destroyed eventually at this flux level. A
dramatic dependence of this efficiency on crystal
quality was observed, contrasting with the
non-phase-matched behaviour reported above. At lower
flux, the efficiency of this cut varies linearly with
thickness and is relatively insensitive to temperature
from 20 to 50° C.

The most perfect crystals show a damage threshold of
about 1 GWcm^{-2} for beam propagation perpendicular to the
(010) plane in a confocal geometry with a section

thickness of 2mm. Examination of crystals of somewhat
lower quality over a period of two years indicated that
no deterioration of either performance or quality of
optical finish occurred in this time. The damage
threshold for a phase-matched cut seems to be lower,
although, to date, only inferior quality phase-matched
cuts have been destroyed in such tests. A value of
~120MW.cm^{-2} obtained for such samples is
unquestionably pessimistic in view of the defects in
the samples used.

REFERENCES.

1. R.J. Twieg and C.W. Dirk, IBM Report, RJ5237154077,
   (1986).
2. F.R. Nash, G.D. Boyd, M. Sargent III, and P.M.
   Bridenbaugh, J. Appl. Phys., 1970, 41, 2564.
3. J. Jerphagnon and S.K. Kurtz, J. Applied Physics,
   1970, 41, 1667.
4. M.V. Hobden, J. Appl. Physics, 1967, 38, 4365.

Electrooptic Properties of a New Organic Pyrazole Crystal

S. Allen

ICI WILTON MATERIALS RESEARCH CENTRE, PO BOX NO. 90, WILTON,
MIDDLESBROUGH, CLEVELAND TS6 8JE, UK

1. INTRODUCTION

Despite the heightened interest over recent years in organic materials for nonlinear optical applications, there have been relatively few reported cases of the full characterisation of the electrooptical properties of organic single crystals. The single aromatic ring compounds POM[1] and MNA[2] have perhaps been the best characterised materials, and recently measurements on the more highly conjugated stilbazolium salt SPCD[3] have been reported. All these measurements have been made at the single wavelength of 633 nm.

We have recently reported[4] the optical, structural and electrooptical characterisation at 633 nm of a new organic compound, 3-(1,1-dicyanoethenyl)-1-phenyl-4,5-dihydro-1H-pyrazole (DCNP). A summary of these results is given in sections 2 and 3 below. This material is found to have a very high electrooptic figure of merit, which may be rationalised in terms of theoretical molecular

137

nonlinearities and the observed crystal structure. The
electrooptic measurements have now been extended to cover
the wavelength range from 630 nm to 1.2 μm. These
measurements are reported in section 4. Significant
dispersion of the electrooptic coefficient is observed.

2. STRUCTURAL AND OPTICAL PROPERTIES OF DCNP

The molecular and crystal structures of DCNP are
shown in Figure 1. Details of synthesis and crystal growth
have been previously reported[4]. Theoretical calculations
of the molecular β-coefficients using a semi-empirical
method[5] have suggested that a family of molecules based on
substituted dihydro-pyrazoles can have very high
nonlinearities. Subsequent powder SHG measurements on the
compound DCNP gave a signal of approximately 100 times
that from urea, using a fundamental wavelength of 1.9 μm.
As shown in Figure 1 the molecules pack in a parallel
fashion within the crystal structure (monoclinic space
group Cc). The molecular planes lie approximately
perpendicular to the monoclinic glide plane, and the
molecular axes are close to the average polar axis. This
axis lies within the crystallographic a-c plane, at an
angle of about 26° to the crystallographic c-axis. The
crystals used in these experiments have been grown from
solutions in acetonitrile, and had well faceted (010)
faces (parallel to the c-glide plane), convenient for the
input and output of light for the experiments described in
the next section.

The crystal has point symmetry m, and so one
principal axis of the refractive index ellipsoid (labelled

'y') is constrained to lie perpendicular to the mirror
plane, along the crystallographic b-axis. The other two

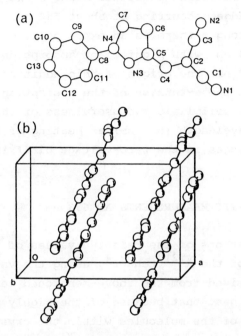

Figure 1 (a) The molecular structure of DCNP and
(b) its crystal structure, viewed down the
crystallographic b-axis.

principal axes lie within the mirror plane, and have been
found using polarising microscopy to be parallel and
perpendicular to the molecular axes, i.e. at 26° and 116°
to the crystallographic c-axis.

The absorption spectra of crystalline samples of DCNP
have been determined for illumination along the y-axis
(defined above) with light polarised along either x or z.
A significant difference is seen between the two spectra,

with an additional feature being present in the E_z curve
(i.e for light polarised along the length of the
molecules). For light polarised along x there is a sharp
absorption edge, occurring at about 580 nm, but for light
polarised along z there is an extra peak, of low intensity
at about 600 nm, with significant absorption occurring out
to about 680 nm. The origin of this additional feature is
unclear, but the extension of the absorption edge for this
polarisation will limit the usefulness of this material at
the He-Ne wavelength (633 nm) as heating of the crystal
leads to changes in the zero-voltage birefringence.

3. ELECTROOPTIC MEASUREMENTS AT 633 nm

Calculations of the relative values of the different
components of the $\chi^{(2)}$ tensor[4], using the values for the
β-tensor derived from the above-mentioned semi-empirical
program, suggest that because of the highly parallel
arrangement of the molecules within the crystal structure
the only significant component is χ_{zzz}. All other
components are expected to be smaller by a factor of at
least 100. Assuming that this will also be the case for
the electrooptic coefficient r much simplifies the
analysis of this crystal.

Samples were prepared for measurement of the large
r_{zzz} coefficient by cutting thin sections (typically 1 - 2
mm thick) out of the larger prismatic crystals. The cuts
were made with a solvent string saw, perpendicular to the
z axis, so that the electric field can be applied along
this axis. The cut crystals were then mounted between
conducting-glass electrodes (separated by spacers of

appropriate thickness so as to avoid stressing the
crystal), with good electrical contact being achieved by
using small quantities of silver paste between electrode
and crystal. The propagation direction for the laser beam
through the crystal was along the optical y axis, normal
to the large (010) facets of the crystals. The input and
output faces were not polished, but reasonable quality of
these faces was obtained by cleavage.

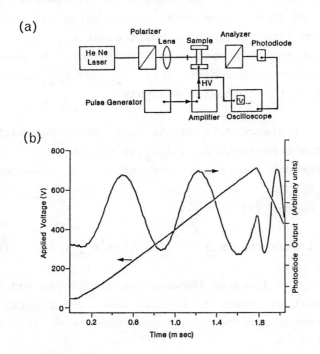

Figure 2 (a) Experimental arrangement for the
measurement of the electrooptical coefficients.
(b) Typical experimental results at 633 nm.

The experimental arrangement for the assessment of
the electrooptic effect is shown in Figure 2. Various

electric fields (in the form of pulses and voltage ramps)
were applied to the crystal using a Philips PM5786 pulse
generator and EOD LA10 linear amplifier, to produce
voltages in the range 54 V to 1500 V. The changes in
intensity of light transmitted through the analyser were
monitored via a fast photodiode on a Gould 1425 digital
storage oscilloscope. The input beam, from a Hughes
3224H-PC 5mW HeNe laser, was polarised at $45°$ to the x and
z axes. In general, the application of an electric field
E_z^o will lead to a change in all three principal refractive
indices and also through the coefficient r_{zxz} to a
rotation of the principal axes x and z about y. However as
described above the coefficients r_{zxx} and r_{zxz} can be
neglected in the following analysis.

Under these conditions the total phase retardation Γ
between the components E_x and E_z of the optical field on
passing through the crystal, of length l is given by[6]

$$\Gamma = \Gamma_o + \delta\Gamma(V)$$

$$= (2\pi l/\lambda)(n_z - n_x)l - (2\pi l/\lambda)(n_z^3 r_{zzz} V/2d) \qquad (1)$$

where d is the distance between the electrodes and V the
applied voltage. When, as in this case, the crystal is
placed between crossed polarisers, the amount of light
transmitted is

$$T = \sin^2(\Gamma/2) \qquad (2)$$

This dependence of T on Γ, and hence on the applied
voltage V, is also shown in Figure 2 for a sample having
dimensions l = 2.8 mm and d = 1.1 mm. The voltage is

increased from 54 V to about 700 V over about 2 msec and more than two complete cycles of the sinusoidal waveform are seen over this range. The short timescales over which the changes in voltage are made eliminates problems due to absorptive heating of the sample, which on longer timescales lead to significant drifting of the signal. The efficiency of an electrooptic modulator at a given wavelength may be defined in terms of its half-wave voltage (V_π), this being the voltage required to switch from full transmission to full extinction of the beam. V_π corresponds to a change in retardation Γ of π radians, and is therefore in this case defined as

$$V_\pi = (\lambda/n_z^3 r_{zzz})\,(d/l) \qquad (3)$$

The half-wave voltage of this particular crystal is determined by the voltage difference between neighbouring maxima and minima in Figure 2. The measured value is about 145 V, which is very low when compared with other available electrooptic materials. Comparison between materials is more readily achieved with the reduced half-wave voltage (V_π^*), which is corrected for the dimensions of the crystal and corresponds to the half-wave voltage of a crystal for which l=d. Correcting the above result using the measured value of 2.55 for (l/d) gives a reduced half-wave voltage of 370 V for DCNP. The corresponding value for MNA is 1.3 kV^2 and that for LiNbO$_3$ is about 3 kV^7. DCNP has a significantly lower reduced half-wave voltage than any of the commonly used inorganic materials and half-wave voltages of less than 50 V will be obtained with only moderate dimensional anisotropy of the crystals.

An alternative electrooptic figure of merit for this material is given by

$$(n_z^3 \cdot r_{zzz})/2 = \lambda/(2V_\pi^*) = (8.6\pm0.2) \times 10^{-10} \text{ m/V} \quad (4)$$

The corresponding figures of merit for MNA and $LiNbO_3$ are 2.7×10^{-10} m/V and 9.5×10^{-11} m/V. To determine the electrooptic coefficient r_{zzz} from this figure of merit it is necessary to know the refractive index n_z. An estimation of n_x and n_z at 633 nm has been made from measurements of the transmission and reflection of two different sample of different thickness. The values obtained are

$$n_x = 1.8 \pm 0.15$$

$$n_z = 2.8 \pm 0.15$$

The high errors are due to variation of transmission measured across the plane of the crystal. The very high birefringence is a result of the anisotropic absorption spectra described above. For z-polarised light there is a large enhancement of the refractive index because of the proximity of the absorption edge. Use of this value for n_z results in an electrooptic coefficient of

$$r_{zzz} = (7.8\pm1.6) \times 10^{-11} \text{ m/V}$$

This corresponds to a second-order hyperpolarisability of

$$\chi_{zzz}(-\omega;\omega,0) = -(2.4\pm0.2) \times 10^{-9} \text{ m/V}$$

4. DISPERSION OF THE ELECTROOPTIC EFFECT

The experiments described above were repeated at a
number of wavelengths in the visible and near infra-red
using a multi-line He-Ne laser (Spindler and Hoyer ML500)
and collimated laser diodes at 750 and 780 nm. It was
found that the problems of drifting of the signal
experienced at 633 nm, caused by heating of the sample by
the absorbed light resulting in a change in refractive
index and crystal dimensions, were much reduced at the
longer wavelengths. The variation of reduced half-wave
voltage with wavelength is shown in Figure 3, with that of
Lithium Niobate also given for comparison. From equation 3
an approximate doubling of V_π^* is expected between 630 and

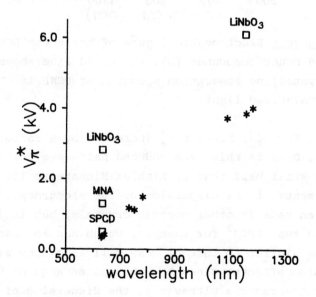

Figure 3 Reduced half-wave voltage (V_π^*) of DCNP (*)
and other compounds (□).

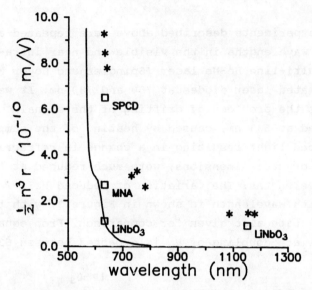

Figure 4 Electrooptic figure of merit for DCNP (*) and other compounds (□). The solid line shows the crystalline absorption spectrum of DCNP in unpolarised light.

1200 nm. For DCNP however V_π^* increases by a factor of more than 10. Despite this, the reduced half-wave voltage of DCNP is still half that of Lithium Niobate at 1100 nm. No measurements of the dispersion of the electrooptic effect have been made in other organic crystals, but it is expected that SPCD[3] for example, which has an absorption spectrum similar to that of DCNP will also show similar dispersion effects. Since the absorption edge of MNA lies further towards the ultraviolet the dispersion of its electrooptic coefficient will probably be smaller over this wavelength range.

The origin of the large increase in V_{π}^{*} can be seen in Figure 4, where the electrooptic Figure of Merit $(n_z^3 \cdot r_{zzz}/2)$ is plotted as a function of wavelength. The position of the absorption edge of DCNP is also shown. A marked resonant enhancement occurs in the lower wavelength region, as this absorption edge is approached. The increase in the Figure of Merit is due to an enhancement of both the coefficient r_{zzz} and the refractive index n_z. Crude estimations of the variation of refractive index were obtained by measuring the transmission and reflection coefficients of thin samples, using polarised laser light at a number of wavelengths. The results are shown in Figure 5, but the accuracy of these measurements is not very high. As expected, the birefringence is reduced on moving away from the proximity of the absorption edge, dropping from about 0.9 at 633 nm to about 0.5 at 1100 nm.

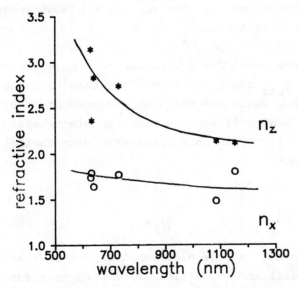

<u>Figure 5</u> Refractive indices n_z (*) and n_x (o) of DCNP. The solid lines are as a guide to the eye only.

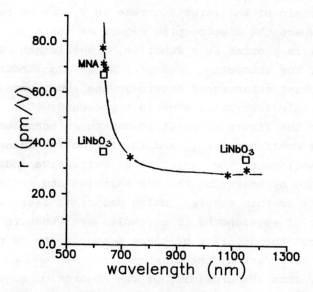

<u>Figure 6</u> Electrooptic coefficients of DCNP (*) and
other compounds (□). The solid line is drawn as a
guide to the eye only.

Using the curve for n_z, values for the electrooptic
coefficient r_{zzz} can be obtained, and these are shown in
Figure 6. Once again the resonant enhancement is clear. At
longer wavelengths it appears that the electrooptic
coefficient of DCNP is only comparable with that of
Lithium Niobate.

5. CONCLUSIONS

Measurements of the electrooptic half-wave voltage,
Figure of Merit and r-coefficient of DCNP as a function of
wavelength show that whereas at 633 nm this compound shows

a significant advantage over Lithium Niobate, at longer wavelengths this advantage is much reduced. At 1100 nm the reduced half wave voltage is still half that of Lithium Niobate, and at laser diode wavelengths around 800 nm the factor in favour of DCNP is about 3. A trade-off must be achieved between highest efficiency and deleterious effects due to absorptive heating of the sample at shorter wavelengths. Nevertheless these measurements underline the potential of organic compounds for the production of a new range of high efficiency electrooptic materials.

6. ACKNOWLEDGEMENTS

This work was carried out with the support of the UK Department of Trade and Industry under a JOERS programme. I am glad to acknowledge the contribution of T D McLean (ICI Electronics Group) in growing the crystals, and P F Gordon and B D Bothwell of ICI Organics Division for the organic synthesis.

7. REFERENCES

1. M. Sigelle and R. Hierle, J. Appl. Phys., 1981, 52, 4199.

2. G. F. Lipscombe, A. F. Garito and R. S. Narang, J. Chem. Phys., 1981, 75, 1509.

3. T. Yoshimura, J. Appl. Phys., 1987, 62, 2028.

4. S. Allen, T. D. McLean, P. F. Gordon, B. D. Bothwell, M. B. Hursthouse and S. A. Karaulov. J. Appl. Phys., 1988, in press.

5. V. J. Docherty, D. Pugh and J. O. Morley, <u>J. Chem.</u>
<u>Soc., Faraday Trans. 2</u>, 1985, <u>81</u>, 1179.
6. A. Yariv and P. Yeh, "Optical Waves in Crystals",
Wiley, New York, 1984, p.511.
7. R. J. Holmes, Y. S. Kim, C. D. Brundle and D. M. Smyth,
<u>Ferroelectrics</u>, 1983, <u>51</u>, 41.

Investigation of Linear and Non-linear Optical Properties of 2-Cyclooctylamino-5-nitropyridine

Ch. Bosshard, K. Sutter, and P Günter

INSTITUTE OF QUANTUM ELECTRONICS, SWISS FEDERAL INSTITUTE OF TECHNOLOGY, ETH HÖNGGERBERG, CH- 8093 ZÜRICH, SWITZERLAND

G. Chapuis

INSTITUTE OF CRYSTALLOGRAPHY, UNIVERSITY OF LAUSANNE, DORIGNY, CH-1015 LAUSANNE, SWITZERLAND

R.J. Twieg and D. Dobrowolski

IBM ALMADEN RESEARCH LABORATORY, SAN JOSE, CA 95120, USA

Abstract

2-Cyclooctylamino-5-nitropyridine (COANP) is a new nonlinear optical crystal with point group symmetry mm2. It has been grown by a temperature difference solution growth technique. Its linear and nonlinear optical properties have been investigated. The refractive indices n_a = 1.699, n_b = 1.847 and n_c = 1.681 (at λ = 550 nm) and the nonlinear optical susceptibility d_{33} (λ_ω = 1.06 μm) have been determined. Type I phase matching is possible at 1.064 μm. A peak conversion efficiency η = $P_{2\omega}/P_\omega$ = 3.6% has been observed with a 0.90 mm thick crystal and a fundamental power P_ω = 560 W.

1.Introduction

Nonlinear optical crystals appear very attractive for applications in image processing and optical communication. In recent years the interest in using organic crystals with charge correlated and highly delocalized π-electron states has increased considerably, since very large nonlinear optical susceptibilities have been measured in some of the materials[1-3] which make them attractive for future cw laser applications. In addition some of the nonlinear optical molecular crystals can be prepared in the form of thin film layers[4], grooves, capillaries or crystal-cored optical fibers[5] which make them particularly interesting for guided wave nonlinear optics[6].

We report on the linear and nonlinear optical properties of COANP, a nonlinear optical crystal with orthorhombic point group symmetry (mm2), large birefringence and nonlinear optical susceptibilities[16]. These properties make COANP a very interesting material for type I phase matched frequency doubling or sum-frequency generation.

2-Cyclooctylamino-5-nitropyridine (COANP) is an amphiphilic molecule. The neutral part is formed by the octyl ring and the polar part by the nitropyridine; see Fig.1. Second-harmonic tests on powders yielded efficiencies 50 times larger than urea.

The synthesis and growth procedure are described in detail in ref.7 and 16.

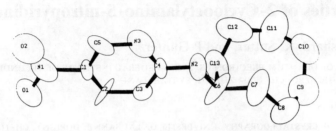

Fig.1: ORTEP plot of a single molecule of COANP with 50% thermal ellipsoides.

2.Structure

The structure of COANP was solved by single crystal x-ray diffraction. Computations were performed by means of the x-ray system (Stewart et al.)[8] and figures drawn with the help of the ORTEP program[9]. The complete resolution of the structure was obtained from the use of direct methods (MULTAN)[10]. Parameters relevant to the measurements and calculations are summarized in table I.

The amphiphilic property gives rise to the packing of the molecules as shown in Fig.2.

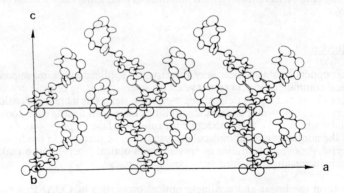

Fig.2: Projection of the COANP structure along the b-axis. The hydrogen bond N(2)-O(2) and the noncovalent bond N(3)-O(1) are schematically represented by open bonds between the N and O atoms.

In the layers perpendicular to the a-axis the polar and neutral parts alternate. Each layer is formed by a series of infinite chains linked by hydrogen bonds. The chains are parallel to the ($\underline{b}$ + $\underline{c}$) direction or alternatively to the symmetry equivalent (-$\underline{b}$ + $\underline{c}$) direction. The oxygen atoms of the nitro group are directed towards the nitrogen atom N(2) to form a hydrogen bond with a N...O distance of 2.9 Å. The N(3)-O(1) interaction is weakly bonding but is not a hydrogen bond. Its N...O distance is 3.9 Å.

3.Optical properties

The optical and nonlinear optical properties of COANP were observed on plates of dimensions up to 9 x 4.5 x 1.3 mm^3 oriented along the crystallographic c, b and a direction respectively, cut with a 0.22 mm diameter diamond wire-saw and polished with diamond paste (0 - 3 μm) in vaseline. The samples were of optical quality showing homogeneous extinction over the whole crystal surface. Fig. 3 shows the absorption spectrum of a 0.90 mm thick crystal for light propagating along the a-axis polarized parallel to the bc-plane. The optical transmission range extends from 470 nm to 1500 nm (50% points), with some absorption bands near 1200 nm and 1400 nm and with a minimum absorption constant α < 1 cm-1 for λ ≈ 1350 nm. The refractive index n_c was measured by means of the immersion technique[11], n_b using a micro-interferometric technique[12]. The birefringence n_a - n_c was determined from the conoscopic figure of a b-plate crystal by counting the interference fringes between optical and crystal axes. The results are shown in table I. Since n_c < n_a < n_b (at 550 nm), COANP is an optically biaxial crystal with the optical axes lying in the (bc) plane.

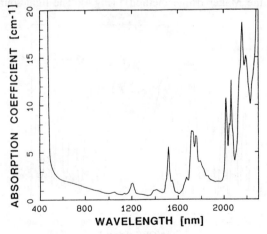

Fig.3: Optical absorption spectrum of a COANP crystal (thickness d = 0.90 mm).

For crystals of point group mm2, the second-harmonic polarization P is given by [13]:

$$P_1 = 2\varepsilon_0 d_{15}E_1E_3$$

$$P_2 = 2\varepsilon_0 d_{24} E_2 E_3 \tag{1}$$
$$P_3 = \varepsilon_0 (d_{31} E_1^2 + d_{32} E_2^2 + d_{33} E_3^2)$$

where d_{ij} are the elements of the second-order polarizability tensor and E_1, E_2 and E_3 the electric field components of the fundamental wave (frequency ω) along the crystallographic a-, b- and c-axis respectively.

For the ideal case of lightly focussed monochromatic Gaussian beams, the second-harmonic intensity, neglecting pump depletion, is given by[13]

$$I_{2\omega} = \frac{2\omega^2}{\varepsilon_0 c^3 \left(n^\omega\right)^2 n^{2\omega}} d_{eff}^2 L^2 \left(I_\omega\right)^2 sinc^2\left(\frac{\Delta k \, L}{2}\right) \tag{2}$$

where we denote $(\sin(x))/x$ by $sinc(x)$. L is the crystal length, d_{eff} the effective nonlinear optical susceptibility and $\Delta k = k_2 - 2k_1$ is the phase mismatch between the fundamental and second-harmonic waves with wave vectors k_1 and k_2 respectively.

For COANP with the refractive indices $n_c < n_a < n_b$ reported above we see that phase matched second-harmonic generation ($\Delta k = 0$) is possible for the configurations using the nonlinear optical polarizabilities d_{31} and d_{32} (type I phase matching) but not for d_{24} and d_{15} (type II phase matching). The nonlinear optical susceptibility d_{33} was measured using the Maker-fringe method[14] (Fig.4). The light source was a Q-

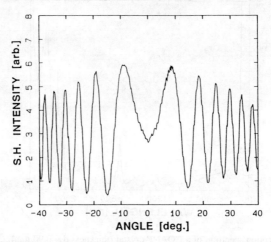

Fig.4: Maker-fringe curve of d_{33} of COANP.

switched Nd:Yag laser operating at $\lambda = 1064$ nm (Specta Physics 3000, peak power $P_\omega \le 2.5$ kW, pulse duration $\tau = 250$ ns, repetition rate 400 Hz). The peak power incident on the crystal was 0.6 kW, the beam waist $w_0 = 0.12$ mm (power density $I_\omega = 1.3$ MW/cm^2). The second-harmonic power was detected by a photomultiplier, gated integrator system and for the reference we used a quartz plate ($d_{11} = 0.4$ pm/V). Details of the computer-controlled experimental set-up have been reported in ref.15. For the measurement we used polished plates oriented perpendicular to the b-axis (thickness: 1.27 mm and 1.24 mm respectively).

The nonlinear optical susceptibility $d_{33} = (10 \pm 2)$pm/V was determined from the envelope of the fringe patterns. The coherence length calculated from our measured refractive index data by using

$$l_c = \frac{\lambda}{4\,(n_\omega - n_{2\omega})} \qquad (3)$$

was $l_c^{33} = 3.5\ \mu m$. From the separation of the fringe minima the same value was determined, showing the good agreement between our linear and nonlinear optics measurements.

Angle tuned phase matched second-harmonic generation for Nd:YAG lasers ($\lambda = 1.06\ \mu m$) has been observed by rotating an a-plate (thickness d = 0.904 mm) by an angle $\theta = (48.9 \pm 0.5)°$ around the c-axis with respect to the a-axis (propagation direction within the crystal $\theta_i = (26.4 \pm 0.2)°$).

The effective nonlinear optical susceptibility for this geometry is $d_{eff} = d_{32}\cos^2\theta_i + d_{31}\sin^2\theta_i$. For a peak intensity $P_\omega = 560$ W ($I_\omega = 1.3$ MW/cm²) the generated second harmonic power was $P_{2\omega} = 20$ W (conversion efficiency 3.6%). Neglecting beam walk-off due to birefringence we estimate from (2) an effective nonlinear susceptibility $d_{eff} = 24$ pm/V.

Summarizing our results COANP crystals of good optical quality having a large nonlinear optical susceptibility were grown. The crystal structure and the linear optical parameters (refractive indices, absorption constant) have been determined, indicating that COANP is a biaxial crystal with point group symmetry mm2. Angle tuned second-harmonic generation of Nd:YAG laser radiation at $\lambda = 1.06\ \mu m$ with a conversion efficiency $\eta = I_{2\omega}/I_\omega = 3.6\%$ (at a power level $P_\omega = 560$ W) has been demonstrated using a 0.9 mm thick a-crystal plate.

Table I: Optical, nonlinear optical and structural data of COANP crystals

Formula:	$C_{13}H_{19}N_3O_2$ (M = 249.32)
Density:	$\rho = 1.24$ g/cm3
Melting Point:	$T_m = 70.9\ °C$
Space group symmetry:	Pca2$_1$ (Z=4)
Point group:	mm2
Unit Cell:	a = 26.281(5) Å, b = 6.655(1) Å, c = 7.630(1) Å V = 1334.5(6) Å^3
Refractive indices: (at $\lambda = 550$ nm)	$n_a = 1.699(4)$, $n_b = 1.847(6)$, $n_c = 1.681(4)$

Nonlinear optical
susceptibilities:
$$d_{33} = (10 \pm 2) \text{ pm/V}$$
$$d_{eff} = d_{32}\cos^2\theta_i + d_{31}\sin^2\theta_i = 24 \text{ pm/V}$$
$$\theta_i = (26.4 \pm 0.2)°$$

Acknowledgement

The authors are very grateful to M.Ehrensperger for crystal growth and J.Hajfler for his expert sample preparation. This work has been supported in part by the Swiss National Science Foundation (NFP 19: Materials for future technology).

References:

[1]   D. J. Williams, Ed.: "Nonlinear optical properties of organic and polymeric materials"(Am. Chem. Soc.Symp. ser.233, Washington D.C., 1983)
[2]   D. J. Williams, Angew. Chem., Int. Ed. Engl. 23,690 (1984)
[3]   R. Twieg, A. Azéma, K. Jain and Y. Y. Cheng, Chem. Phys.Lett. 25, 202 (1986)
[4]   I. Ledoux, D. Josse, P. Vidakovic and J. Zyss, Optical Engineering 25, 202 (1986)
[5]   B. K. Nayar in: "Nonlinear optical properties of organic and polymeric materials", D. J. Williams, Ed.(Am. Chem. Soc. Symp. ser.233, Washington D.C.(1983), p.153
[6]   G. Stegeman in: "Integrated Optics", H. P. Nolting and R. Ulrich, Ed (Springer Verlag,Berlin, Heidelberg,(1985) p.178
[7]   H. Arend, R. Perret, H.Wüest and P.Kerkoc, J. Cryst. Growth 74,321 (1986)
[8]   J. A. Stewart, G. O. Kruger, H. L. Ammon, C. W. Dickinson and S. R. Hall, Computer Science Center, Univ. of Maryland
[9]   C. K. Johnson, "ORTEP: Fortran Thermal Ellipsoid Plot Program for Crystal Structure Illustrations", ORNL-3794, 2nd version, Oak Ridge, Tennessee(1971)
[10]   P. Main, S. J. Fiske, S. E. Hull, J.-P. Declercq, G. Germain and M. M. Woolfson, "MULTAN-A system of computer programs for the automatic solution of crystal structures"
[11]   N. H. Hartshorne and A. Stuart, "Crystals and the polarizing microscope" (E. Arnold Ltd, London) 1970
[12]   J. Gahm, Zeiss Mitteilunngen 2, 389 (1962) and 3, 3 (1963)
[13]   See, e.g., A. Yariv,"Quantum Electronics"(John Wiley and Sons, New York, (1975)
[14]   See, e.g., S. K. Kurtz in:"Quantum Electronics", H. Rabin and C. L. Tang, Eds.(Academic Press, New York, 1975), p.209
[15]   J. C. Baumert, J. Hoffnagle and P. Günter, SPIE Vol 492 Europ. Conf. on Opt. SystemsandApplications, 1984 (Amsterdam)
[16]   P. Günter, Ch. Bosshard, K. Sutter, H. Arend, G. Chapuis, R. J. Twieg, D. Dobrowolski,Appl. Phys. Lett. 50(9), 486(1987)

Organic Crystals for Non-linear Optics: Molecular Engineering of Non-centrosymmetric Crystal Structures

J.F. Nicoud

INSTITUT DE PHYSIQUE ET CHIMIE DES MATÉRIAUX DE STRASBOURG, INSTITUT CHARLES SADRON-CRM, 6 RUE BOUSSINGAULT, F-67083 STRASBOURG CEDEX, FRANCE

1 INTRODUCTION

Organic materials able to lead to frequency conversion of laser light are currently under intensive investigation[1]. This is particularly due to the development of optical communications and optical signal processing. These materials must present efficient non-linear optical (NLO) properties, and with this target we have investigated some strategies of organic synthesis, in order to get organic crystals having enhanced quadratic electric susceptibilities. The case of second harmonic generation (SHG) has been more specificaly studied. The high non-linear efficiencies of these molecular crystals result from a double optimization. The first step is the design of an "active" molecule having high hyperpolarizability in relation to its transparency range. The second step is the preparation of crystals having a non-centrosymmetric structure, without which no quadratic effect would occur. The point is to get an optimized packing of the active molecules, hopefully according to the most efficient crystal classes that allow phase-matching conditions[2,3].

In this paper we will recall our previous results in the field of amino-nitro-aromatic derivatives. Then we will present some recent results concerning the use of chirality as an efficient strategy for the tentative control of non-centrosymmetry in organic crystals. In addition to the classical nitro-phenyl or nitro-pyridine derivatives, some new conjugated chains have been studied, such as chalcone and 2-benzylidene-1,3-indanedione derivatives. The non-linear optical properties of these new materials have been tested by SHG on crystalline powders, at 1.06 μm or 1.32 μm .

157

Efficient nitroaromatic NLO derivatives

4-nitroaniline (4NA) and 2-amino-5-nitropyridine (ANP) are well known molecules for their satisfactory hyperpolarizability. Their transparency range in the solid state is typically 0.5-2.0 μm, with a slight advantage for the pyridine derivatives, owing to the blue shift attributed to the nitrogen heteroatom[2]. These materials have been extensively studied for possible use with 0.9-1.5 μm lasers. Our aim was to prepare crystalline compounds having the required non-centrosymmetric crystal packing for $\chi^{(2)}$ effects. Starting from a given structure organic chemistry allows the preparation of a variety of derivatives, but this empirical method leads statistically to 29% acentric crystals, and among these only a few have the desired NLO efficiency. This implies waste of time and material, so we adopted a strategy using chirality which reduces time loss. The use of enantiomerically pure derivatives ensures that the crystallization will occur in one of the 11 enantiomorphous point groups that are a subset of the 21 non-centrosymmetric point groups. We have to keep in mind that this method leads to a 100% probability of getting an acentric crystal structure, but not necessarily an optimized packing. We got indeed chiral materials showing nearly zero SHG efficiency; we think that in exploring a new class of hyperpolarizable molecules , chirality is a good means of investigation. Chirality can be used also with additional effects increasing the chances of preparing materials with suitable crystal structure. We first made use of hydrogen-bonding interactions, easily established between the nitro and hydroxyl groups. We prepared accordingly a number of 4NA and ANP derivatives starting from chiral amino-alcohols. The most efficient among them are recalled in Table 1. Only NPP has been chosen for further studies. From the analysis of its P2$_1$ crystal structure, it appears that the molecular packing is nearly optimized for quadratic phase-matched effects. The other materials of Table 1 have comparable SHG efficiencies, but their crystal structures are not yet established. We then investigated achiral amino-alcohol or amino-acid derivatives, thinking that intermolecular hydrogen-bonds could lead alone to a good acentric packing. Table 2 shows that this strategy may be successful notably with 3-aminopropanol derivatives the crystal structures of which are still unknown.

Following another strategy, we studied the effect of a highly polar group grafted onto the charge-transfer polarized conjugated backbone of the achiral molecule[3]. The polar cyano group was chosen first in testing this strategy leading to the discovery of two simple materials: N-(4-nitrophenyl)-N-methylaminopropionitrile (NPPN) and N-(4-nitrophenyl)-N-methylaminoacetonitrile (NPAN) which have high SHG efficiencies (Table 2). Only the NPAN

Table 1 Chiral amino-alcohols 4NA and ANP derivatives

D	Z	R	R'	$I^{2\omega}$/urea(1.06μ)	
CH₃ structure, HOH₂C–CH–NH–	CH	H	H	22	
	N	H	H	92	NPA
CH₂–Ph structure, HOH₂C–CH–NH–	N	H	H	130	NPPA
	CH	H	H	150	NPP
pyrrolidine structure, N–CH₂OH	CH	CN	H	22	CNPP
	CH	H	OH	140	HNPP

Table 2 Achiral 4NA derivatives

D	R	R'	$I^{2\omega}$/urea (1.06μ)	
HOH₂C(CH₂–NH–)(CH₂)	H	H	80	APNP
	CF₃	H	30	
	H	CF₃	80	
HOOC(CH₂–CH₂–NH–)	H	H	115	BANP
NC–CH₂–N–	H	H	140	NPAN
NC(CH₂)(CH₂)–N–(H₃C)(H₃C)	H	H	85	NPPN

crystal structure is known: the crystals are orthorhombic,
point group mm2, space group Fdd2[4]. As for NPP, a
nearly optimized molecular orientation for phase-matched
$\chi^{(2)}$ effects is evidenced in NPAN[5].

Newly investigated materials

Some enone derivatives are already known as non-linearly
active materials[6]. We prepared a variety of chalcone
derivatives, using the chiral prolinol group as a donor,
and a nitro or cyano group as an acceptor. In two cases
we get positive powder tests, whereas a previous study[6]
reports a series of cyano derivatives without activity.
This success due to chirality prompted us to prepare also
achiral materials; among them we found some new compounds
showing significant SHG activity (Table 3). Due to the
red colour of some derivatives, the SHG powder test have
been done by using 1.32 μm laser light. We completed the
search of enones by preparing the 4-prolinol derivative
of 2-benzylidene-1,3-indanedione (Figure 1), a deep red
material showing high SHG efficiency at 1.32 μm. On the
other hand, we tried to get new efficient materials in
the pyridine-1-oxide family. Recently we met success by
discovering 3-methyl-4-nitropyridine-1-oxide (POM) as one
of the NLO crystals allowing the easiest crystal growth
conditions[7,8]. The high hyperpolarizabity of 4-dime-
thylaminostilbazole-N-oxide (DASNO) has been measured

Table 3 Polarized chalcones

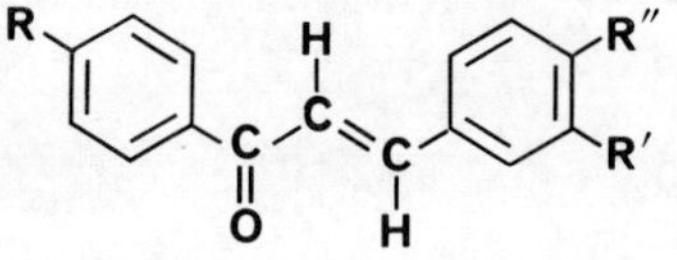

U = urea
POM = 13 x U
NPAN = 140 x U

R	R'	R"	$I^{2\omega}$ (1.32μ)
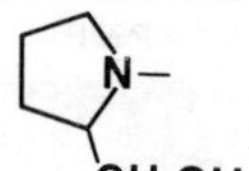 N— CH$_2$OH (pyrrolidine)	H	NO$_2$	> U
CN	H	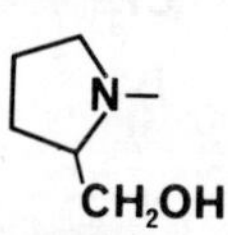 N— CH$_2$OH (pyrrolidine)	> U
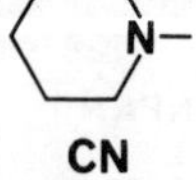 N— (piperidine)	H	NO$_2$	≃ POM
CN	OCH$_3$	OCH$_3$	> U ; < POM
CN	H	NMe$_2$	≃ NPAN

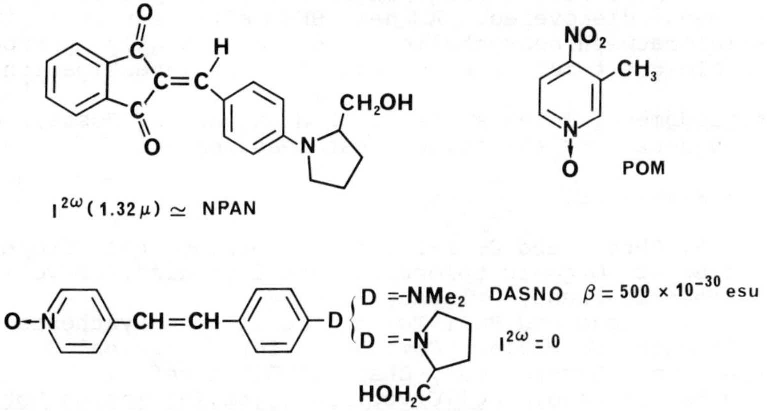

$I^{2\omega}(1.32\mu) \simeq$ NPAN

Figure 1

(β = 500 x 10^{-30} esu)[7]. We however failed in our effort to get an acentric DASNO derivative; the chiral prolinol derivative has even zero SHG efficiency (Figure 1). We then attempted to control the bulk dipolar alignment of DASNO molecules through inclusion complexation, by using a chiral host. When a solution of DASNO and deoxycholic acid (DCA) in absolute EtOH is slowly evaporated, bright yellow crystals are formed, showing significant SHG efficiency at 1.06 μm. The guest and host powders have respectively zero efficiency in the same conditions. Further studies are in progress in the analysis of this new inclusion material. Here, its relatively high SHG efficiency, in comparison with those early reported in that field,[9,10] may result from the compensation of the inevitable dilution of the guest molecule by choosing one having a large hyperpolarizability. We wish to point out a possible advantage of inclusion complexation. The fact that the guest molecule is packed alone could suppress the bathochromic shift in the electronic absorption spectrum of the crystal, mainly due to exciton-splitting and dipole-dipole interactions, along with the charge-resonance interactions in the first excited singlet state[11].

2 CONCLUSION

We continued our work in molecular and crystal engineering for non-linear optical materials. We focused our attention on obtaining organic crystals for $\chi^{(2)}$ effects, having non-centrosymmetric packing. We largely used chirality as an efficient strategy. Some new charge-

transfer compounds showing significant SHG powder tests
have been discovered. A new SHG efficient inclusion
material between deoxycholic acid and a highly hyperpolarizable guest molecule is under further investigation.

Acknowledgment. I thank Dr. I. Ledoux and D. Josse, of
CNET-Bagneux, for the powder test measurements.

3 REFERENCES

1. D.S. Chemla and J. Zyss, "Nonlinear Optical Proper-
    ties of Organic Molecules and Crystals", 2 vols.,
    Academic Press, Orlando, FL, 1987.
2. J.F. Nicoud and R.J. Twieg, "Design and Synthesis of
    Organic Molecular Compounds for Efficient Second
    Harmonic Generation", Chap. II.3, in ref. 1.
3. J.F. Nicoud, Mol.Cryst.Liq.Cryst.Inc.Nonlin.Opt.,
    1988, 156, 257.
4. P. Vidakovic, M. Coquillay and F.Salin,J.Opt.Soc.Am.
    B., 1987, 4(6), 998.
5. M. Barzoukas, D. Josse, P. Fremaux, J. Zyss, J.F.
    Nicoud and J.O. Morley, J.Opt.Soc.Am.B., 1987, 4(6),
    977.
6. R.J. Twieg and K. Jain, in "Nonlinear Optical
    Properties of Organic and Polymeric Materials", D.J.
    Williams ed., ACS Symp.Ser.N°233, Washington 1983.
7. J. Zyss, D.S. Chemla and J.F. Nicoud, J.Chem.Phys.,
    1981, 74(9), 4800.
8. J. Badan, R. Hierle, A. Périgaud and P. Vidakovic, "
    Growth and Characterisation of Molecular Crystals",
    Chap.II.4, in ref. 1.
9. S. Tomaru, S. Zembutsu, M. Kawachi and M. Kobayashi,
    J.Incl.Phenom., 1984, 2, 885.
10. D.F. Eaton, A.G. Anderson, W. Tam and Y. Wang,
    J.Am.Chem.Soc., 1987, 109, 1886.
11. F. Brisse, G. Durocher, S. Gauthier, D. Gravel, R.
    Marquès, C. Vergelati and B. Zelent, J.Am.Chem.Soc.,
    1986, 108, 6579.

The Examination of Novel Organic Materials for Second Harmonic Generation

P.J. Davis, S.R. Hall*, R.J. Jones, P.V. Kolinsky*, and A.J. Weinel

GEC HIRST RESEARCH CENTRE, EAST LANE, WEMBLEY, MIDDLESEX HA9 7PP, UK

M.B. Hursthouse and S.A. Karaulov

DEPARTMENT OF CHEMISTRY, QUEEN MARY COLLEGE, MILE END ROAD, LONDON E1 4NS, UK

INTRODUCTION

Organic materials of certain types have been shown to possess high non-linear optical coefficients[1]. From the examples discovered to date a certain number of empirical rules have been drawn up that permit the design of materials that may potentially give the same non-linear optical effects. In the case of materials for $\chi^{(2)}$ effects these can be summarised on the molecular scale by the structure (1). The $\chi^{(2)}$ effect is drastically affected by the relative positions that the active molecular fragments (1) take up on the macroscopic scale.

DONOR〜〜〜〜ACCEPTOR

(1) π-electron pathway

Molecular Requirements for good $\chi^{(2)}$

The major problem with the design of crystalline organic molecular materials is that it is not possible to accurately predict crystal structure and hence relative molecular alignment. The search for new organic materials for second order non-linear optical effects is therefore an empirical one.

In this paper the ungraded powder S.H.G. results on a series of materials closely related to "DAN"(2) and "NPP" (3) are described.

163

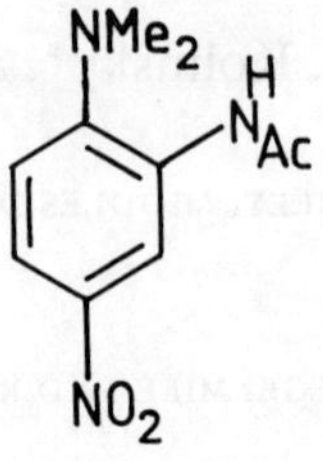

EXPERIMENTAL   **(2)** 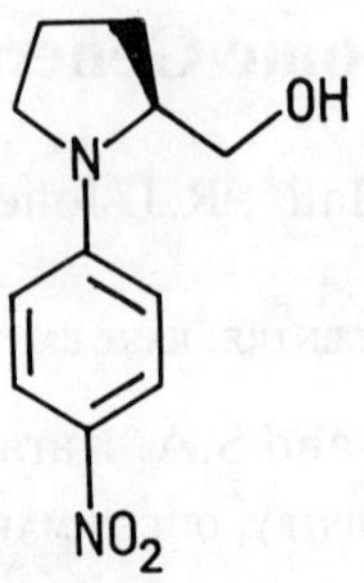   **(3)**

All of the materials were made using well established organic chemical procedures. Every material for which a result has been given passed elemental microanalysis.

The laser testing carried out used a Q-switched Neodymium/YAG laser (25 ns FWHM, 1.064 μm) as the fundamental input. The organic powder (ungraded) was packed into a sealed cell set at an angle of 45° to the input beam and the quantity of reflected second harmonic light (0.532 μm) at an angle of 90° to the input beam was measured. As a reference a sample of urea (AnalaR) was subjected to the same treatment. The result obtained from this system for NPP(3) is shown so that a true comparison can be made with other published results.

RESULTS AND DISCUSSION

The materials synthesised were all chosen as close analogues of NPP(3) or DAN (2), materials already well recognised for their high $\chi^{(2)}$ coefficients. In particular the effect of varying the group X in structure (4) was examined.

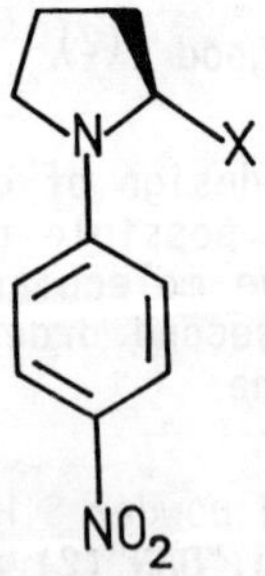

(4)

From the results of ungraded powder second harmonic generation given in Table 1 certain conclusions can be drawn:

1. The-OH group in NPP appears to be the best of the groups tried for the disruption of dipole-dipole interactions between nitroaniline groups. The only derivative that shows real promise is the material (13).

2. The replacement of the nitro aniline framework by amino-nitro-ethylene does not give a good $\chi^{(2)}$ material. The crystal structure of (10), the direct analogue of NPP, shows that the molecules favour pairing which then negates the polar character of each individual molecule itself (see Appendix One).

3. The most promising material of those studied is (13) and the crystal structure is given in Appendix One. The material (13) was designed as a close analogue of NPP and DAN with the aim of increasing the melting point with the acetamide grouping. In fact (13) has a melting point (112°C) which is below that of DAN (166°C) and NPP (115°C).

REFERENCE

"Non linear Optical Properties of Organic Molecules and Crystals", D.S. Chemla and J Zyss Eds., Acadmeic Press, 1987

Table 1 Results of S.H.G. Testing on Ungraded Powders of the Structure shown below

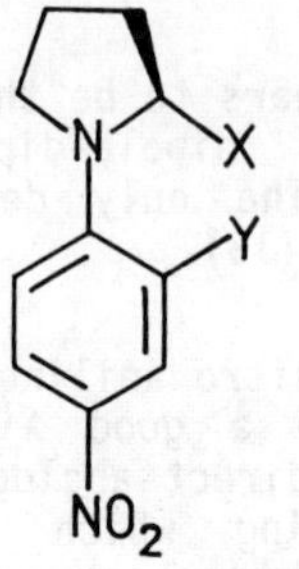

Structure		Number	MPT (°C)	SHG (urea = 1.0)
X	Y			
CH_2NH_2	H	5	89.5-90.5	0.68
CH_2NH_3Cl	H	6	>203(dec)	1.7
CH_2CN	H	7	164-165	0.02
CH_2OT_S	H	8	97-99	0.01
CH_2NPhth	H	9	245.246	15.5
CO_2H	H	14	138-139	7.0
CO_2Me	H	15	79-79.5	26
$CONH_2$	H	16	202-205	17.0
$CONHNH_2$	H	17	209-211	1.66
CONHEt	H	18	139-141	0.13
CH_2OH	H	3	115-116	355
CH_2OH	NHAc	13	110-112	204

Table 1 (Contd)

Structure	Number	MPT (°C)	SHG (Urea=1)
CO$_2$H CO$_2$Me	19 20	143-144.5 84-86	36 8.0
	10	59-60	0.0
	11	94-95	0.0
	12	112-114	0.5

APPENDIX ONE

Crystal data, intensity data, collection parameters
and details of refinement for (13)

Crystal data:

Mol. formula	C(13) H(17) O(4) N(3)
a,A	7.717
b,A	20.462
c,A	8.517
alpha, o	90.00
beta, o	90.00
gamma	90.00
V, A(3)	1.345.09
System	orthorhombic
space group	P2(1)2(1)2(1)
Z	4
radiation	Mo K(alpha) w.l. 0.71069 A

Data collection:

theta min/max	1.5,25
temperature	R.t.
total data unique	1382
total data observed	915
significant test	Fo3*sigma(Fo)

Refinement:

No. of parameters	237
absorption correction	psi-scan
weighting scheme	2002/[sigma**2(Fo)+0.0F.0]
final R=	0.062
final RW=	0.050

Crystal Structure of Compound (10)

Push–Pull Polyenes and Carotenoids for Non-linear Optics

M. Blanchard-Desce and J.-M. Lehn*

CHIMIE DES INTERACTIONS MOLÉCULAIRES, COLLÈGE DE FRANCE, 11 PLACE M. BERTHELOT, 75005 PARIS, FRANCE

I. Ledoux and J. Zyss

CENTRE NATIONAL D'ETUDES DES TÉLÉCOMMUNICATIONS, 196 AVENUE H. RAVERA, 92220 BAGNEUX, FRANCE

DESIGN

Modification of natural carotenoids[1] provides direct access to the molecular engineering of polyolefinic chains yielding molecules of interest for the design of molecular devices or of new materials[2]. For instance combination of an electron donor group at one end, and of an electron acceptor group at the other end of the polyenic chain yields push-pull carotenoids. These compounds may exhibit interesting electrochromic[3], semi-conducting[3] or non-linear optical properties[5,6].

In this paper we wish to report the synthesis of seven series of such push-pull polyenes (1)-(3), (5)-(6) and carotenoids (4), (7) and point to their interest as organic materials for optical devices[7]. The donors chosen were either the classical p-dimethylaminophenyl group or the benzodithia group, the latter unit being also present in the tetrathiafulvalenes which yield organic metals. The acceptor groups comprise aldehyde (a), dicyanomethylene (b), pyridine (c), N-methylpyridinium (d), p-cyanophenyl (e) and p-nitrophenyl (f).

SYNTHESIS OF COMPOUNDS (1)-(7)

The aldehydes (2a), (3a) and (4a) were prepared by condensation of the dialdehydes (8), (9) and (10)[8] respectively with <u>one</u> equivalent of the Wittig or Wittig-Horner reagents generated from the corresponding precursors (11) or (12)[9] in liquid-liquid[10] (Wittig) or solid-liquid[11] (Wittig-Horner) phase-transfer conditions at room temperature. The aldehydes (5a), (6a) and (7a) were synthesized similarly with the Wittig reagent generated from (13)[12].

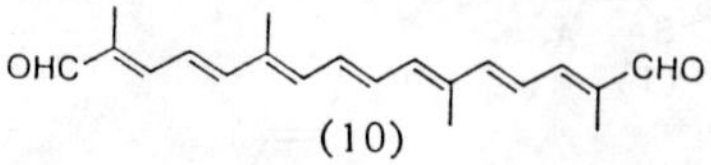

(10)

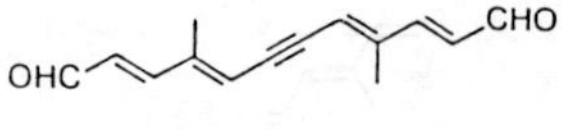

(9)

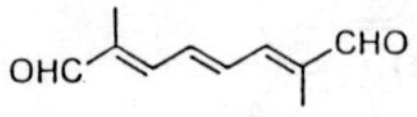

(8)

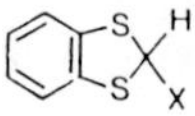

(11) : X = PPh$_3^+$ BF$_4^-$

(12) : X = PO(OEt)$_2$

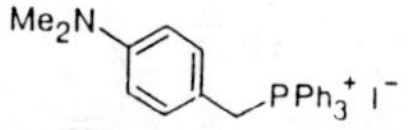

(13)

Reaction of (1a)[13]-(7a) with malonitrile in refluxing ethanol in presence of a catalytic amount of piperidine yields the dicyanomethylene derivatives. Wittig-Horner condensation of 4(-CH$_2$POPh$_2$)-pyridine[2] with the aldehydes (1a)-(4a) in solid-liquid phase transfer conditions[11] at room temperature afforded the 4-pyridyl derivatives. The methyliodides (1d)-(4d) were obtained by treatment of the corresponding pyridine derivatives with neat methyliodide at 20 °C.

The reagents (p-CN)C$_6$H$_4$CH$_2$PO(OEt)$_2$[14] and (pNO$_2$)C$_6$H$_4$CH$_2$PO(Et)$_2$[14] were condensed with the aldehydes (1a), (2a) and (3a) in solid-liquid phase transfer conditions[11] to give respectively the cyanophenyl and nitrophenyl derivatives.

OPTICAL PROPERTIES OF COMPOUNDS (1)-(7)

The compounds (1)-(7) exhibit broad and intense absorption band in the visible region characteristic of internal charge transfer (Table 1) . One may expect the conjugated chain to play the role of a molecular wire connecting the D and A groups as in the caroviologens[2], push-pull oligo-acetylenes[15] or dissymetrical polyenes[16].

Table 1. Electronic Absorption Data : λ_{max} (nm) and [log ε] values for compounds (1)-(7), (a)-(f)[a].

	(1)	(2)	(3)	(4)	(5)	(6)	(7)
(a)	372	456	466	500	450	461	498
	[4.23]	[4.66]	[4.68]	[4.88]	[4.66]	[4.73]	[4,92]
(b)	446	562	540	588	560	538	588
	[4.66]	[4.75]	[4.58]	[4.83]	[4.74]	[4.68]	[4.86]
(c)	394	457	452	505			
	[4.60]	[4.77]	[4.76]				
(d[b])	476	534	493	540			
	[4.54]	[4.66]	[4.68]				
(e)	410	465	457				
	[4.58]	[4.80]	[5.30]				
(f)	452	495	467				
	[4.40]	[4.60]	[4.88]				

[a] In chloroform at 20 °C.
[b] In DMSO solution at 20 °C.

Each series (a)-(f) shows a bathochromic effect with increasing number of conjugated double bonds as well as hypsochromic effet of the triple bond with regard to the double bond. Moreover each compound displays a pronounced and complex solvatochromic behavior. These results suggest that these compounds should present electrochromic as well as second-order non-linear optical properties.

We investigated the SHG efficiencies of the series of push-pull polyenes (1)-(7) by means of the Kurtz powder technique[17]. The powder samples were irradiated at 1.06μm and 1.32μm with an home made Q-switched YAG : Nd^{3+} laser. The SHG-powder efficiencies are listed in Table 2.

Table 2. Second Harmonic Generation Efficiencies of powdered samples of compounds (1)-(7), (a)-(f)[a]

	(1)	(2)	(3)	(4)	(5)	(6)	(7)
(a)	0 0	f 0	f 0	0 0	0,5	0	0 0
(b)	f 0	ε 0	0 0	0 0	0 0	0 0	0 0
(c)	f 0	5 0	10 30	12 13			
(d)	0 0	ε 0	ε 0	0 0			
(e)	f 0	f 0	f 0				
(f)	1 13	0 0	0 0				

a For each compound the second harmonic signal is given
relative to that of urea for irradiation at 1.06 μm (left)
and 1.32 μm (right); f : two-photons fluorescence; ε : SHG
signal between those of KDP and urea; 0 : no detectable
signal.

Compounds of series (a), (b), (c), (d) and (f) gave SHG
effects, (5a), (2c), (3c), (4c) and (1f) displaying by far
the most intense signals. The compounds of series (c)
exhibit large SHG powder efficiencies along with an
interesting chain length effect.

CONCLUSION

Taking into account the possible annihilation of SHG by
bulk effects, the present results, although qualitative,
indicate that push-pull polyenes and carotenoids (1)-(7)
are good candidates for further optical studies such as
EFISH experiments (allowing quantitative measurement of β)
as well as incorporation into organized assemblies (e.g.
Langmuir-Blodgett films). For instance, determination of β
would yield comparison of the donor strength of the
benzodithia donor group with the p-dimethylaminophenyl
one; as well as ranking of the different acceptor groups[18].
Moreover it would allow investigation of the dependence of
β on the conjugation length. It has been established both
empirically[19] and theoretically[20] that an extension of the
conjugation path between the donor and acceptor groups
results in a substantial increase in the β value. However,
up to now, there were no experimental data available for
series of push-pull polyenes of length increasing up to
nine conjugated double bonds. Preliminary results
indicate that some compounds are at least 20 times more
effective than p-nitroaniline[21].

REFERENCES

1.   'Carotenoids', O. Isler, Birkhaüser Verlag, Basel,
      1971.
2.   T.S. Arrhenius, M. BLanchard-Desce, M. Dvolaïtsky,
      J.-M. Lehn and J. Malthête, Proc. Natl. Acad. Sci.
      U.S.A., 1986, 83, 5355.

3.   J.R. Platt, J. Chem. Phys., 1961, 34, 862.
4.   J. Simon and J.-J. André, 'Molecular Semicon-
      ductors', Springer-Verlag, Berlin, 1985.
5.   D.J. Williams, Angew. Chem. Int. Ed. Engl., 1984, 23,
      690.
6.   'Nonlinear Optical Properties of Organic Molecules
      and Crystals', D.S. Chemla and J. Zyss eds., Academic
      Press, Orlando, 1987.
7.   J. Zyss, J. Mol. Electr., 1985, 1, 25; A.M. Glass,
      Science, 1984, 226, 657.
8.   O. Isler, H. Gutman, H. Lindlar, M. Montavon, R.
      Ruëgg and P. Zeller, Helv. Chim. Acta, 1956, 39, 463.
9.   J. Nakayama, Synthesis, 1975, 38; I. Degani and R.
      Fochi, ibid, 1976, 471; K. Akiba, K. Ishikawa and
      N.Inamoto, Bull. Chem. Soc. Jpn., 1978, 51, 2674.
10.   G. Märkl and A. Merz, Synthesis, 1973, 295.
11.   R. Baker and R.J. Sims, Synthesis, 1981, 117.
12.   H. Bredereck, G. Simchen and W. Griebenow, Chem.
      Ber., 1973, 106, 3732.
13.   Z.-i. Yoshida, T. Kawase, H. Awaji, I. Sugimoto, T.
      Sugimoto and S. Yoneda, Tetrahedron Lett., 1983,
      3469.
14.   F. Kagan, R.D. Birkenmeyer and R.E. Strube, J. Am.
      Chem. Soc., 1959, 81, 3026.
15.   A.E. Stiegman, V.M. Miskowski, J.W. Perry and D.R.
      Coulter, J. Am. Chem. Soc., 1987, 109, 5884.
16.   F. Effenberger, H. Schlosser, P. Baüerle, S. Maier,
      H. Port and H.C. Wolf, Angew. Chem. Int. Ed. Engl.,
      1988, 27, 281.
17.   S.K. Kurtz and T.T. Perry, J. Appl. Phys., 1968, 39,
      3798.
18.   H.E. Katz, K.D. Singer, J.E. Sohn, C.W. Dirk, L.A.
      King and H.M. Gordon, J. Am. Chem. Soc., 1987, 109,
      6561.
19.   A. Dulcic, C. Flytzanis, C.L. Tang, D. Pépin, M.
      Fétizon and Y. Hoppilliard, J. Chem. Phys., 1981, 74,
      1559.
20.   J.O. Morley, V.J. Docherty and D. Pugh, J. Chem. Soc.
      Perkin Trans.2, 1987, 1351.
21.   M. Barzoukas, work in progress.

Linear and Non-linear Optical Properties of Thiophene Oligomers

D. Fichou and F. Garnier

CNRS, ER 241-LABORATOIRE DES MATÉRIAUX MOLÉCULAIRES 2-8, RUE HENRY DUNANT, 94320 THIAIS, FRANCE

F. Charra, F. Kajzar, and J. Messier

CENTRE D'ETUDES NUCLÉAIRES DE SACLAY, 91191 GIF-SUR-YVETTE CEDEX, FRANCE

INTRODUCTION

One-dimensional semiconducting conjugated polymers such as polyacetylenes and polydiacetylenes proved to show intense and fast third-order optical nonlinearities as a consequence of the large pi-electron delocalization[1]. Recently, nonlinear optical processes have also been characterized in aromatic heterocyclic polymers such as polythiophene[2,3].

In an attempt to check the influence of the conjugation length of polythiophene upon its optical properties, we synthesized a series of alpha-conjugated thiophene oligomers α-nT, with n = 3, 4, 5, 6 and 8. To this respect, oligomers as model compounds present two major advantages:
1 - A well-defined chemical structure with:
   .no mislinking in β-position on the thiophene units,
   .a precise and controlled conjugation length,
   .the possibility to observe different solid-state conformations.
2 - Low molecular weights, allowing a better processability.
These two combined advantages would allow a refined interpretation of the dependence of the electronic properties of polythiophene upon the chain length from both theoretical and experimental view-points.

We present in this paper preliminary results on the linear and nonlinear (cubic susceptibilities $\chi^{(3)}$) optical properties of thiophene oligomers in thin films.

176

EXPERIMENTAL

α-nT oligomers are prepared in high yields according to known procedures[4,5], via organo-metallic coupling of thiophene units starting from the commercial bithiophene α-2T and terthiophene α-3T.

1 - Synthesis of α-nT when n = 4, 6 and 8:

$$H \overset{}{-}\underset{S}{\boxed{}}_{2,3,4} \overset{}{-} H \xrightarrow{LDA} H \overset{}{-}\underset{S}{\boxed{}}_{2,3,4} \overset{}{-} Li \xrightarrow[x2]{CuCl_2} H \overset{}{-}\underset{S}{\boxed{}}_{4,6,8} \overset{}{-} H$$

where LDA = lithium diisopropylamide.

2 - Synthesis of α-nT when n = 5:

$$Br \overset{}{-}\underset{S}{\boxed{}}_{3} \overset{}{-} Br + 2\ BrMg \overset{}{-}\underset{S}{\boxed{}} \overset{}{-} H \xrightarrow{NiCl_2 dppp} H \overset{}{-}\underset{S}{\boxed{}}_{5} \overset{}{-} H$$

where $NiCl_2(dppp)$ (dppp = $Ph_2P(CH_2)_3PPh_2$) is a nickel-phosphine complex acting as an efficient cross-coupling catalyst.

Thin films of α-nT are very easily prepared by simple vacuum evaporation onto glass slides at a pressure of 10^{-5} to 10^{-6} torr by heating the powdered material in a tungsten crucible. Since the α-nT oligomers are susceptible to polymerize thermally, the sublimation is carried out within a few seconds, at a rate of about 100 nm.s^{-1}. The films used in this work have thicknesses in the range 2000-9000 Å.

The cubic susceptibilities $\chi^{(3)}$ $(-3\omega; \omega, \omega, \omega)$ of the α-nT thin films have been determined from third harmonic generation (THG) measurements, using the classical Maker fringes method by rotating the films around their vertical axis. The THG intensities are corrected of self-absorption for each sample. The experimental setup used here has been described elsewhere[6].

RESULTS AND DISCUSSION

ABSORPTION SPECTRA

The absorption spectra of α-5T, α-6T and α-8T thin films are shown in respectively Fig.1a,1c and 1e.Spectra of α-5T and α-6T in solution in benzene are also presented, except for α-8T which is totally unsoluble in usual organic solvents. As it can be observed for α-5T and α-6T,

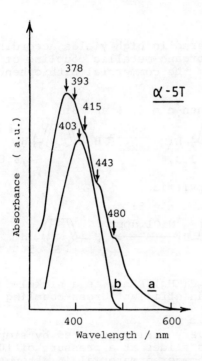

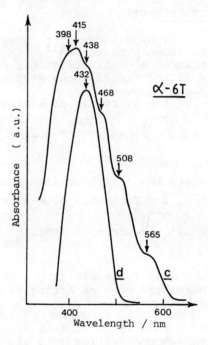

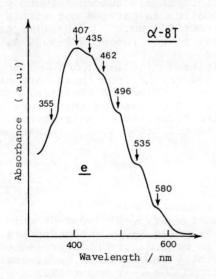

Figure 1 - Absorption spectra of α-5T, α-6T and α-8T in respectively thin films (a, c and e) and benzene solutions (b and d).

absorption in solution (1b and 1d) and solid-state films (1a and 1c) is quite different. It must be mentioned that this difference does not arise from a change of the chemical composition, as evidenced by the absorption spectra of the redissolved films.

In addition to spectral extensions on both the short and long wavelength sides as compare to the solutions, the spectra of the films exhibit a fine structure, with pronounced shoulders of decreasing intensities on the long wavelength side of the absorption band. This step-wise absorption of solid-state α-nT oligomers in thin films may be attributed to privileged molecular conformers due to the rotation of each repeat thiophene unit around the long axis of the α-nT molecule. To every shoulder of the spectra would then correspond a "rotamer" with well-defined angles (φ_i) between the successive rings, the average angle of each rotamer thus controlling the inter-ring conjugation rate and as a consequence the absorption wavelength. These various conformational structures would have been frozen-in during the sublimation process at the surface of the glass slides. The following scheme represents α-6T in a s-trans planar conformation ($\varphi = 0$) and shows the five angles which may vary to yield the observed rotamers:

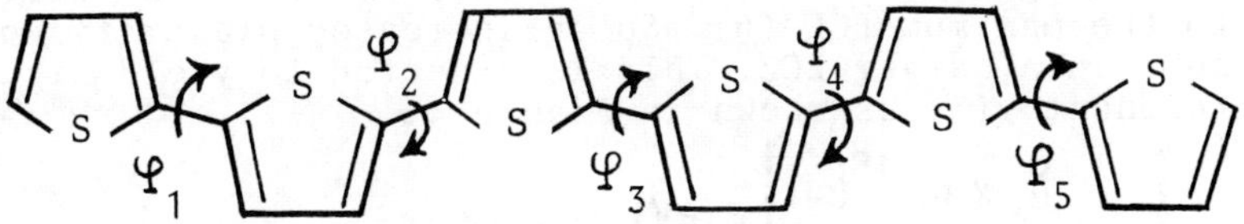

CNDO CALCULATIONS

We have performed self-consistent field CNDO computations (including mono-excited states and configuration interactions) of the excited states energies and transition dipole moments of α-nT oligomers. The calculated transition dipole moments between excited states show that in most cases only two (u,g) or three of them are relevant in the linear and nonlinear optical properties. Figure 2 depicts the variations of the energy E_u of the first one-photon allowed transition with the number of thiophene units n.

The influence of the rotation angle φ between successive rings has also been computed, assuming that the inter-ring bond lengths is not affected. A comparison with the dependence upon n shows that each rotational conformation can be roughly characterized by an equivalent conjugation length (ECL) which is obtained as a number of totally

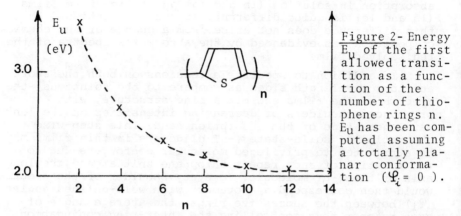

Figure 2- Energy E_u of the first allowed transition as a function of the number of thiophene rings n. E_u has been computed assuming a totally planar conformation ($\varphi_i = 0$).

planar (φ =0) repeat units by reporting the calculated or measured one-photon transition energy in Figure 2. The calculated dependence of E_u, E_g and γ_o (second hyper polarizability at zero frequency) as a function of φ are shown in Figure 3. In the case of α-6T for example, the maximum absorption of the film corresponds to an ECL of 3. The spectrum (Fig.1c) also shows that very few species correspond to the maximum ECL (6). So an increased planarity would result in a larger ECL. This is expected to yield higher non-resonant χ(3) as shown in Figure 4.

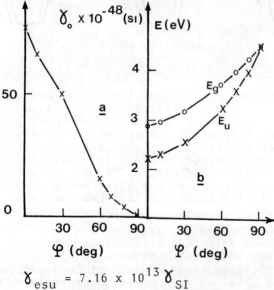

Figure 3 - Évolution of (a) the second hyper polarizability at zero frequency γ_o of α-6T and (b) the energies of the two first excited states E_u and E_g of α-6T as a function of the rotation angles φ_i. The five angles are arbitrarily equal for the calculations.

$$\gamma_{esu} = 7.16 \times 10^{13} \gamma_{SI}$$

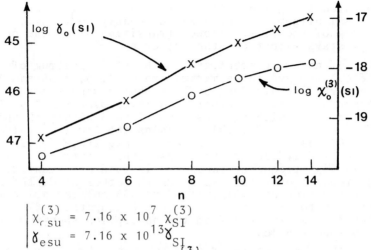

$$\left| \begin{array}{l} \chi^{(3)}_{r\,su} = 7.16 \times 10^7 \; \chi^{(3)}_{SI} \\ \gamma_{esu} = 7.16 \times 10^{13} \gamma_{SI} \end{array} \right.$$

<u>Figure 4</u> - Calculated γ_o and $\chi_o^{(3)}$ of α-6T as a function of the number of repeat units n.

THIRD HARMONIC GENERATION

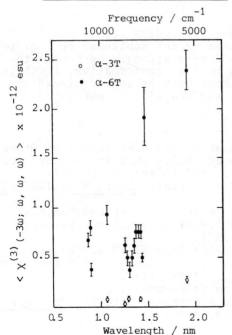

The cubic susceptibilities $\chi^{(3)}(-3\omega;\omega,\omega,\omega)$ of α-3T and α-6T thin films as a function of incident laser wavelength, are shown' in Figure 5. It can first be noticed that the $\chi^{(3)}$ values of α-3T are almost one order of magnitude smaller than those of α-6T all over the fundamental frequency range. This result is in good agreement with our theoretical predictions at zero frequency (see Fig.4) and emphasizes the importance of the conjugation length upon $\chi^{(3)}$ as

<u>Figure 5</u> - Cubic susceptibilities $\chi^{(3)}$ of α-3T and α-6T thin films as a function of the fundamental wavelength.

compared to E_u (see Fig.2); the cubic optical nonlinearity of α-nT oligomers increases noticeably up to n=14 while the energy of the first allowed transition (i.e. the optical gap) stays almost constant after n=8.

If now we concentrate on the $\chi^{(3)}$ spectrum of α-6T, one can observe three maxima around 1.06, 1.35 and 1.91 µm. The first two are due to three photon resonances and correspond to α-6T conformers absorbing over the 350-450 nm region, i.e. with a modest ECL (2 to 3 planar thiophene units, from Fig.2). This explains the low $\chi^{(3)}$ values in the 10^{-12}esu range, two orders of magnitude smaller than those of polydiacetylene. These experimental $\chi^{(3)}$ in the 0.8-1.5 µm frequency range are in good agreement with the theoretical ones for n=2-3. On the other hand, the higher $\chi^{(3)}$ observed at 1.91 µm (THG at 635 nm) may arise from a three photon resonance in highly conjugated α-6T conformers (ECL=6), as expected from the absorption spectrum of the film (see Fig.1c).

The $\chi^{(3)}$ values of α-5T have also been determined and are smaller than those of α-6T all over the 0.8-1.9 µm wavelength range. At 1.907 µm, $\chi^{(3)}$=1.88 x 10^{-12}esu for α-5T and $\chi^{(3)}$=2.38 x 10^{-12}esu for α-6T.

CONCLUSION

Thin films of thiophene oligomers with n=3,4,5,6,8 have been prepared. Their linear and nonlinear optical properties have been shown to depend strongly upon both the conjugation length and the spatial conformation. Electrical properties of these films have been presented elsewhere[7], and more detailled optical studies will be published in the near future.

REFERENCES

1. See for example, Nonlinear Optical Properties of Organic Molecules and Crystals, ed. D.S.Chemla and J.Zyss, Academic Press, 1987, Vol.2, and references therein.
2. A.J.Heeger, D.Moses and M.Sinclair, Synth.Met.,1987, 17, 343.
3. P.N.Prasad in Nonlinear Optical of Polymers, ed.A.J. Heeger, J.Orenstein and D.R.Ulrich, Mat.Res.Soc.Symp. Proc., 1988, Vol.109, p.271.
4. J.Kagan and S.K.Arora, Heterocycles, 1983, 20, 1937.
5. K.Tamao, S.Kodama, I.Nakajima, M.Kumada and Y.Kobayashi, J.Chem.Soc., Chem.Commun., 1985, 1194.
6. F.Kajzar and J.Messier in reference 1, p.51.
7. D.Fichou, G.Horowitz and F.Garnier, Int.Conf.Sci.Tech. Synth.Met., Santa Fe, USA, 1988.

Derivatives of the Cyclized Amide Form of 2-Hydroxyquinoline and Related Molecules as Potential Second-order Electro-optically Active Materials

C.L. Honeybourne, R.J. Ewen, and K.J. Atkins

MOLECULAR ELECTRONICS AND SURFACES GROUP, BRISTOL POLYTECHNIC, BRISTOL BS16 1QY, UK

A number of derivatives of urea, benzene, pyridine and pseudolinear dyes(1), in which donor-acceptor pairs of substituents are selectively positioned to give strong mutual interactions, exhibit low-lying excited singlet electronic states having large transition moments(2). A further prerequisite(3) is that large changes in the molecular dipole moment accompany the large transition moments in the case of the major diagonal contributions to the ß-tensor. This set of circumstances can lead to good figures of merit with respect to second harmonic generation(SHG) provided that the packing of the molecules or ions in the solid state is non-centrosymmetric.

The use of derivatives of 2-hydroxyquinoline(I) and related heterocycles based on a naphthenoid framework offers the following advantages:
(i)   Due to tautomerism, (I) and related hydroxy heterocycles(see Figure 1) have a urea-like constituent in their structure.
(ii)   There is a choice of locations for one or more donor-acceptor pairs of substituents.
(iii)   There are a number of locations for the placement alkyl groups either to inhibit centrosymmetric packing or to facilitate the deposition of Langmuir-Blodgett films(4).
(iv)   The presence of one or more in-ring N-atoms permits the synthesis of =N→O derivatives for adjustment of the ground-state dipole moment(5).
(v)   The lack of rotational symmetry in the 1-aza 2-aza,1,2-diaza and 1,3-diaza derivatives of naphthalene will enhance the likelihood of non-centrosymmetric packing.

183

(vi)   Co-ordination to metal ions, including the
       generation of optically active(chiral or dis-
       symmetric)complexes or complex ions is possible.

In order to confine ourselves to a limited and clearly
targeted programme of chemical synthesis, we have used
simple bonding arguments and deductions made on the
basis of transition moments and dipole moment changes
to assess the likely relative magnitudes of the major
diagonal contributions to the ß-tensor(3).

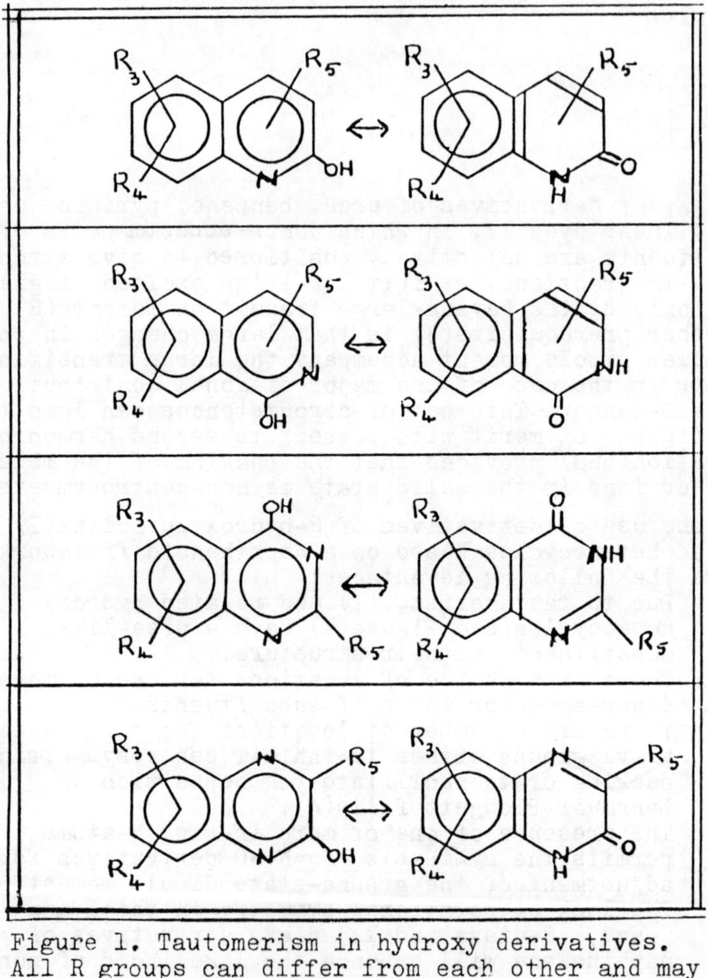

Figure 1. Tautomerism in hydroxy derivatives.
All R groups can differ from each other and may
be alkyl, donor(e.g.NH_2) or acceptor (e.g.NO_2).

We seek structures that can, in common with MNA, give
a mesoionic,quinonoid depiction of a single electron
excitation. In Figure 2 we present some molecules that
can behave in the latter way, confining ourselves to
those systems with the acceptor substituent(A) on the
carbon ring. There is a similar set with A on the hetero
atomic ring.

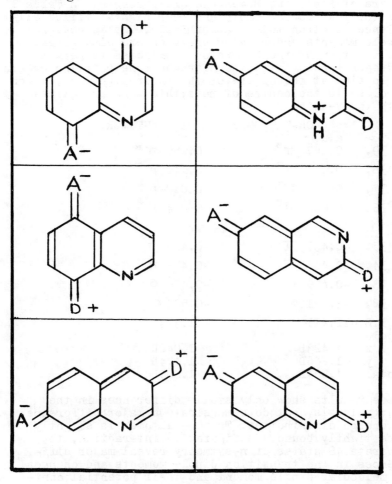

Figure 2. Location of donor(D)-acceptor(A) pairs
in order to obtain a meso-ionic quinonoid depiction
of one-electron excitation. D and A may be inter -
changed.

Theoretical aspects of 2-hydroxy-6-nitroquinoline.

We select 2-hydroxy-6-nitroquinoline as an example of
a system that has (A) and (D) groups in a favourable
location. There are two tautomeric forms of this com-
pound(2H6NQ)(see Figure 1). The results of our CNDO/2
calculations, using the equations and original parameters of Pople and Beveridge are presented in Table 1
for the four lowest empty and four highest filled all-
valence electron molecular orbitals. The molecular
dipole moments and binding energies are also given.
The location of the OH proton is selected to permit
hydrogen bonding with the nitrogen lone pair.The origin
is located at the centre of the inter-ring bond, which
is close to the centre of mass;this bond is colinear
with the y-axis.

	OH-TAUTOMER	OXO-TAUTOMER
	a.u.	a.u.
39.	0.167 π^*	0.157 π^*
38.	0.141 π^*	0.140 π^*
37.	0.067 π^*	0.048 π^*
36.	0.022 π^*	0.024 π^*
35.	-0.437 π	-0.416 π
34.	-0.466 π	-0.473 π
33.	-0.475 σ	-0.476 σ
32.	-0.491 σ	-0.500 σ
B.E.	-11.411	-11.374
Dipole:		
x	4.423D	1.042D
y	1.206D	1.474D
Total	4.584D	1.805D

These results show only small differences in the
energy levels, but do demonstrate a major difference
in charge distribution. The OH-tautomer is slightly
more tightly bound. Configuration interaction, to
generate 48 states of π-symmetry reveal major differences in the transition dipole moments and changes
in molecular dipole moment and their potential contributions to the ß-tensor. We have not yet sought a set
of parameters to optimise the fit between calculated
and theoretical values of dipole moments and ß. We
have used a well established set of parameters purely

for comparative purposes only at this stage. Using the
numbering system 1-8 for the occupied π-levels and 9-14
for the empty π-levels, we find that the three lowest
lying states,I,II and III make the major contributions
to ß-DIAG in both the static and dynamic(w = 1.17 eV)
regimes. These states are composed almost exclusively
of the 7-9, 7-10, 8 -9 and 8-10 one-electron excitations
in the case of the OH-tautomer. In the OXO-tautomer,
there is a rather more extensive mixing of configurations: 7-9, 7-10, 8-9 and 8-10 are again important, but
8-11,8-12,6-9 and 7-12 make small but significant
contributions. This highlights the fact that the four-
level model is quite inappropriate. Pugh and Morley
have already stressed the importance of utilising very
extensive configuration interaction in some classes of
materials. Using equation 53 of Ref.3 we obtain the
estimates of ß-DIAG(xxx) and ß-DIAG(yyy) within the
π-approximation shown below. We conclude that ß for the
OXO-tautomer is higher than that of the OH-tautomer,
mainly by virtue of the lower ground-state electric
dipole moment of the OXO-tautomer.

	OH-TAUTOMER	OXO-TAUTOMER
State I	0.78(7-9)+0.61(8-10)	0.83(8-9)+ 0.46(7-10)
	(SUM STATE)	+0.18(7-9 + 6-9 + 7-12)
	3.866 eV	3.840 eV
State II	0.97(8-9)+0.24(7-10)	0.88(8-10) + 0.37(7-9)
	(SUM STATE)	+0.17(8-9 +8-11 +8-12)
	4.185 eV	4.219 eV
State III	0.76(8-10)-0.63(7-9)	0.78(7-10) - 0.46(8-9)
	(DIFFERENCE STATE)	+0.29(8-11 - 7-9)
	4.965 eV	4.648 eV

ß-DIAG(O):

xxx	7.49	10.90
yyy	0.76	0.94

ß-DIAG(w):

xxx	10.86	w = 1.17eV	16.30
yyy	1.10		1.47

Table 1. Details of estimates of ß-DIAG($\times 10^{30}$/cgsesu)
within the π-states approximation using 48
one-electron configurations for the two
tautomeric forms of 2-OH-6-NO_2-quinoline.

The one-electron excitation involving the principal
shift of charge in the OXO tautomer is 8-10. The molecular orbital coefficients for the nitro-group (before
excitation; after excitation) are -0.038, 0.090, 0.097;
0.419, -0.298, -0.296. The major contribution to ß-DIAG
is given by State III which has a large coefficient for
the 8-10 excitation.

In the OH tautomer, there are similar shifts in charge
towards the nitro-group given by the excitations 7-10,
8-10 and 8-11(in decreasing order). Although State III
again makes the largest contribution to ß-DIAG, State I
and State II also make significant contributions. It is
not possible, in the case of the OH tautomer, to identify a unique excitation having a large transition moment
coupled with a large change in molecular dipole moment.

REFERENCES.

1. J.Zyss, J.Molec.Electron., 1 25 (1985).

2. Non-linear Properties of Organic Molecules and
   Crystals, Edts.,D.S.Chemla and J.Zyss, Academic
   Press, Orlando, 1987.

3. D.Pugh and J.O.Morley, Ref.2, Vol.I, p.193.

4. A.Barraud and M.Vandevyver, Ref.2, Vol.I, p.357.

5. J.F.Nicoud and R.J.Twieg, Ref.2, Vol.I, p.227.

6. J.A.Pople and D.L.Beveridge in "Approximate
   Molecular Orbital Theory", McGraw-Hill,New York,1970.

Second and Third Order Non-linear Optical Properties of End-capped Acetylenic Oligomers

J.W. Perry, A.E. Stiegman, S.R. Marder, and D.R. Coulter

JET PROPULSION LABORATORY, CALIFORNIA INSTITUTE OF TECHNOLOGY, 4800 OAK GROVE DRIVE, PASADENA, CA 91109, USA

The second and third order nonlinear optical, NLO, properties of a series of end-capped acetylenic oligomers have been investigated. These molecules are of interest as model compounds for developing relationships between molecular structure and NLO properties as well as for their potential for applications to NLO devices. Symmetric and asymmetric end-capped acetylenes have been studied. A series of symmetric acetylenes have been studied in solution by using third harmonic generation, THG. The effects of chain length and the nature of the end-capping group on the third order susceptibilities have been investigated. New asymmetric acetylenes with electron donor and acceptor end-groups have been synthesized. These materials have been screened for second harmonic generation, SHG, using the Kurtz powder method.

The potential of organic materials for nonlinear optics has recently come under intense investigation.[1-5] Although polymeric and molecular materials with substantial optical nonlinearities have been developed, broad device applications of these materials have not yet been realized. This is largely due to the fact that many material properties or conditions must be met simultaneously for such applications. The required properties[6] are dependent on the specific device application but would generally include large nonlinearity, fast nonlinear response, high optical quality (eg. low scattering loss), low optical absorption, fabricability into various forms including thin films, good mechanical and thermal stability and for $\chi^{(2)}$ applications, macroscopic noncentrosymmetry. It has proven quite difficult to achieve all the required properties in a simple molecular or polymeric material and therefore there has been substantial recent interest in exploiting novel molecular engineering strategies whereby specific structural features are introduced in order to impart the range of desired properties into a polymeric material. One such strategy is to introduce chromophores or segments with large optical nonlinearity, either as guest-host systems or as covalently attached pendant or main chain groups, into a polymer structure with desirable processability and physical properties. Significant progress has been achieved using this approach for $\chi^{(2)}$ polymers[7] and reports on prototype devices have appeared.[8] Recent work[9-13] has indicated that substantial third order nonlinearities may be achieved with finite sized conjugated molecules or segments. Thus, it is important to develop an understanding of the relation between nonlinear susceptibility and molecular structure for the wide range of possible types of finite conjugated structures.

This work focusses on the NLO properties of oligmeric end-capped acetylenes. These molecules are well suited to a study of the effects of end-groups because of the availability of well defined synthetic methods. The acetylenic linkage is of interest as a prototypical conjugated segment which can readily conjugate with aromatic or other π electron systems. Additionally, because of the rigid linear structures of these molecules and their large polarizability anisotropy, they are of interest for potential liquid crystal phases or applications.

The symmetric and asymmetric end-capped diacetylenes were synthesized using copper catalyzed (Hay catalyst) oxidative coupling reactions of the appropriate terminal acetylenes.[14] The asymmetric end-capped monoacetylenes were prepared using acetylide substitution of para-substituted halobenzenes. For the p-amino,p'-nitrodiphenylacetylenes, the electronic spectroscopic[15] and liquid crystal properties[16] (for the N-substituted diacetylenes) have been previously reported. The full details of the synthesis as well as a complete x-ray crystallographic study will be published shortly. The synthesis and linear spectroscopy of the other donor-acceptor diacetylenes will be reported elsewhere.

A wide range of asymmetric acetylenes have been prepared as shown in Table 1. We have prepared in this series of compounds molecules with donors and acceptors of varying strength, resulting in materials whose color ranges from yellow to dark red, as well as molecules of varying length of the acetylenic linkage. For example, for the case of p-amino, p'-nitrodiphenylacetylenes the number of linking acetylenes varies from one to three. We have also been interested in exploring the potential of organometallic moieties in NLO materials and because of recent developments using ferrocene organometallic groups as donors, we were motivated to synthesize several ferrocene substituted acetylenes shown in Figure 1. This series of molecules provides an interesting range of electronic and structural variation and while our intention is to determine the molecular hyperpolarizabilities of this series of compounds, thus far we have only screened several of them for macroscopic second order nonlinearity by using the Kurtz powder SHG technique.[17]

Powder SHG measurements have been made using 1064 nm and 1907 nm radiation as the fundamental for SHG. The SHG measurements were made on unsized powder samples which had been lightly ground and placed between glass cover slips. A two-channel optical system was used to provide intensity normalization of the SHG signals on every laser shot. Powder SHG efficiencies were determined relative to urea powder (50 - 100 μm particle size) for 1064 and 1907 nm.

Table 2 summarizes the powder SHG efficiencies for the compounds studied thus far. The powder SHG efficiency of 1 has been reported previously[4] and our measurement is in agreement with the reported value. The values for 2 and 3 are significantly lower, although one might expect larger β values for these molecules than for 1. The powder SHG results suggest unfavorable crystal packing for 2 and 3. In fact, our crystallographic studies clearly indicate that 2 and 3 have centrosymmetric crystal structures whereas 1 is noncentrosymmetric. The observation of weak but finite SHG from 2

<u>Table 1.</u>   Asymmetric substituted diphenylacetylenes and ferrocenylphenylacetylenes.

$$R_1\text{-}\left(\bigcirc\right)\text{-}(C{\equiv}C)_n\text{-}\left(\bigcirc\right)\text{-}R_2$$

#	R1	R2	n
2	NH_2-	$-NO_2$	1
3	NH_2-	$-NO_2$	2
	NH_2-	$-NO_2$	3
	CH_3O-	$-NO_2$	2
4	CH_3S-	$-NO_2$	2
	NH_2-	$-CN$	2
	CH_3S-	$-CN$	2
	NH_2-	$-CO_2CH_3$	2

$$\left(\text{Fe}\right)\text{-}(C{\equiv}C)_n\text{-}\left(\bigcirc\right)\text{-}R$$

#	R	n
	$-NO_2$	1
	$-NO_2$	2
5	$-CN$	1

and **3** is intriguing given the usual interpretation that noncentrosymmetry would be required. However, these observations may be explained by large surface SHG effects or by inversion symmetry breaking defects in the crystals and perhaps involving two-photon resonances as discussed by Hochstrasser et al.[18] The SHG efficiency of the organometallic compound **5** is significant although a factor of about 15 times lower than the previously reported ferrocene compound Z-ferrocenyl,p-nitrophenylethylene.[19] The result does support the notion that the ferrocene group is a good electron donor and that its compounds have some tendency

<u>Table 2.</u> Powder SHG efficiencies for some asymmetric end-capped acetylenes. SHG efficiencies are reported relative to urea at 1064 and 1907 nm.

#	Compound	SHG eff. 1064nm	SHG eff. 1907nm
1	$NH_2C_6H_4C_6H_4NO_2$	2.0	1.9
2	$NH_2C_6H_4CCC_6H_4NO_2$		0.4
3	$NH_2C_6H_4(CC)_2C_6H_4NO_2$	0.05	0.07
4	$CH_3SC_6H_4(CC)_2C_6H_4NO_2$	0.6	0.2
5	$Fe(C_5H_5)_2CCC_6H_4CN$		3.7

towards noncentrosymmetric crystallization. Powder SHG
efficiencies of the other compounds in the series will be reported
shortly.

Third-order susceptibilities of several symmetric end-capped
acetylenes have been determined in toluene solution using wedged
cell THG techniques.[20,21] THG $\chi^{(3)}$s were measured using 1064 and
1907 nm fundamental radiation. Some salient aspects of the
experimental system, the wedged THG cell, and the data analysis are
described in another paper in these proceedings.[22]

Typical experimental THG fringes are shown in Figure 1 for
diphenylbutadiene in toluene. The data are fit to a simple
fringing expression:[21]

$$I(3\omega) = A \sin^2(\Delta\psi/2) \qquad (1)$$

From the fitted parameters we obtain the third harmonic intensity
maximum and the coherence length, l_c. $\chi^{(3)}$ for the solutions are
determined relative to toluene using the solution and toluene THG
fringe parameters and the expression:

$$\chi^{(3)} = 1/l_c \; [(1-R^{1/2})(\chi^{(3)}l_c)_g + R^{1/2}(\chi^{(3)}l_c)_r] \qquad (2)$$

where $R = A/A_r$ and the subscripts g and r refer to the window
material and the reference liquid, respectively.

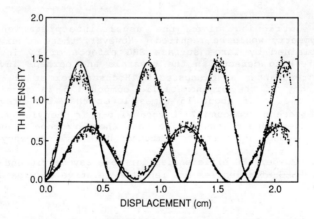

Figure 1   THG wedge cell fringes from 0.988M
diphenylbutadiyne in toluene (larger signal) and neat toluene using
1907 nm fundamental wavelength. TH intensity is in arbitrary
units. The wedge cell angle was 0.264°.

The solute susceptibility is estimated using an ideal two-component solution expression:

$$\chi^{(3)}_2 = (\chi^{(3)}_{obs} - \chi^{(3)}_1)/x_2 + \chi^{(3)}_1 \qquad (3)$$

where 2 refers to the solute and x is the mole fraction. The $\chi^{(3)}$ calculated in this way corresponds to the neat solute with the local field factor of the solution.

Table 3 summarizes the THG $\chi^{(3)}$s determined for the molecules studied at 1064 and 1907 nm. Two main structural features have been explored here. One is the effect of the conjugated chain length and the other is the effect of the end-capping group on $\chi^{(3)}$. One can see that on going from trimethylsilylacetylene, **6**, to bistriethylsilyldiacetylene, **7**, to bistriethylsilyltetraacetylene, **8**, there is a significant increase in $\chi^{(3)}$ with increasing chain length. For **7** and **8** the consistency of the values of $\chi^{(3)}$ at 1064 and 1907 nm suggests that the $\chi^{(3)}$s for these molecules are essentially nonresonant, well below three-photon resonance with the lowest allowed optical transition from the ground state. The values of $\chi^{(3)}$ for **6,7** and **8** do not scale simply according to the expected power law based on the conjugated chain length or number of repeating units.[10-12] This could be taken as an indication of a contribution of the silyl end groups, which are relatively weak electron donating groups. In going to phenyl end-capping groups in **9** and **10** we again have weak electron donating groups but now the π-electrons of the phenyls can conjugate with the acetylene π orbitals normal to the molecular plane, thus providing a route for enhanced electron delocalizaton. Note that the $\chi^{(3)}$ for **9** with one triple bond is intermediate between that for **7** and **8**. On going to **10** with two triple bonds $\chi^{(3)}$ increases significantly, becoming roughly equal to the value for **8** at 1907 nm. It is important to note that for the phenyl terminated molecules there is a significant enhancement in $\chi^{(3)}$ on going from 1907 to 1064 nm fundamental wavelength. It is reasonable to expect that these molecules will show significant dispersion in $\chi^{(3)}$ over this range as the allowed electronic absorption approaches the

Table 3. Third harmonic generation $\chi^{(3)}$ for symmetric end-capped acetylenes. $\chi^{(3)}$ is given for 1064 and 1907 nm fundamental wavelengths in 10^{-14} esu. Fc is ferrocenyl.

#	Compound	1064 nm	1907 nm
6	$(CH_3)_3SiCCH$		7
7	$(Et)_3Si(CC)_2Si(Et)_3$	11	12
8	$(Et)_3Si(CC)_4Si(Et)_3$	97	95
9	$C_6H_5CCC_6H_5$	67	46
10	$C_6H_5(CC)_2C_6H_5$	180	92
11	$Fc(CC)_2Fc$		225
12	$t-C_6H_5(CHCH)C_6H_5$	122	61
13	$t,t-C_6H_5(CHCH)_2C_6H_5$		154

third harmonic wavelength, 355 nm, for the 1064 nm fundamental. In
these cases, three-photon pre-resonance enhancement is anticipated.
The electron donating ability of the ferrocenyl moiety has been
noted earlier. The bisferrocenyl compound, **11**, provides a case
where there are relatively good electron-donating π-conjugated end
groups symmetrically terminating a diacetylene group. The $\chi^{(3)}$ for
this compound at 1907 nm is enhanced over the phenyl substituted
diacetylene by more than a factor of two. The ferrocenyl compound
is colored, however, with a low lying absorption band at 438 nm and
the $\chi^{(3)}$ value may be influenced by three photon pre-resonance
enhancement, even though the third harmonic at 636 nm is in the
transparent region of the solution spectrum. $\chi^{(3)}$ for **11** is
reasonably large , an order of magnitude larger than the THG $\chi^{(3)}$
at 1907 nm for CS_2.[23] Clearly, substantial work is needed to
evaluate the potential of metallocenes and other organometallics
for $\chi^{(3)}$ NLO properties.   THG studies on other metallocene
compounds are in progress in our laboratory.
   For a simple comparison with the phenyl capped acetylenes, we
also measured the THG $\chi^{(3)}$s of stilbene, **12**, and diphenylbutadiene,
13. Again $\chi^{(3)}$ is larger for the longer conjugated molecule.
$\chi^{(3)}$s are higher than for the corresponding acetylene compounds
consistent with greater electron delocalization and lower lying
electronic absorption in the double bond linked molecules.
   In conclusion, some preliminary results on the second order NLO
properties of asymmetric end-capped acetylenes have been reported
and only moderate powder SHG efficiencies have been realized thus
far. It is likely that structural modifications made to promote
favorable crystal packing will result in substantially enhanced
nonlinearities. A study of the third order susceptibilities of
some symmetric end-capped acetylenic oligomers was also presented
and significant nonlinearity has been observed in relatively small
conjugated molecules. The trends observed provide some insight
into the design of finite-sized conjugated segments with enhanced
third-order nonlinearity.

   ACKNOWLEDGEMENTS
   The research described in this paper was performed by the Jet
Propulsion Laboratory, California Institute of Technology as part
of its Innovative Space Technology Center which is supported by the
Strategic Defense Initiative Organization through an agreement with
the National Aeronautics and Space Administration. SRM thanks the
National Research Council and the National Aeronautics and Space
Administration for an NRC-Resident Research Associateship at the
Jet Propulsion Laboratory.

   REFERENCES

1.   'Nonlinear Optical Properties of Organic and Polymeric
      Materials,' D. J. Williams, ed., ACS Symp. Ser. 233, American
      Chemical Society, Washington, D. C., 1983.
2.   a)'Molecular and Polymeric Optoelectronic Materials:
      Fundamentals and Applications,' G. Khanarian, ed., Proc. SPIE

682 (1987); b)'Advances in Nonlinear Polymers and Inorganic Crystals, Liquid Crystals and Laser Media,' S. Musikant, ed., Proc. SPIE 824 (1988).

3. 'Nonlinear Optical Properties of Polymers,' A. J. Heeger, J. Orenstein and D. R. Ulrich, ed., Materials Research Society Symposium Proceedings Vol. 109, M. R. S., Pittsburgh, 1988.

4. 'Nonlinear Optical Properties of Organic Molecules and Crystals,' Vols. 1 and 2, D. S. Chemla and J. Zyss, ed., Academic Press, Orlando, 1987.

5. 'Nonlinear Optical and Electroactive Polymers,' P. N. Prasad and D. R. Ulrich, ed., Plenum Press, New York, 1988.

6. G. I. Stegeman, R. Zanoni and C. T. Seaton, in ref. 3, p. 53.

7. R. DeMartino, D. Haas, G. Khanarian, T. Leslie, H. T. Man, J. Riggs, M. Sansone, J. Stamatoff, C. Teng, and H. Yoon, in ref. 3, p. 65.

8. J. I. Thackara, G. F. Lipscomb, R. S. Lytel and A. J. Ticknor, in ref. 3, p. 19.

9. J. Ducuing, 'International School of Physics, E. Fermi LXIV-Nonlinear Spectroscopy,' North Holland, Amsterdam,1977, p 276.

10. D. N. Beratan, J. N. Onuchic and J. W. Perry, *J. Phys. Chem.,* 1987, *91*, 2696.

11. A. Garito, J. R. Heflin, K. Y. Wong and O. Zamani-Khamiri, in ref. 3, p. 91.

12. B. M. Pierce, in ref. 3, p. 109.

13. S. H. Stevenson, D. S. Donald and G. R. Meredith, in ref. 3, p. 103.

14. A. S. Hay, *J. Org. Chem.,* 1962, *27*, 3320.

15. A. E. Stiegman, V. M. Miskowski, J. W. Perry, and D. R. Coulter, *J. Am. Chem. Soc.,* 1987, *109*, 5884.

16. C. Fouquey, J.-M. Lehn and J. Malthete, *J. Chem Soc., Chem. Commun.,* 1987, 1424.

17. S. K. Kurtz and T. T. Perry, *J. Appl. Phys.,* 1968, *39*, 3798.

18. B Dick, R. M. Hochstrasser and H. P. Trommsdorff, in ref. 4, Vol. 2, ch. III-6, p. 159.

19. M. L. H. Green, S. R. Marder, M. E. Thompson, J. A. Bandy, D. Bloor, P. V. Kolinsky and R. J. Jones, *Nature*, 1987, 330.

20. G. R. Meredith, B. Buchhalter and C. Hanzlik, *J. Chem. Phys.,* 1983, *78*, 1543.

21. F. Kajzar and J. Messier, *J. Opt. Soc. Am. B*, 1987, *4*, 1040.

22. S. R. Marder, J. W. Perry, F. Klavetter, and R. H. Grubbs, in these proceedings.

23. G. R. Meredith, B. Buchhalter and C. Hanzlik, *J. Chem. Phys.,* 1983, *78*, 1533.

New Developments in Diacetylene and Polydiacetylene Chemistry

J.R. Allan, G.H.W. Milburn, A.J. Shand, J. Tsibouklis, and A.R. Werninck

DEPARTMENT OF APPLIED CHEMICAL SCIENCES, NAPIER POLYTECHNIC OF EDINBURGH, IO COLINTON ROAD, EDINBURGH EH IO 5DT, UK

1 INTRODUCTION

Interest in diacetylenes and polydiacetylenes was heightened after Wegner[1,2] discovered that particular diacetylenes polymerised in the solid state retaining crystallinity from monomer to polymer. At Napier Polytechnic of Edinburgh, investigation into crystal structures of diacetylenes has continued since the Department of Applied Chemical Sciences commenced work in this field[3-8]. The "rigid rod" part of diacetylene molecules is the type of structure element that is commonly found in liquid crystals[9] and Napier Polytechnic has been active in producing diacetylenes with liquid crystal properties[10-13].

Production of diacetylenes which do not have a centre of symmetry in conjunction with a considerable extension of delocalised π-orbitals in both monomers and polymers may lead to diacetylenes and polydiacetylenes that exhibit relatively large non-linear optical coefficients, $\chi^{(2)}$ and $\chi^{(3)}$[14]. At Napier Polytechnic, two series of diacetylenes are currently being produced to test this idea.[10,15,16].

Diacetylene molecules can be made to form polymers in which the diacetylene unit, $-C{\equiv}C-C{\equiv}C-$, remains intact for possible subsequent cross-linking.[17] At Napier Polytechnic, a less common approach was taken by incorporating 1st series transition metal atoms into a polymeric chain containing the diacetylene unit.[18]

A brief discussion of the results obtained in the light of the aforementioned points follows under the Titles:

I   Crystallography of Diacetylenes
II   Liquid Crystal Diacetylenes and Polydiacetylenes
III Diacetylenes with high non-linear optical coefficients
IV   Metal-complex polymers of diacetylenes.

I - Crystallography of Diacetylenes

Early work at Napier Polytechnic was concerned with the sulphonate ester series.[3-6]

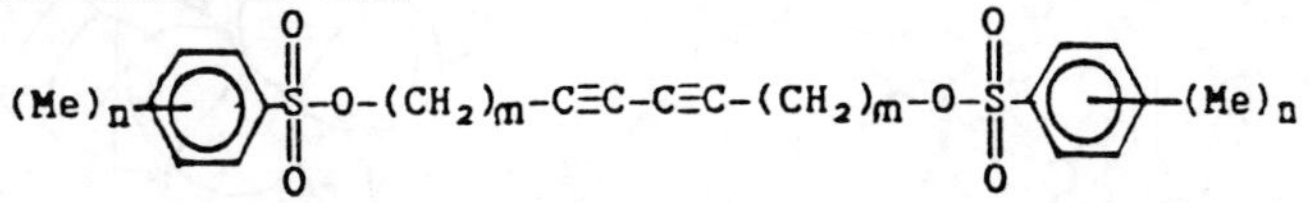

$$(Me)_n-\langle\bigcirc\rangle-\overset{O}{\underset{O}{\overset{\|}{\underset{\|}{S}}}}-O-(CH_2)_m-C{\equiv}C-C{\equiv}C-(CH_2)_m-O-\overset{O}{\underset{O}{\overset{\|}{\underset{\|}{S}}}}-\langle\bigcirc\rangle-(Me)_n$$

$1 \leq n \leq 5$ $1 \leq m \leq 4$

but more recent work has involved carboxylate esters and other diacetylenes from other projects[7,8].

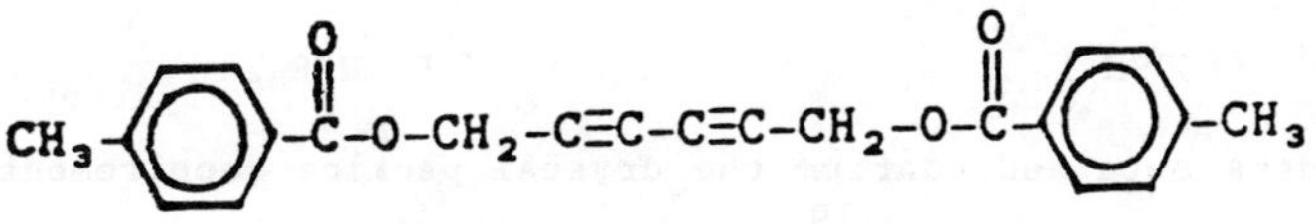

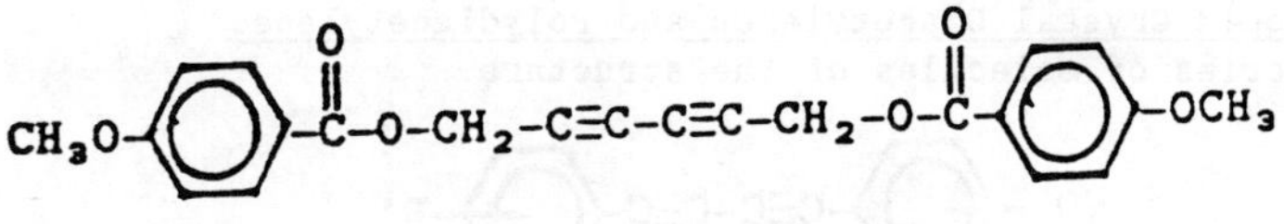

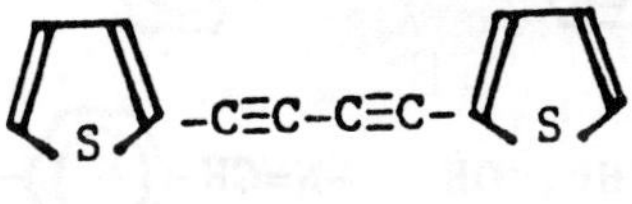

$$CH_3-\langle\bigcirc\rangle-\overset{O}{\overset{\|}{C}}-O-CH_2-C{\equiv}C-C{\equiv}C-CH_2-O-\overset{O}{\overset{\|}{C}}-\langle\bigcirc\rangle-CH_3$$

$$CH_3O-\langle\bigcirc\rangle-\overset{O}{\overset{\|}{C}}-O-CH_2-C{\equiv}C-C{\equiv}C-CH_2-O-\overset{O}{\overset{\|}{C}}-\langle\bigcirc\rangle-OCH_3$$

$$\langle\underset{S}{\bigcirc}\rangle-C{\equiv}C-C{\equiv}C-\langle\underset{S}{\bigcirc}\rangle$$

A

Packing diagrams of (A) are shown in Figures 1 and 2.

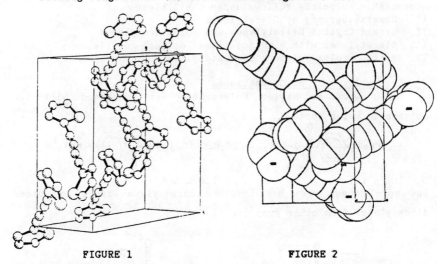

FIGURE 1   FIGURE 2

The data obtained confirm the crystal packing requirements for solid state polymerisation.[19]

II - Liquid Crystal Diacetylenes and Polydiacetylenes

A series of molecules of the structure.

$$NO_2 - \text{⬡} - C \equiv C - C \equiv C - \text{⬡} - R'$$

B

R' = NH$_2$, OMe, Br, OH   −N=CH−⬡−OMe   −N=CH−⬠O

−N=CH−⬡−OH, OMe   −N=CH−⬡−OH

−N=CH−⬡−O−CH$_2$−O

Several members of this family have displayed liquid crystallinity[10,11] and d.s.c. data for one example is shown in Figure 3 which is quite typical.

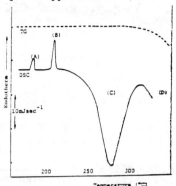

FIGURE 3

Polymerisation can occur in the liquid crystal state and the polymer formed is a nematic liquid crystal.[10]
With the aid of collaborators in Hungary, a different series of liquid crystal diacetylenes, some based on cholesterol, has been made.[13] Some examples of these are:

$$(ChOOCO(CH_2)_m-COO-CH_2-C\equiv C-)_2$$

Ch =

CH-CH₃-(CH₂)₂-CH with CH₃ and CH₃ branches

$$1 \leq m \leq 6$$

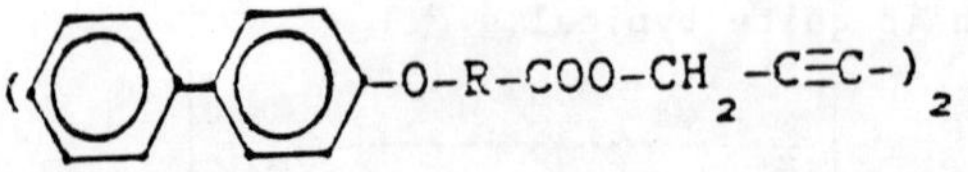

$$(\langle\bigcirc\rangle\!-\!\langle\bigcirc\rangle)\!-\!O\!-\!R\!-\!COO\!-\!CH_2\!-\!C\!\equiv\!C\!-)_2$$

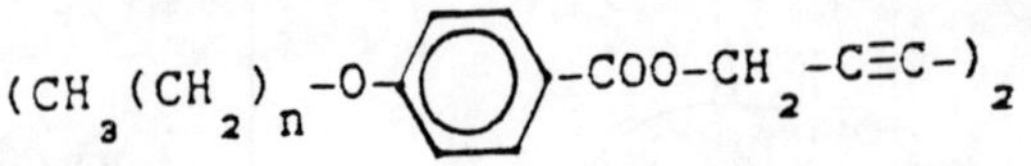

$$(CH_3(CH_2)_n\!-\!O\!-\!\langle\bigcirc\rangle\!-\!COO\!-\!CH_2\!-\!C\!\equiv\!C\!-)_2$$

$$0 \leq n \leq 12$$

As in the other series, several of these polymerise in the liquid crystal state to liquid crystal polymer.

III - Diacetylenes with High Non-Linear Optical Coefficients

The production of diacetylenes with high non-linear optical coefficients has followed 2 routes at Napier Polytechnic. The liquid crystalline series (B) has shown some promise and results are shown in Table 1.

Since it is not possible to predict whether any particular diacetylene of the unsymmetrical type shown above will crystallise to give non-centrosymmetric crystals, a requirement for high values of $\chi^{(2)}$,[14] diacetylenes were synthesised which contain chiral centres. This series is represented by the general structure:

(R'CONH)   where R = $CH(CH_3)_2$ and R' = CH_3

\
H----C➝R   = CH_3 = Ph
/
$CO_2CH_2C\equiv C)_2$ = CH_2Ph = $4\text{-}NO_2C_6H_4$

Several of these materials polymerise in the solid state by the action of light and γ-irradiation to give highly coloured polymers.[16] The non-linear optical properties are currently under investigation.

IV - Metal Complex Polymers of Diacetylenes

These materials are to be discussed elsewhere[13]

Butadiyne   SHG

1. NO_2–⟨○⟩–C≡C–C≡C–⟨○⟩–Br   0.01

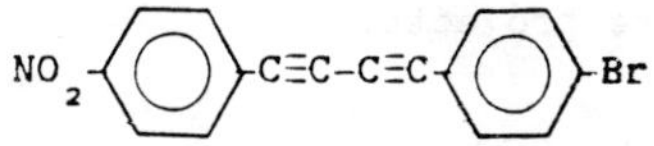

2. NO_2–⟨○⟩–C≡C–C≡C–⟨○⟩   0.02
                              MeO

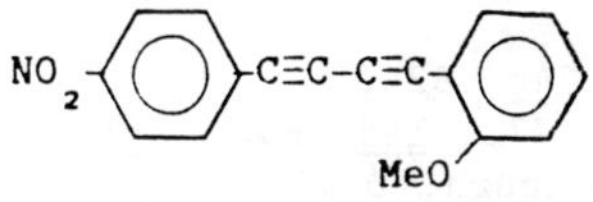

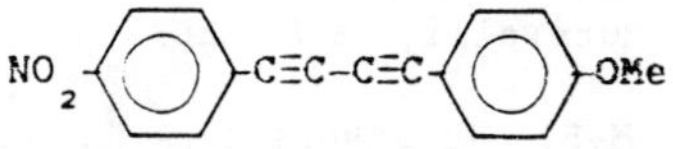

3. NO_2–⟨○⟩–C≡C–C≡C–⟨○⟩–OMe   0.03

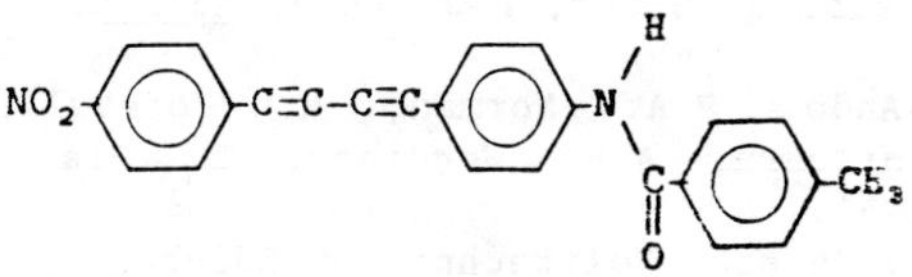

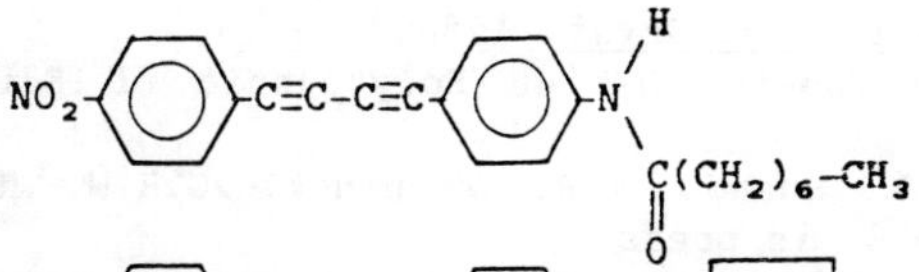

4. NO_2–⟨○⟩–C≡C–C≡C–⟨○⟩–N(H)–C(O)–⟨○⟩–CH_3 0.01

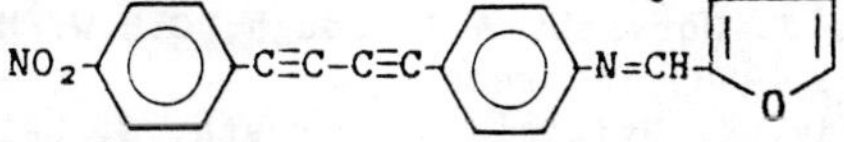

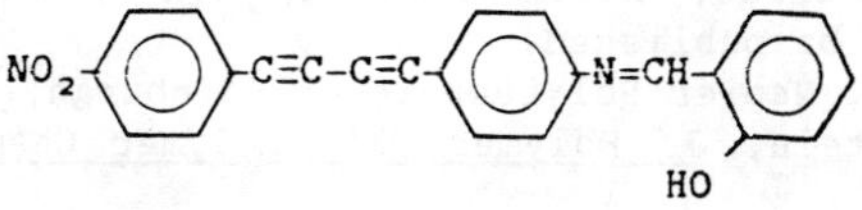

5. NO_2–⟨○⟩–C≡C–C≡C–⟨○⟩–N(H)–C(O)$(CH_2)_6$–CH_3 0.07

6. NO_2–⟨○⟩–C≡C–C≡C–⟨○⟩–N=CH–furan(O)   0.10

7. NO_2–⟨○⟩–C≡C–C≡C–⟨○⟩–N=CH–⟨○⟩   0.33
                                        HO

Standard

1. 2-methyl-4-nitroaniline   4.32

2. 3-nitroaniline   0.62

3. 3-bromonitrobenzene   0.02

Table 1

Conclusion
    Diacetylenes and polydiacetylenes have been explored along a
broad front at Napier Polytechnic and the work done has indicated
the direction of extension of these projects.

REFERENCES

1. G. Wegner, Z Naturforsch, 1969, 24b, 824.
2. G. Wegner, Die Makromol Chemie, 1971, 145, 85.
3. A.R. Werninck, E. Blair, H.W. Milburn, D.J. Ando, D. Bloor, M.
    Motevalli, M.B. Hursthouse, Acta Cryst, 1985, C41, 227.
4. R.J. Day, D.J. Ando, D. Bloor, P.A. Norman, E. Blair, A.R.
    Werninck, H.W. Milburn, M. Motevalli, M.B. Hursthouse, Acta
    Cryst., 1985, C41, 1456.
5. M. Motevalli, P.A. Norman, M.B. Hursthouse, A.R. Werninck,
    H.W. Milburn, E. Blair, D. Bloor, P.J. Ando, Acta Cryst, 1986,
    C42, 1049.
6. D. Bloor, D.J. Ando, P.A. Norman, M. Motevalli, M.B.
    Hursthouse, H.W. Milburn, A.R. Werninck, E. Blair, Acta.
    Cryst, 1986, C42, 1051.
7. A. Lough, PhD Thesis, Napier Polytechnic of Edinburgh, 1988.
8. M.J. Barrow, A. Lough, G.H.W. Milburn, A.R. Werninck, Acta
    Cryst, 1988, unpublished results.
9. B Bahadur, Mol. Cryst. Liq. Cryst, 1984, 109, 3.
10. J. Tsibouklis, PhD Thesis, Napier Polytechnic of Fdinburgh,
    1988.
11. J. Tsibouklis, A.J. Shand, A.R. Werninck, G.H.W. Milburn,
    Liquid Crystals, 1988, in press.
12. M.J. Barrow, G. Hardy, J. Horvath, A.J. Lough, G.H.W. Milburn,
    K. Nyitrai, Acta Cryst, 1988, in press.
13. G.H.W. Milburn, G. Hardy, K. Nyitrai, J. Horvath, G. Balazs,
    J. Varga, A.J. Shand, New Polymeric Materials, 1988, in press.
14. J. Zyss, J. of Non-Crystalline Solids, 1982, 211.
15. J. Tsibouklis, A.J. Shand, A.R. Werninck, G.H.W. Milburn,,
    Chemtronics, 1988, to be published.
16. E. Bolton, PhD thesis, Napier Polytechnic of Edinburgh, 1988.
17. Y. Ozcayir, A. Blumstein, J. Polymer Sci. Polymer Chem. ed.,
    1986, 24, 1217.
18. L. Macindoe, PhD Thesis, Napier Polytechnic of Edinburgh, 1988.
19. D. Bloor, "Developments in Crystalline Polymers - I"
    ed. D.C. Bassett, Applied Science, 1982.

Third Harmonic Generation in Organic Dye Solutions

A. Penzkofer* and W. Leupacher

NATURWISSENSCHAFTLICHE FAKULTÄT II-PHYSIK, UNIVERSITÄT REGENSBERG, D-8400 REGENSBURG, FRG

1 INTRODUCTION

Light at the third harmonic frequency, $\nu_3 = 3\nu_L$, may be generated by the direct third-order nonlinear interaction $\nu_L + \nu_L + \nu_L \rightarrow \nu_3$ due to the third-order nonlinear susceptibility $\chi_{THG}^{(3)}$, or it may be generated by cascading the second harmonic generation, $\nu_L + \nu_L \rightarrow \nu_2$, and the frequency mixing, $\nu_2 + \nu_L \rightarrow \nu_3$. The cascading interaction is due to the second-order nonlinear optical susceptibilities $\chi_{SHG}^{(2)}$ and $\chi_{FM}^{(2)}$. Phase-matching, $\Delta k = 0$, is necessary for efficient light generation at the third harmonic frequency. Two nonlinear media in series are necessary for phase-matching both the second harmonic generation and the frequency mixing. The various generation schemes of phase-matched third harmonic light generation are summarized in Table 1.

In this paper the efficient phase-matched third harmonic generation in some organic dye solutions is studied. A picosecond Nd-phosphate glass laser is used as pump source. The third-order nonlinear susceptibilities and hyperpolarizabilities are determined. The limiting factors of the third-harmonic conversion efficiency at high pump pulse intensities are discussed. The third harmonic generation is resonantly enhanced by two-photon absorption (TPA, S_0-S_1 absorption peak between fundamental and third harmonic frequency). The phase-matching at a certain dye concentration is achieved by the anomalous dispersion of the refractive index of the dye above the S_1-absorption band.

Table 1   Schemes of phase-matched light generation at third harmonic frequency. TPA = two-photon absorption. IC = inversion
center.

Medium	Phase-matching	Process	Resonance	Reference
Metal vapors	puffer gas	direct	TPA	1-3
Inert gases	puffer gas	direct	-	4
Organic dye				
solutions	solvent	direct	TPA	5-11
vapors	puffer gas	direct	TPA	12
Birefringent crystals				
with IC	birefringence	direct	-	13-15
without IC	birefringence	direct and cascading	-	13,14 16,17
without IC	birefringence	cascading in two crystals	-	18,19
Liquid crystals and layered materials	reciprocal lattice vector	direct	-	20,21

2 RESULTS

Determination of nonlinear susceptibilities

As long as other nonlinear optical processes and
pump pulse depletion may be neglected the third harmonic energy conversion efficiency $\eta_E = W_3/W_L$ is given by[22]

$$\eta_E = \frac{\kappa}{3^{3/2}} \; |\chi_{THG}^{(3)}|^2 I_{0L}^2 \tag{1}$$

with

$$\kappa = \frac{4\pi^2 \nu_3^2 \{ \exp(-3\alpha_L \ell) + \exp(-\alpha_3 \ell) - 2\exp[-(\alpha_3 + 3\alpha_L)\ell/2]\cos(\Delta k \ell) \}}{n_3 n_L^3 c_0^4 \varepsilon_0^2 [(\alpha_3 - 3\alpha_L)^2/4 + \Delta k^2]} \tag{2}$$

α_L and α_3 are the linear absorption coefficients at ν_L
and ν_3, respectively. n_L and n_3 are the corresponding
refractive indices. ℓ is the sample length, c_0 is the
vacuum light velocity, and ε_0 is the permittivity. The

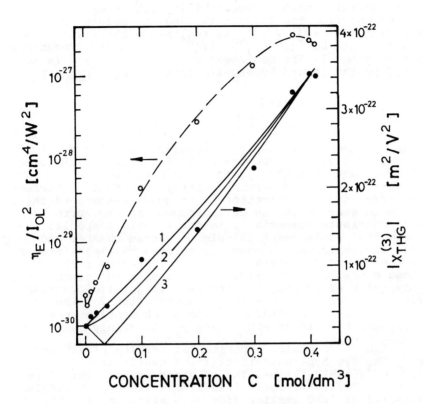

Figure 1 Third harmonic energy conversion efficiency and third-order nonlinear susceptibility for methylene blue in methanol. Solid curves are calculated for $\chi_D^{(3)} > 0$ (1), imaginary (2), and <0 (3).

wave-vector mismatch is given by $\Delta k = 6\pi\nu_L (n_3 - n_L)/c_0$. A temporal and spatial Gaussian input pulse shape is assumed [intensity $I_L = I_{0L} \exp(-t^2/t_0^2 - r^2/r_0^2)$]. An effective interaction length may be defined by

$$\ell_{eff} = \frac{\exp(-3\alpha_L \ell/2) + \exp(-\alpha_3 \ell)}{[(\alpha_3 - 3\alpha_L)^2/4 + \Delta k^2]^{1/2}} \tag{3}$$

The third order susceptibility $\chi_{THG}^{(3)}$ comprises contributions from the solvent (S) and the solute (D), i.e. $\chi_{THG}^{(3)} = \chi_S^{(3)} + \chi_D^{(3)}$. $\chi_S^{(3)}$ is real since the solvent is transparent, but $\chi_D^{(3)} = \chi_D^{(3)'} - i\chi_D^{(3)''}$ is complex (resonance contributions). The nonlinear susceptibility $\chi^{(3)}$ is related to the second hyperpolarizability by

$$\chi^{(3)} = NL^{(4)}\gamma^{(3)}/\varepsilon_0 \tag{4}$$

N is the number density of molecules and $L^{(4)} = (n_3^2+2) \times (n_L^2+2)^3/81$ is the Lorentz-local field correction factor. The real and imaginary parts of $\chi_D^{(3)}$ may be resolved by measuring the third harmonic energy conversion efficiency versus dye concentration.[9] For non-phasematched third harmonic generation the cell windows and the surrounding air contribute essentially to the signal. A special experimental arrangement (sample in vacuum chamber and cell window thickness equal to an even multiple integer of the coherence length $\ell_{coh} = \pi/\Delta k$) is necessary to avoid these contributions.[22] Fig.1 shows the third harmonic conversion efficiency and the resulting third-order nonlinear susceptibility versus concentration for the dye methylene blue in methanol.[9] The S_1 absorption peak of methylene blue is at 650 nm and $\chi_D^{(3)}$ is mainly real.

Table 2 contains experimental results of $\chi^{(3)}$ and $\gamma^{(3)}$. The dye hyperpolarizabilities center around 10^{-59} cm^{4}V^{-3} ($\simeq 10^{-34}$ esu). The solvents are far out of resonance. Their hyperpolarizabilities are approximately a factor of 1000 smaller (for discussion see Ref.9).

Efficient Phase-Matched Third Harmonic Generation

For some dyes phase-matched collinear third harmonic generation of Nd:glass laser pulses is possible at a fixed concentration C_{PM} due to the anomalous refractive index dispersion above the S_1 absorption band. High conversion efficiencies require long effective interaction lengths (small linear absorptions α_3, see Eq.3). The ℓ_{eff} values at C_{PM} are given in Table 2. Fig.2 shows the absorption cross-section spectrum of the dye PYC in hexafluoroisopropanol. The absorption minimum of this dye is at 375 nm and does not coincide with $\lambda_3 = 351.3$ nm. Nd-silicate glass lasers may be frequency tuned near to the absorption minimum.[23]

The third harmonic conversion efficiency η_E versus pump pulse peak intensity is plotted in Fig.3 for the

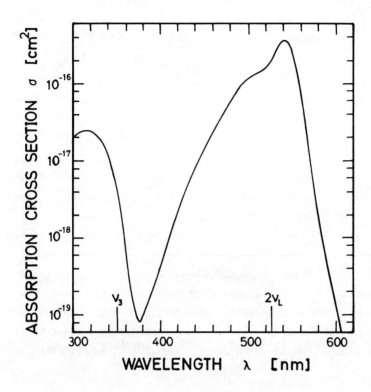

Figure 2   Absorption cross-section spectrum of 0.0825
            molar PYC in HFIP.

dye PYC in hexafluoroisopropanol. At high pump pulse intensities the conversion efficiency saturates. η_E-values
at $I_{OL} = 2 \times 10^{11}$ W/cm² are listed in Table 2.

Limitation Of Conversion Efficiency

        At high pump pulse intensities the two-photon ab-
sorption dynamics (two-photon absorption, excited-state
absorption, amplified spontaneous emission, refractive
index changes) and the self-phase modulation reduce the
third harmonic conversion efficiency.[11] Some dependences
of the conversion efficiency on material parameters are
summarized in Table 3.[1,2,11]

Table 2   Dye and solvent parameters and THG results

Dye	Solvent	C [mol/dm³]	ℓ_{eff} [μm]	$\|\chi^{(3)}_{THG}\|$ [m²V⁻²]	$\|\gamma^{(3)}_{THG}\|$ [cm⁴V⁻³]	η_E a)
Rhodamine 6G	ME	0.3 b)	2.8	8×10^{-22}	1.4×10^{-59}	
Fuchsin	ME	0.25 b)	7.9	5.1×10^{-22}	1.2×10^{-59}	
Methylene blue	ME	0.37 c)	13	3.2×10^{-22}	4×10^{-60}	
Safranine T	HFIP	0.33 c)	48	1.7×10^{-22}	3.3×10^{-60}	1×10^{-4}
PYC	HFIP	0.0825 c)	113	2×10^{-22}	1.7×10^{-59}	2×10^{-4}
HMICI	HFIP	0.08 c)	160	2.48×10^{-22}	2×10^{-59}	4×10^{-4}
-	ME	24.73 b)	2.7	2.4×10^{-23}	6×10^{-63}	1×10^{-7}
-	HFIP	9.46 b)	5.1	1.4×10^{-23}	1×10^{-62}	1.3×10^{-7}

a: $I_{OL} = 2\times10^{11}$ W/cm². b: not phase-matchable. c: phase-matched concentration C_{PM}. ME = methanol. HFIP = hexafluoroisopropanol. PYC = 1,3,1',3'-tetramethyl-2,2'-dioxopyrimido-6,6'-carbocyanine hydrogen sulphate. HMICI = 1,3,3,1',3',3'-hexamethylindocarbocyanine iodide. $\chi^{(3)}$ (esu)=$(9\times10^{8}/4\pi)\chi^{(3)}$ (SI). $\gamma^{(3)}$ (esu)=$8.088\times10^{24}\gamma^{(3)}$ (SI).

3   CONCLUSIONS

The highest conversion efficiency obtained was 4×10^{-4}. Efficiencies η_E up to the percent region are expected for dyes with extremely low α_3 values and moderate excited state absorption cross-sections.

REFERENCES

1. J.F. Reintges, 'Nonlinear Optical Parametric Processes in Liquids', Academic Press, Orlando, 1984.
2. J.F. Reintges, in 'Laser Handbook', edited by M. Bass and M.L. Stitch, North-Holland, Amsterdam, 1985, Vol.5, Chapter 1.
3. C.R. Vidal in 'Tunable Lasers', edited by F.L. Mollenauer and J.C. White, Springer, Berlin, 1987, p. 57.
4. A.H. Kung, J.F. Young and S.E. Harris, Appl. Phys. Lett., 1973, 22, 301.

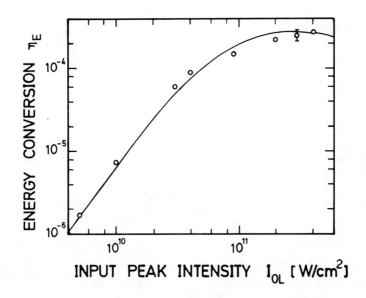

INPUT PEAK INTENSITY I_{OL} [W/cm^2]

Figure 3   Third harmonic energy conversion efficiency of
           0.0825 molar PYC in HFIP. Sample length 0.1 mm.

5. P.P. Bey, J.F. Guiliani and H. Rabin, IEEE J. Quant. Electron.,
   1971, QE-7, 86.
6. R.K. Chang and L.K. Galbraith, Phys. Rev., 1968, 171, 993.
7. J.C. Diels and F.P. Schäfer, Appl. Phys., 1974, 5, 197.
8. L.I. Al'Perovich, T.B. Baveav and V.V. Shabalov, Sov. J. Appl.
   Spectrosc., 1977, 26, 196.
9. W. Leupacher and A. Penzkofer, Appl. Phys., 1985, B36, 25.
10. W. Leupacher, A. Penzkofer, B. Runde and K.H. Drexhage, Appl.
    Phys., 1987, B44, 133.
11. A. Penzkofer and W. Leupacher, Opt. Quant. Electron., 1988, 20,
    222.
12. V.F. Lukinykh, S.A. Myslivets, A.K. Popov, and V.V. Slabko,
    Appl. Phys., 1985, B38, 143.
13. P.D. Maker and R.W. Terhune, Phys. Rev., 1965, 137A, 801.
14. S.A. Akhmanov, L.B. Meisner, S.T. Parinov, S.M. Saltiel and
    V.G. Tunkin, Sov. Phys. JETP, 1977, 46, 898.
15. A. Penzkofer, F. Ossig and P. Qiu, Appl. Phys. B, to be
    published.
16. C.C. Wang and E.L. Baardsen, Appl. Phys. Lett., 1969, 15, 396.
17. P. Qiu and A. Penzkofer, Appl. Phys., 1988, B45, 225.

Table 3   Limitation dependences of third harmonic generation[11]

Linear absorption, α_L $\qquad\qquad$ $\eta_E \propto |\chi_{THG}^{(3)}|^2 I_{OL}^2 \alpha_L^{-2}$

Linear absorption, α_3 $\qquad\qquad$ $\eta_E \propto |\chi_{THG}^{(3)}|^2 I_{OL}^2 \alpha_3^{-2}$

TPA of $2\nu_L$, $\sigma_{LL}^{(2)}$ $\qquad\qquad$ $\eta_E \propto |\chi_{THG}^{(3)}|^2 (\sigma_{LL}^{(2)})^{-2}$

Excited state absorption, σ_{ex}

  without ground state depletion

  $(\alpha_{ex} = N_{ex}\sigma_{ex} \propto \sigma_{ex}\sigma_{LL}^{(2)} I_{OL}^2)$ $\qquad$ $\eta_E \propto |\chi_{THG}^{(3)}|^2 (\sigma_{ex}\sigma_{LL}^{(2)} I_{OL})^{-2}$

  with ground state depletion

  $(\alpha_{ex} = N_D\sigma_{ex})$ $\qquad\qquad$ $\eta_E \propto |\chi_{THG}^{(3)}|^2 I_{OL}^2 \sigma_{ex}^{-2}$

Refractive index dispersion

  $(D = \partial n_3/\partial\nu - \partial n_L/\partial\nu)$ $\qquad$ $\eta_E \propto |\chi_{THG}^{(3)}|^2 I_{OL}^2 D^{-2}$

Refractive index change due to

excited state population, Δn $\qquad$ Change of C_{PM}

Nonlinear refractive index, n_2 $\qquad$ Change of C_{PM}

---

18. R. Piston, Laser Focus, 1978, 14/7, 66.
19. D. Eimerl, IEEE J. Quant. Electron., 1987, QE-23, 575.
20. J.W. Shelton and Y.R. Shen, Phys. Rev. Lett., 1971, 26, 538.
21. N. Bloembergen and A.J. Sievers, Appl. Phys. Lett., 1970, 17, 483.
22. M. Thalhammer and A. Penzkofer, Appl. Phys., 1983, B32, 137.
23. H. Schillinger and A. Penzkofer, Opt. Commun., to be publ.
24. R.W. Minck, R.W. Terhune and C.C. Wang, Appl. Opt., 1966, 5, 1595.

Optical Bistability in Non-linear Absorbing Molecules

S. Speiser* and F.L. Chisena

ENGINEERED MATERIALS SECTOR, ALLIED SIGNAL INC., MORRISTOWN, NJ 07960,
USA

1 INTRODUCTION

Optical bistability is characterized by two different light transmission states of an optical system for a given input light intensity.[2,3] In order to observe optical bistability, a nonlinear optical medium and an optical feedback are required. A nonlinear absorbing medium in an optical resonator was the first configuration for which the existence of optical bistability was theoretically predicted by Szoke et al.,[4] and by McCall.[5]

Although difficult to implement experimentally, absorptive bistability has been a major subject of theoretical research involving the elucidation of the mechanisms and the characterization of the relevant microscopic phenomena. The main part of this theoretical work, initiated by Bonifacio and Lugiato,[6] was carried out for a nonlinear medium consisting of an ensemble of two-level systems, thus exhibiting saturation of resonant absorption. In this context, analysis of the steady-state and temporal behavior of the stability conditions were carried out,[7-9] using, for the most part, the mean-field approximation, which ignores propagation effects. Recent efforts concentrated on media exhibiting dynamically increasing absorption.[10]

Recently we have analyzed the steady-state characteristics of absorptive bistability assuming particular microscopic models for the optical nonlinearity. In particular we considered nonlinear molecular absorbers, such as organic laser dyes. The nonlinearity originates from unique intensity dependence of the complex index of refraction, manifested also in the photoquenching of molecular fluorescence.[11] The main analytical tool is the recently developed complex nonlinear eikonal approximation.[12-14]

Optical bistability in an organic dye has already been observed.[15-17] The present paper deals with many aspects of similar systems; in particular, the relations between molecular parameters and the conditions for observing optical bistability are investigated.

2 SUMMARY OF EARLIER RESULTS[17]

Optical bistability has been observed in highly concentrated
fluorescein dye solution and in thin (~1μm) doped polymeric films.
For fluorescein, at concentrations larger than 10^{-5} mole/l dye
dimers are formed. The dimer-monomer equilibrium constant is
10^5l/mole so that most of the dye species are in the dimer form. At
480 nm the dimer absorption cross section is 10^{-18}cm²/molecule,
while that for the dye monomer molecule is 7.6×10^{-17}cm²/molecule.
Upon laser excitation dimers dissociate to form monomers thus
providing a highly nonlinear laser induced absorption. The high
nonlinear absorption coefficient can be utilized for optically
bistable response of the dye system.

Optical bistability was observed by placing dye solutions or dye
thin films inside a Fabry-Perot resonator and exciting it with 480 nm
dye laser pulses of 10 ns duration. The effect is more pronounced in
10^{-4} mole/l fluorescein than in 10^{-6}mole/l fluorescein in which
dimer formation is not that efficient.

3 OPTICAL BISTABILITY IN FLUORESCEIN SALTS

In order to observe optical bistability in a molecular absorber
due to either absorption saturation or excited state absorption one
has to look for system free of higher molecular aggregates.[18-20]

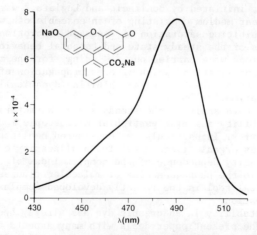

Fig. 1   Absorption spectrum for disodium fluorescein in ethanol
solution. The same spectrum is observed in PMMA doped thin film

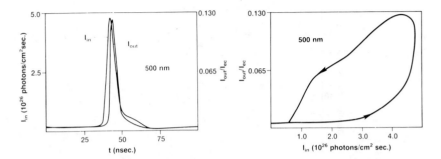

Fig. 2   Input and output intensity 500 nm pulses (left graph)   for
9.45 x 10^{-5} mole/l disodium fluorescein/ethanol solution with the
corresponding optical bistability hysteresis loop (right graph)

Such systems are: disodium fluorescein, eosin Y and erythrosin B
where dimer formation is known to be inefficient (K = 5.01 l/mole) at
the concentration range employed here.[19]
        The absorption spectrum of disodium fluorescein is shown in Fig.
1, no significant deviations from Beer's law are observed at
concentrations lower than 10^{-3} mole/l, or in the PMMA doped thin
film samples. Similar spectra were obtained for eosin Y and
erythrosin B. Figure 2 shows a typical input and output 500 nm laser
pulses for 9.45 x 10^{-5} mole/l disodium fluorescein ethanol solution
together with the associated optical bistability loop. The output
intensity I_{out} is normalized to that of an empty cavity I_{ec}
excited by the same laser input pulse. It is evident that in this
case optical bistability is characterized by a lag of the output
pulse with respect to the input pulse and by an expected
counterclockwise loop.
        The optical bistability response of a molecular absorber is
expected to depend on the optical density of the medium.[14] The best
way to examine this dependence is by measuring the optical
bistability spectrum. This can be done by simply measuring the area
of the optical bistability loop as a function of the input laser
frequency at a constant laser peak power and for a given sample
concentration. The results for disodium fluorescein are shown in
Fig 3. Optical bistability is observed at a much narrow spectral
range than that spanned by the absorption band. This result is in
agreement with theoretical calculation of the spectral response of
optical bistability of molecular absorbers.[15] The corresponding
bistability response of erythrosin B is shown in Figure 4 with
similar behavior for eosin Y both in solution and in this films. Such
a bistability loop is characteristic of intensity-induced increasing

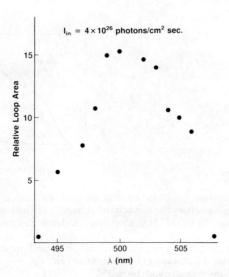

Fig. 3   Spectrum of the optical bistability area for disodium fluorescein

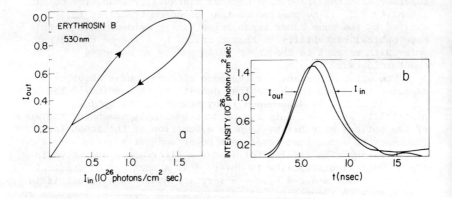

Fig. 4   Input and output intensity 530 nm pulses (b) for 10^{-6} mole/l erythrosin B/ethanol solution with the corresponding optical bistability hysteresis loop (a)

absorption.[2] This is to be expected as S_1 absorption for disodium
fluorescein peaks at 530 nm and is more intense than S_0
absorption.[14] Thus excitation to the red of the absorption peak
results in primarily S_1 absorption which increases with increasing
pump intensity.[11] The fact that the bistability loop is reversed
due to change in the electronic absorption mechanism suggests a
purely electronic route for inducing optical bistability. This is the
first example of purely increasing pump electronic absorption optical
bistability. Samples of PMMA thin films doped with disodium
fluorescein exhibit a similar response. It should be noted that this
type of increasing absorption bistability is due to intensity induced
changes in the level populations and requires the resonator for
feedback.[12,13] This is a different situation compared to
dynamically increasing absorption in semiconductors where cavity-less
optical bistability is observed.[10] The bistability spectral
behavior for eosin Y and erythrosin B is similar to that observed for
disodium fluorescein.

4 CONCLUSION

Our results on optical bistability in fluorescein dyes show the
suitability of dye systems for nonlinear optical applications. The
optical nonlinearites involved are the same as those that determine
fluorescein performance in four wave mixing and real time holography
experiments.[22,23] However, as was demonstrated here, depending on
the conditions, fluorescein may dimerize and thus give rise to a
novel type of optical nonlinearity.[21] This type of nonlinearity is
a kind of photochromism which was shown to be instrumental in the
bistable response of photochromic systems.[24] For disodium
fluorescein, eosin Y and erythrosin B we have shown evidence for
optical bistability due to either absorption saturation or to excited
state absorption, depending on the excitation wavelength. This is
indicative of mainly electronic mechanism for inducing the
bistability response, which is different from previous examples of
optical bistability in dyes, which involve thermally induced
nonlinearity.[16,17] Our study increases the scope of applications of
organic materials for nonlinear optics.[25] Progress in this field is
considered essential in order to realize optical hardware needed for
optical data processing.[26]

REFERENCES
1. Permanent address: Department of Chemistry, Technion-Israel
   Institute of Technology, Haifa 32000, Israel.
2. J.A. Goldstone, 'Optical bistability.' In 'Laser Handbook',
   Vol. 4, M.L. Stitch, M. Bass (eds) (North-Holland, Amsterdam
   1985).
3. H.M. Gibbs, 'Optical Bistability: Controlling Light with
   Light' (Academic, New York 1985).

4. A. Szoke, V. Dannell, J. Goldhar, N.A. Kurnit, Appl. Phys. Lett., 1969, 15, 376.
5. S.L. McCall, Phys. Rev. A, 1974, 9, 1515.
6. R. Bonifacio, L.A. Lugiato, Phys. Rev. A, 1978, 18, 1129.
7. G.P. Agrawal, H.J. Carmichael, Phys. Rev. A, 1979, 19, 2074.
8. R. Roy, M.S. Zubairy, Phys. Rev. A, 1980, 21, 274.
9. J.A. Goldstone, E.M. Garmire, IEEE J. Quantum Electron, 1983, QE-19, 208.
10. D.A.B. Miller, J. Opt. Soc. Am.B, 1984, 1, 857.
11. S. Speiser, N. Shakkour, Appl. Phys. B, 1985, 38, 191 and references therein.
12. M. Orenstein, S. Speiser, J. Katriel, Opt. Commun., 1984, 48, 367.
13. M. Orenstein, S. Speiser, J. Katriel, IEEE J. Quantum Electron., 1985, QE-21, 1513.
14. M. Orenstein, J. Katriel, S. Speiser, Phys. Rev. A, 1987, 35, 1192.
15. Z.F. Zhu, E.M. Garmire, IEEE J. Quantum Electron., 1983, 19, 1495.
16. M.C. Rushford, H.M. Gibbs, J.L. Jewell, N. Peyghambarian, D.A. Weinberger, C.F. Li, 'Optical Bistability 2,' ed. by C.M. Bowden, S.L. McCall (Plenum, New York 1983) pp. 345-352.
17. S. Speiser, F.L. Chisena, Appl. Phys. B, 1988, 45, 137.
18. R.W. Chambers, T. Kajiwara, D.R. Kearns, J. Phys. Chem., 1974, 78, 380.
19. I. Lopez Arbeloa, Part 1, J. Chem. Soc., Faraday Trans., 2 77, 1725; Part 2, J. Chem. Soc., Faraday Trans. 2 77, 1735 (1981).
20. W.E. Ford, J. Photochem., 1987, 37, 189.
21. S. Speiser, V.H. Houlding, J.T. Yardley, Appl. Phys. B, 1988, 45, 237.
22. T. Todorov, L. Nikolova, N. Tumova, V. Dragotinova, Opt. Electron. 13, 209 (1981); IEEE J. Quant. Electron., 1986, QE-22, 1262.
23. M.A. Kramer, W.R. Tompkin, R.W. Boyd, Phys. Rev. A, 1986, 34, 2026.
24. C.J.G. Kirkby, R. Cush, I. Bennion, Opt. Commun., 1985, 56, 288.
25. D.S. Chemla, J. Zyss, 'Nonlinear Optical Properties of Organic Molecules and Crystals' (Academic, New York 1987).
26. A.R. Tanguay, Jr., Opt. Eng., 1985, 24, 2.

Organometallics

The Synthesis of Ferrocenyl Compounds with Second-order Optical Non-linearities

J. A. Bandy, H.E. Bunting, M.L.H. Green*, S.R. Marder, and M.E. Thompson

INORGANIC CHEMISTRY LABORATORY, SOUTH PARKS ROAD, OXFORD OX I 3QR, UK

D. Bloor

DEPARTMENT OF PHYSICS, QUEEN MARY COLLEGE, MILE END ROAD, LONDON E I 4NS, UK

P.V. Kolinsky and R.J. Jones

GEC RESEARCH LTD., HIRST RESEARCH CENTRE, EAST LANE, WEMBLEY, MIDDLESEX HA9 7PP, UK

The development of new materials with large optical nonlinearities is an exciting discipline with applications in telecommunications, optical information processing and optical computing.[1-3] Two important manifestations of light interacting with a nonlinear medium are the process of second harmonic generation (SHG) which is relevant to new laser technology, and the electro-optic effect which has applications in telecommunications and in integrated optics.

There has been a tremendous effort devoted the synthesis of organic materials with large second-order optical nonlinearities. It is now well established that large second-order optical nonlinearities are associated with structures that have large differences between ground state and excited state dipole moments as well as large transition dipole moments.[4] Molecules with π-donor-acceptor interactions are promising candidates to fulfil the above requirements. The afore mentioned criteria will ensure large molecular second-order nonlinearities; however it is imperative that the molecular dipole reside in a noncentrosymmetric environment, if the molecular nonlinear polarization is to lead to an observable effect in the bulk material.

In comparison to the large effort focussed on the synthesis of organic compounds with large SHG efficiencies, little attention has been paid to organometallic compounds. Until recently, there have been only a few reports of organometallic compounds exhibiting second-order optical nonlinearities and the SHG efficiencies have been modest.[5-7] The diversity of oxidation states as well as ligand environments of organometallic and coordination complexes can impart excellent acceptor or donor properties upon the metal centre. Further, the ligand sphere of organometallic and coordination complexes can often be easily varied in a systematic manner to give specific properties to the complex which may improve the electronic or crystallographic factors leading to very large coefficients of $\chi^{(2)}$. On this basis, we initiated an exploratory synthesis programme to prepare organometallic complexes exhibiting large second-order nonlinearities. Here we present results which suggest that complexes incorporating the well studied ferrocenyl moiety can

219

undergo very efficient second harmonic generation as evidenced by the Kurtz powder test.

Deprotonation of $\{(\eta-C_5H_5)Fe(\eta-C_5H_4)CH_2P(C_6H_5)_3\}$ I,[8] **1**, with n-butyllithium, followed by addition of 4-nitrobenzaldehyde, yielded dark purple crystals of (Z)-[1-ferrocenyl-2-(4-nitrophenyl) ethylene],[9] **2**, and the previously reported (E)-[1-ferrocenyl-2-(4-nitrophenyl)ethylene], **3**. Compounds **2** and **3** were readily separated by chromatography on alumina. Solutions and crystals of **2** appear to be stable indefinitely in air. Compound **2** was characterized by microanalysis and n.m.r. spectra which indicated the structure shown in Figure 1(a). The UV-visible spectrum of **2** was markedly affected by solvent. In heptane there were three absorptions at 320, 406, and 466 nm whereas in N,N'- dimethylformamide there were only two absorptions at 340 and 492 nm.

In view of this solvatochromic behavior, (Z)-[1-ferrocenyl-2-(4-nitrophenyl)ethylene], **2**, was screened for SHG efficiencies using the Kurtz powder technique on unsized particles.[10] Because of the intense optical absorption of the crystals of **2** the fundamental light from a Nd:YAG laser operating at 1.064 μm, was Raman shifted to 1.907 μm with a high pressure hydrogen cell. This resulted in frequency doubled light at 0.953 μm which had an intensity 62 times that of the urea reference standard. The origin of this large nonlinearity may largely be due to disparate contributions of both canonical forms of (Z)-[1-ferrocenyl-2-(4-nitrophenyl)ethylene], **2**, shown in Figure 1, to the ground and excited states of the molecule, leading to a large $\Delta(\mu_{gs}-\mu_{es})$, which is consistent with the observed solvatochromic behavior.

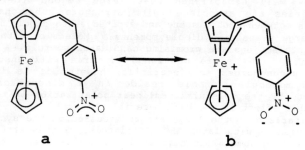

a   b

Figure 1. Two canonical forms of (Z)-[1-ferrocenyl-2-(4-nitrophenyl) ethylene].

A single crystal X-ray diffraction study of **2** was performed on a crystal obtained from slow evaporation of a saturated solution of 2-propanol. Crystals of **2** obtained in this manner crystallize in the non-centrosymmetric space group F2dd. The molecular structure is shown in Figure 2. We note that the nitrophenyl ring and the η-C_5H_4 ring are not coplanar. This minimizes unfavourable steric interactions between protons on these groups but decreases the conjugation between the groups as well. From the molecular packing diagram shown in Figure 3, it is clear that the the molecular dipole of **2** will be preserved to an substantial extent in the crystalline

material.   Indeed, the the tilting of the molecular dipole should
favor efficient phased matched second harmonic generation.

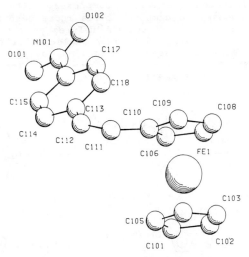

Figure 2. Molecular structure of **2** with hydrogen atoms omitted for
clarity.

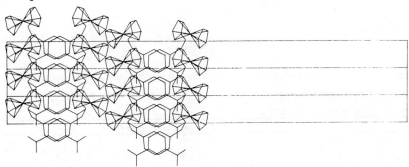

Figure 3. Crystal packing diagram for **2** with only half of the
molecules in the unit cell shown for clarity.

It is interesting to note that complex **2** undergoes a reversible
one electron oxidation, in water, as determined by cyclic
voltametric studies.   Thus the iron center which was originally an
electron rich donor (Fe(II)), in a neutral, 18 electron complex has
become after oxidation an electron deficient acceptor (Fe(III)), in
a cationic, 17 electron complex.   This raises two new important
questions: 1) can electrochemistry be used as a switching mechanism
for controlling nonlinear optical properties and 2) will
paramagnetic species (such as the oxidized form of **2**) have unusual
nonlinear optical properties?

　　　　To gain insight into the potential scope of ferrocene complexes as nonlinear optical materials, we set out to synthesize a series of substituted ferrocenyl ethylene complexes. The complexes were prepared analogously to compound **2** via route described by Pauson[8] (Figure 4). Our choice of tetrahydrofuran as the solvent and n-butyllithium as the base led to reasonable yields of both "Z" and "E" isomers. As can be seen from Table 1, a variety of complexes can be prepared.[11] Using a similar procedure, Toma et. al. have prepared compounds **3**, **5** and **7**.[12]

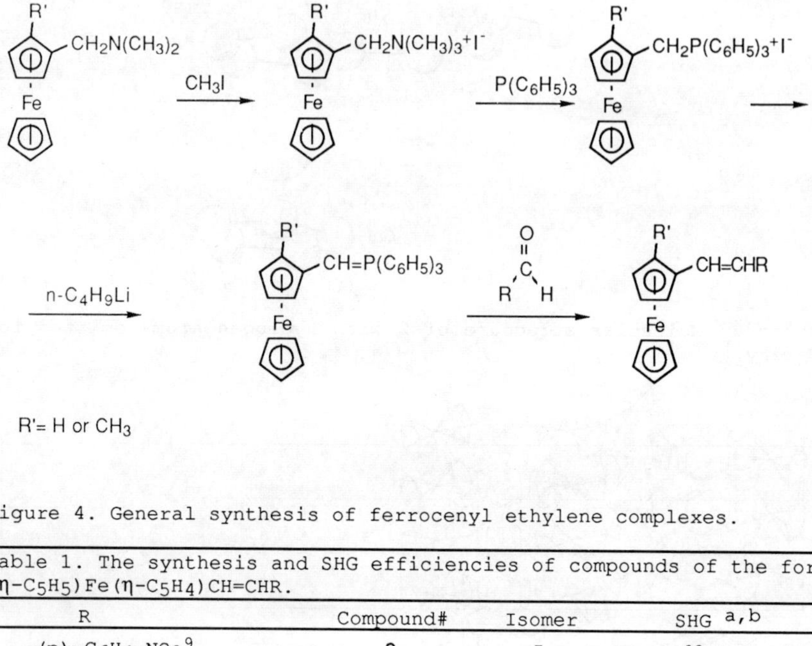

R'= H or CH$_3$

Figure 4. General synthesis of ferrocenyl ethylene complexes.

Table 1. The synthesis and SHG efficiencies of compounds of the form $(\eta-C_5H_5)Fe(\eta-C_5H_4)CH=CHR$.

R	Compound#	Isomer	SHG [a,b]
-(p)-C$_6$H$_4$-NO$_2$[9]	2	Z	62
-(p)-C$_6$H$_4$-NO$_2$[12]	3	E	0
-(p)-C$_6$H$_4$-CN[11]	4	Z	0.95
-(p)-C$_6$H$_4$-CN[12]	5	E	0.87
-(p)-C$_6$H$_4$-CHO[11]	6	Z	0
-(p)-C$_6$H$_4$-CHO[12]	7	E	0.72
-(E)-CH=CH-(p)-C$_6$H$_4$-NO$_2$[11]	8	Z	0
-(E)-CH=CH-(p)-C$_6$H$_4$-NO$_2$[11]	9	E	0.04
-C=CH-CH=C(NO$_2$)O[11]	10	E	0.03

a-All measurements were performed at 1.907 μm.
b-The SHG signal is the magnitude of the signal at 0.953 μm for the compound of interest relative to a urea reference standard measured under the same conditions.

These compounds were also screened for SHG efficiencies by the Kurtz powder test and the results are summarized in Table 1. Although none of the complexes exhibited SHG efficiencies of the order of **2**, several had efficiencies comparable to urea. Furthermore over half of the compounds gave non-zero efficiencies. This result is encouraging in that it implies that this general class of compounds does not have an overriding propensity to crystallize in centrosymmetric space groups.

We decided to introduce asymmetry and *optical activity* in the molecules by substituting the cyclopentadienyl ring which contains the acceptor moiety with an additional methyl group. It was our hope that by analogy to approaches used in organic systems this would lead to a more desirable alignment of the molecular dipole in the crystal lattice. Using a procedure developed by Sokolov, the compound (s)-$(\eta-C_5H_5)Fe(\eta-C_5H_3CH_3)CH_2N(CH_3)_2$,[13] **12**, was synthesized and carried on the to a series of ferrocenyl ethylene complexes as shown in Figure 4. Compound **12** was isolated in 82% enantiomeric excess (ee) as determined by n.m.r. spectra recorded with chiral shift reagents. Thus the ferrocenyl ethylene complexes isolated were assumed to be isolated in ~82% ee, as racemization was not expected in this system. The "Z" isomers were formed in much lower yields in this series of compounds as compared to the series of compound shown in Table 1. As a result, only the "E" isomers have been screened for SHG efficiencies. The results which are shown in Table 2[14] are very encouraging. Every compound exhibited larger efficiencies than the analogous compound in Table 1. Compound **17** for example has a SHG efficiency over 500 times that of compound **10**. Thus in the the compounds described here we have attempted to engineer into the molecules both the desired electronic and crystallographic properties desirable for second-order optical nonlinearities.

Table 2. The synthesis and SHG efficiencies of compounds of the form (s)-$(\eta-C_5H_5)Fe(\eta-C_5H_3CH_3)CH=CHR$ (~ 82% ee of "s" isomer).[14]

R	Compound#	Isomer	SHG [a,b]
-(p)-C_6H_4-NO$_2$	13	E	8.0
-(p)-C_6H_4-CN	14	E	1.2
-(p)-C_6H_4-CHO	15	E	2.5
-(E)-CH=CH-(p)-C_6H_4-NO$_2$	16	E	6.4
-C=CH-CH=C(NO$_2$)O	17	E	17.0

a-All measurements were performed at 1.907 μm.
b-The SHG signal is the magnitude of the signal at 0.953 μm for the compound of interest relative to a urea reference standard measured under the same conditions.

In these preliminary studies we have demonstrated that organometallic compounds can have large optical nonlinearities.

Clearly, further synthetic, crystallographic and optical studies are
needed if we are to fully understand and exploit the large optical
nonlinearities of organometallic compounds.

Acknowledgments:

    We thank Genetics International Inc. for financial support (to
S.R.M.), the S.E.R.C. for support (to M.E.T.), and the S.E.R.C. and
the Royal Society for support (to D.B.)

References and Footnotes:

1.   D.J. Williams, Angew. Chem. Int. Ed. Engl., 1984, 23, 690.
2.   D.J. Williams, 'Nonlinear Optical Properties of Organic and
      Polymeric Materials, ACS Sypm. Ser., 233', American
      Chemical Society, Washington, 1983.
3.   D.S. Chemla and J. Zyss, 'Nonlinear Optical Properties of
      Organic Molecules and Crystals'. Academic Press, Orlando,
      1987, Volumes 1 and 2.
4.   C.C. Teng and A.F. Garito, Phys. Rev. B, 1983, 28, 6766.
5.   C.C. Frazier, M.A. Harvey, M.P. Cockerham, E.A. Chauchard
      and C.H. Lee, J. Phys. Chem., 1986, 90, 5703.
6.   D.F. Eaton, A.G. Anderson, W. Tam and Y. Yang, J. Amer.
      Chem. Soc., 1987, 109, 1886.
7.   J.C. Calabrese and W. Tam, Chem. Phys. Letters, 1987,
      133, 245.
8.   P.L. Pauson and W.E. Watts, J. Chem. Soc., 1963, 2990.
9.   M.L.H. Green, S.R. Marder, M.E. Thompson, J.A. Bandy, D.
      Bloor, P. V. Kolinsky and R. J. Jones, Nature, 1987, 330,
      360.
10.   S.K. Kurtz and T.T. Perry, J. Appl. Phys., 1968, 39, 3798.
11.   M.L.H. Green, S.R. Marder, D. Bloor, M.E. Thompson, P. V.
      Kolinsky and R. J. Jones, J. Organomet. Chem., submitted for
      publication.
12.   S. Toma, A. Gaplovsky and P. Elecko, Chem. Papers, 1985,
      39, 115.
13.   V.I. Sokolov, L.L. Troitskaya and O.A. Reutov, J.
      Organomet. Chem., 1979, 182, 537.
14.   H.E. Bunting, M.L.H. Green, S.R. Marder, D. Bloor, P.
      V. Kolinsky and R. J. Jones, to be submitted to J. Chem.
      Soc., Chem. Commun..

The Synthesis of Organometallic Compounds with Second-order Optical Non-linearities

J. A. Bandy, H.E. Bunting, M.H. Garcia, M.L.H. Green,
S.R. Marder, and M.E. Thompson

INORGANIC CHEMISTRY LABORATORY, SOUTH PARKS ROAD, OXFORD OX I 3QR, UK

D. Bloor

DEPARTMENT OF PHYSICS, QUEEN MARY COLLEGE, MILE END ROAD, LONDON E I
4NS, UK

P.V. Kolinsky and R.J. Jones

GEC RESEARCH LTD., HIRST RESEARCH CENTRE, EAST LANE, WEMBLEY, MIDDLESEX
HA9 7PP, UK

Organometallic and coordination chemists have for many years
observed that metal-organic compounds can have very intense metal
to ligand or ligand to metal charge transfer bands in the uv-
visible spectrum. These intense charge transfer bands, to a great
extent arise from excellent overlap of the the diffuse metal d-
orbitals and the ligand π orbitals.[1] In addition, metal centres
can be made extremely electron rich or electron deficient by tuning
their oxidation states or the ligand environments. Thus transition
metal compounds are often very strong oxidizing or reducing agents.
Over the past ten years a pattern has emerged for the design of
organic materials with large second-order optical nonlinearities,
in which the molecules have large changes of dipole moments between
ground and excited states and large transition dipole moments.[2]
Using these molecular criteria, it seems clear that transition
metal organic compounds can indeed have very large molecular
hyperpolarizabilities (β). Transition metals are also known to
stabilize unusual and unstable organic fragments such as carbenes,
carbynes and cyclobutadienes to mention a few; therefore new
classes of materials may be investigated. In addition, it may be
relatively easy to fine tune either electronic or crystallographic
properties by simple replacement of a labile ligand.

For these and other reasons we have begun to explore the
possibility of using organometallic compounds in the field of
nonlinear optics.[3-5] In this paper we will present some preliminary
results on compounds in which the transition metal can behave as an
acceptor.

We have examined a class of compounds in which two iron atoms
are bound to a formally cationic carbon with one organic
substituent. Casey et al. has described the synthesis and
reactivity of this class of compounds generally referred to as
diiron alkenylidyne complexes.[6-10] The bridging carbon, with its
formally vacant p-orbital is expected to exhibit excellent acceptor
properties. We hoped therefore, that di-iron alkenylidyne

complexes such as the previously reported {(η-C$_5$H$_5$)$_2$Fe$_2$(CO)$_2$(μ-CO)(μ-(E)-C-CH=CH-C$_6$H$_4$-(p)-N(CH$_3$)$_2$)}$^+$BF$_4^-$ [7,8], **1**, would exhibit large second-order optical nonlinearities, by virtue of the presence of the excellent donor and acceptor moiety joined by an extended π system as shown in Figure 1. Both the ^{13}C n.m.r. spectra and the variable temperature dynamic ^{1}H n.m.r. spectra of **1**, indicate that the canonical formulation **1b** contributes substantially to the ground state of the molecule. [9]

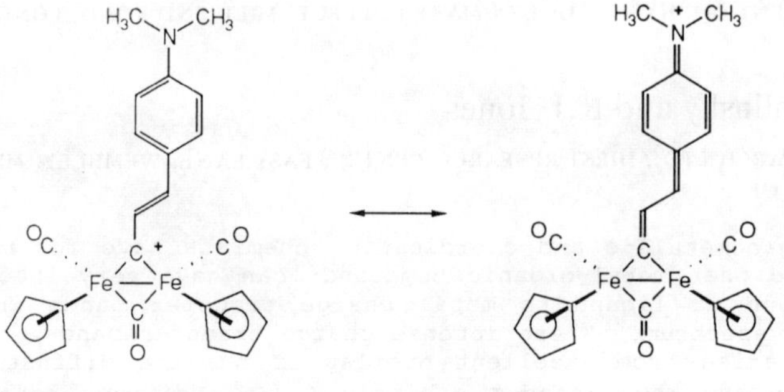

Figure 1. Two canonical representations of {(η-C$_5$H$_5$)$_2$Fe$_2$(CO)$_2$(μ-CO)(μ-(E)-C-CH=CH-C$_6$H$_4$-(p)-N(CH$_3$)$_2$)}$^+$BF$_4^-$.

We therefore synthesized a series of diiron alkenylidyne complexes using the the procedure described by Casey [7,8] which involves a condensation reaction of {(η-C$_5$H$_5$)$_2$Fe$_2$(CO)$_2$(μ-CO)(μ-C-CH$_3$)}$^+$BF$_4^-$ [11], with appropriately substituted aldehydes (Figure 2), such as N,N'-dimethylaminobenzaldehyde in the case of **1**. Unsized powders of these compounds were tested for second harmonic generation efficiencies using the Kurtz powder technique. [12] Since most of the compounds range in color from red to purple due to the low energy charge transfer band, light from a Nd:YAG laser operating at 1.064 μm was Raman shifted to 1.907 μm with a high pressure hydrogen gas cell. The results of this work are summarized in Table 1.

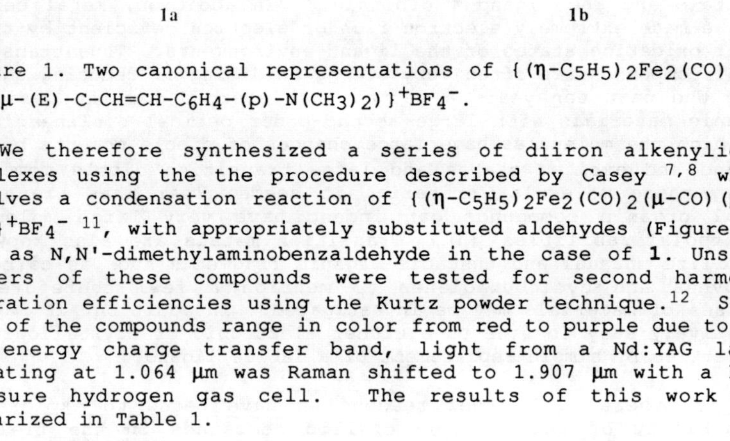

Figure 2. Synthesis of diiron alkenylidyne complexes.

Table 1. Second Harmonic Generation powder efficiencies of compounds of the form $\{(\eta\text{-}C_5H_5)_2Fe_2(CO)_2(\mu\text{-}CO)(\mu\text{-}(E)\text{-}C\text{-}CH=CH\text{-}R)\}^+BF_4^-$.

R	Compound#	SHG [a,b]
$-C_6H_5$ [8]	3	0
$-C_6H_4\text{-}(p)\text{-}Cl$	4	0
$-C_6H_4\text{-}(p)\text{-}N(CH_3)_2$ [7,8]	1	0.77
$-(E)\text{-}CH=CH\text{-}C_6H_5$	5	0
$-(E)\text{-}CH=CH\text{-}C_6H_4\text{-}(p)\text{-}N(CH_3)_2$	6	0

a- All measurements were performed at 1.907 μm.
b- The SHG signal is the magnitude of the signal at 0.953 μm for the compound of interest relative to a urea reference standard measured under the same conditions

Interestingly, only compound **1** exhibits macroscopic optical nonlinearities, with a powder second harmonic generation efficiency 0.77 times that of the urea reference standard. To gain a better understanding of the structural factors giving rise to the nonlinear optical properties of **1**, a single crystal X-ray analysis was performed. A suitable crystal was obtained by vapour diffusion of diethyl ether into a nearly saturated solution of **1** in acetone. The structure was solved in the *centrosymmetric* space group $P2_1/n$.[13] Apparently, the deviation from centrosymmetry for the packing of the molecules of **1** in the crystal lattice was sufficiently minute that it was undetectable crystallographically, yet large enough to give rise to appreciable second harmonic generation. These observations imply that **1** may have an extremely large molecular hyperpolarizability (β) and that upon modification of the ligands to improve the alignment of the molecular dipole in the crystal lattice, large second harmonic generation efficiencies may be expected.

The molecular structure of **1** is shown in Figure 3 and 4 and selected bond lengths and bond angles are given in Table 2. These data reveal the large contribution of canonical formulation **1b** to the solid state structure. In **1**, we observe localization of double bond character in the phenyl ring and complete delocalization between $C(1)$ and $C(4)$. The latter was not seen in $\{(\eta\text{-}C_5H_5)_2Fe_2(CO)_2(\mu\text{-}CO)(\mu\text{-}(E)\text{-}C\text{-}CH=CH\text{-}C_6H_4\text{-}(p)\text{-}CH_3)\}^+BF_4^-$ [7,8], **7**, where the analogous vinyl group retains substantial double bond character. Comparisons of dimensions of the phenyl groups are impossible since in the case of **7** these were constrained to 1.395 Å during the refinement. Another similar structure which has been reported is $\{[(\eta\text{-}C_5H_5)_2Fe_2(CO)_2(\mu\text{-}CO)]_2(\mu\text{-}C_5H_3)\}^+BF_4^-$ [17], **8**, where we do observe the same delocalization along the $\mu\text{-}C_5H_3$ group.

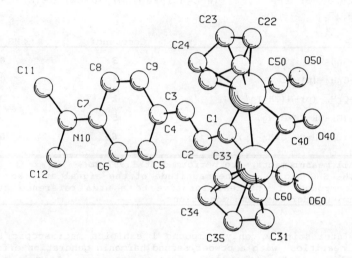

Figure 3. Molecular structure of **1**, hydrogen atoms and BF$_4$⁻ omitted for clarity.

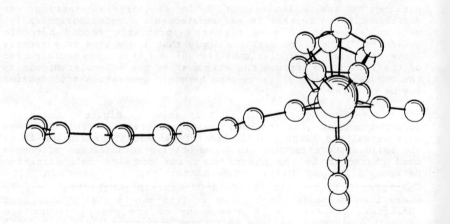

Figure 4. View of **1** along plane of aromatic ring.

Table 2. Selected bond lengths (Å) and angles(°) for **1**.

Fe(1) - Fe(2)	2.505(1)	
Fe(1) - C(1)	1.861(5)	
Fe(1) - C(21)	2.108(7)	
Fe(1) - C(22)	2.075(6)	
Fe(1) - C(23)	2.065(6)	
Fe(1) - C(24)	2.115(6)	
Fe(1) - C(25)	2.125(7)	
Fe(1) - C(40)	1.941(5)	
Fe(1) - Fe(50)	1.764(6)	

Fe(2) - C(1)	1.895(5)
Fe(2) - C(31)	2.089(5)
Fe(2) - C(32)	2.121(6)
Fe(2) - C(33)	2.122(6)
Fe(2) - C(34)	2.116(6)
Fe(2) - C(35)	2.080(6)
Fe(2) - C(40)	1.921(5)
Fe(2) - C(50)	1.758(6)

C(1) - C(2)	1.385(6)
C(2) - C(3)	1.373(7)
C(3) - C(4)	1.414(6)
C(4) - C(5)	1.420(7)
C(5) - C(6)	1.372(7)
C(6) - C(7)	1.414(7)
C(7) - C(8)	1.422(7)
C(8) - C(9)	1.355(7)
C(9) - C(4)	1.413(7)
C(7) - N(10)	1.344(6)
N(10) - C(11)	1.440(7)
N(10) - C(12)	1.467(7)

Fe(1) - C(1) - Fe(2)	84.5(2)
Fe(1) - C(1) - C(2)	140.9(4)
Fe(2) - C(1) - C(2)	143.6(4)
C(1) - C(2) - C(3)	122.8(5)
C(2) - C(3) - C(4)	127.9(5)
C(3) - C(4) - C(5)	124.1(5)
C(3) - C(4) - C(9)	119.3(5)
C(5) - C(4) - C(9)	116.5(4)
C(4) - C(5) - C(6)	121.1(5)
C(5) - C(6) - C(7)	121.3(5)
C(6) - C(7) - C(8)	117.5(5)
C(6) - C(7) - N(10)	121.5(5)
C(8) - C(7) - N(10)	120.9(5)
C(7) - C(8) - C(9)	120.3(5)
C(8) - C(9) - C(4)	122.9(2)

C(21) - C(22)	1.363(8)
C(22) - C(23)	1.396(8)
C(23) - C(24)	1.383(8)
C(24) - C(25)	1.367(8)
C(25) - C(21)	1.354(8)

C(25) - C(21) - C(22)	109.4(7)
C(21) - C(22) - C(23)	107.3(6)
C(22) - C(23) - C(24)	106.9(7)
C(23) - C(24) - C(25)	108.3(7)
C(24) - C(25) - C(21)	108.1(7)

C(31) - C(32)	1.379(7)
C(32) - C(33)	1.385(7)
C(33) - C(34)	1.391(7)
C(34) - C(35)	1.391(7)
C(35) - C(31)	1.411(7)

C(35) - C(31) - C(32)	107.8(6)
C(31) - C(32) - C(33)	108.5(6)
C(32) - C(33) - C(34)	108.3(6)
C(33) - C(34) - C(35)	108.0(6)
C(34) - C(35) - C(31)	107.4(6)

C(40) - O(40)	1.169(6)
C(50) - O(50)	1.134(7)
C(60) - O(60)	1.123(7)

Fe(1) - C(40) - Fe(2)	80.9(2)
Fe(1) - C(40) - O(40)	138.1(4)
Fe(2) - C(40) - O(40)	141.0(4)
Fe(1) - C(50) - O(50)	179.6(6)
Fe(2) - C(60) - O(60)	176.4(6)

It may have been expected that the planes defined by Fe(1), Fe(2), C(1) (plane 1) and by the C(4)-C(9) inclusive (plane 2) would be coplanar with the vinyl group C(1), C(2), C(3) (plane 3) to maximize orbital overlap and electron delocalization. However, this is not observed; the angle between planes 2 and 3 is 12.4°. This is very similar to the (p)-tolyl analogue **8** [8], where the angle between the vinyl group and the tolyl ring is found to be 11.0°.

This bend in **1** is not due to the crystal packing as there are no close intermolecular interactions. The closest distance between ions is between C(21) and one of the fluorines in the BF_4 (3.142 Å)

The results here suggest that compound **1** has an extremely large molecular second-order hyperpolarizability (β) however due unfavourable orientations of the molecular dipole in the crystal lattice, the large β does not manifest itself by the Kurtz powder test. It therefore necessary to attempt to alter the crystal packing by incorporation of hydrogen bonding ligands, optical activity or simple variation of the counter ion. Studies to address these issues are currently in progress. It is safe to say that these problems are not insurmountable and that the intrinsically large molecular hyperpolarizabilities of organometallic compounds make this class of molecules worthy of future investigation.

Acknowledgments:

We thank Genetics International Inc. for financial support (to S.R.M.), the S.E.R.C. for support (to M.E.T.), and the S.E.R.C. and the Royal Society for support (to D.B.)

References and Footnotes:

1.   G.L. Geoffroy and M.S. Wrighton 'Organometallic Photochemistry', Academic Press, New York, 1979.
2.   D.J. Williams, 'Nonlinear Optical Properties of Organic and Polymeric Materials, ACS Sypm. Ser., 233', American Chemical Society, Washington, 1983.
3.   M.L.H. Green, S.R. Marder, M.E. Thompson, J.A. Bandy, D. Bloor, P. V. Kolinsky and R. J. Jones, Nature, 1987, 330, 360.
4.   M.L.H. Green, S.R. Marder, D. Bloor, M.E. Thompson, P.V. Kolinsky and R.J. Jones, J. Organomet. Chem., submitted for publication.
5.   H.E. Bunting, M.L.H Green, S.R. Marder, D. Bloor, P.V. Kolinsky and R.J. Jones, to be submitted to J. Chem. Soc, Chem. Commun..
6.   C.P. Casey and S.R. Marder, Organometallics, 1985, 4, 411.
7.   C.P. Casey, M.S. Konings, R.E.Palermo, and R E. Colborn, J. Am. Chem. Soc., 1985, 107, 5296.
8.   C.P. Casey, M.S. Konings and S.R. Marder, Polyhedron, in press.
9.   C.P. Casey, M. S. Konings, S.R. Marder and Y. Takezawa, Organometallics, in press.
10.   C. P. Casey, M. S. Konings and S. R. Marder, J. Organomet. Chem., in press.
11.   M. Nitay, W. Priester and M. Rosenblum, J. Am. Chem. Soc., 1978, 100, 3620.
12.   S.K. Kurtz and T.T. Perry, J. Appl. Phys., 1968, 39, 3798.

13. Crystal data: $C_{24}H_{22}Fe_2NO_3BF_4$, M=570.9, monoclinic, space group $P2_1/n$, a = 10.702(1), b = 13.349(1), c = 16.932(2), β=93.42(1)$^\circ$, U = 2414.6Å^3, Z = 4, D_c = 1.57 Mgm^{-3}, λ(Mo-Kα_1) = 0.70930 Å, μ(Mo-Kα_1) = 12.57 cm^{-1}, F(000) = 1160. Crystal dimensions: 0.05 x 0.20 x 0.525 mm. At convergence R = 0.042, R_w = 0.049 for 2417 reflections [I > 3σ(I)] in the range 1 < θ < 25°. All non-hydrogen atoms were refined anisotropically. Hydrogens atoms were included in the calculated positions riding on their attached carbon atoms. The cyclopentadienyl rings and disordered BF_4^- ion were refined subject to soft vibrational restraints.[14,15] All calculations were performed on the VAX 11/750 computer of the Chemical Crystallography Laboratory, Oxford using the Oxford CRYSTALS package.[16] R = Σ(|F_o| - |F_c|)/Σ|F_o|, R_w = [Σw(|F_o| - |F_c|)^2/Σw|F_o|^2]$^{1/2}$.

14. J.S. Rollet in, 'Crystallographic Computing', F. R. Ahmed, ed, Munksgaard, Copenhagen, 1969, p. 169.
15. J. Waser, <u>Acta</u>. <u>Crystallogr</u>., 1963, <u>16</u>, 1091.
16. D.J. Watkin, J.R. Carruther, P.W. Betteridge, 'CRYSTALS User Guide', Chemical Crystallography Laboratory, University of Oxford, Oxford, 1985.
17. C.P. Casey, M.S. Konings and K.J. Haller, J. <u>Organomet</u>. <u>Chem</u>., 1986, <u>301</u>, C55.

n_2 Measurements on Some Metallocenes

C.S. Winter, S.N. Oliver, and J.D. Rush

BRITISH TELECOM RESEARCH LABORATORIES, MARTLESHAM HEATH, IPSWICH IP5 7RE, UK

SUMMARY

The molecular hyperpolarisabilities of molten ruthenocene, hafnocene and ferrocene have been measured and compared with a liquid ferrocene derivative. The value of the third order coefficient, γ, was found to be two to three times that of nitrobenzene for 10 ns duration pulses at 1.06 μm. The observed n_2 may be electronic in origin rather than rotational.

1 INTRODUCTION

Although the theory of the molecular origin of the large second order hyperpolarisabilities in organic materials is reasonably well understood, and many interesting materials have been examined[1], there is much less known about materials with large third order hyperpolarisabilities. Most of the systems studied have been molecules with large second order coefficients or conjugated polymers. We have started the study of a largely neglected group of materials - organo-metallics. Although the results of some powder studies have been published[2,3], little work has been done on third order phenomena. Organo-metallics are of interest because, in many cases, they contain low lying energy states that could be used to resonance enhance the non-linearity to obtain responses large enough for device consideration. We report here the results of a study of a number of metallocenes using the optical power limiter technique developed by Soileau[4-6].

2 EXPERIMENTAL.

Ruthenocene, hafnocene dichloride, ferrocene and zirconocene dichloride were used as supplied by Aldrich Chemical Company. The materials were all gently degassed under vacuum. The molten phase was obtained by heating the metallocene above its melting point in a sealed cuvette under helium. A liquid derivative of ferrocene, bis(trimethylsilyl)ferrocene (BTMSF), that melts at 16°C was also prepared[7]. All four materials were measured in 5mm path length cuvettes.

The technique used to measure the n_2 coefficients of the materials was based on the optical power limiter developed by Soileau *et al*[4-6]. In this technique a laser beam with a gaussian spatial profile is focused in to the medium under study, the transmitted light is then refocused onto a pinhole and thence onto a detector (see figure 1). When the input power exceeds a critical level self-focusing of the beam into a filament occurs which moves the focal point of the system and defocuses the output at the plane of the pin-hole. The power observed by the detector is then pinned at a constant level; an example of this is displayed in figure 2 where the results for nitrobenzene and bis(trimethylsilyl) ferrocene are presented. This pinning arises because the pin-hole allows only the on-axis component to be measured; if the pin-hole were to be removed, all the power transmitted through the sample would be measured at the detector. In this case a break in the plot of input power to output power is still observed, although the output is not pinned, since the intensity in the filament is so high that non-linear absorption occurs.

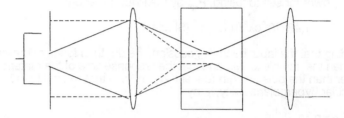

Figure 1 : Schematic of the experimental apparatus showing the arrangement of lenses, sample, pin-hole and detector. The effect of beam self-focusing is shown by the dotted line.

For materials with a simple non-linearity corresponding to the real part of $\chi^{(3)}$ the observed limiting is *power* dependent[4]. For materials with more complex non-linearities the onset of power limiting is *intensity* dependent. The two can be distinguished by varying the focal length of the focusing lens. With a gaussian beam the focal spot size is directly proportional to the focal length of the lens, thus the intensity at the focal point is proportional to the inverse square of the focal length of the lens. Hence, if the focal length of the first lens is doubled and the phenomenon is intensity dependent then the critical power increases four fold, whereas if it is a power dependent phenomenon there is no change in the critical power.

Measurements were carried out using a Quantel 585 laser which delivers 10 ns pulses at 1.06 μm in a TEM$_{00}$ mode. The pulse repetition rate was 2 Hz to minimise thermal effects in the sample. Pulse energies were measured using a pyroelectric energy ratiometer which permitted the monitoring of individual pulses. The detector also contains a fast silicon diode to enable pulse shape monitoring using a Tektronix storage oscilloscope. The spatial profile was monitored using a spiricon linear

234   *Organic Materials for Non-linear Optics*

photodiode array, which permits the observation of the spatial profile of a
single pulse. The laser was attenuated using a half-wave plate/polariser
combination which minimises beam walk during attenuation. Singlet 'best
form' lenses were used to focus the 7mm diameter beam onto the sample.
Each experimental point consists of the average of 10 pulses at a given
setting of the half-wave plate.

3 RESULTS AND DISCUSSION

For a material with a positive n_2 the increase in the refractive index at the high
intensity centre of the beam compared with the wings causes the beam to
self-focus. When the self-focusing is greater than the beam diffraction the
beam self-focuses into a filament (or, if pulsed radiation is used, a time-
dependent movement of the focal spot occurs). The theory developed by
Marburger[8] predicts that for a spatially gaussian beam this occurs at a certain
critical power. n_2, the non-linear refractive index, is related to the observed
critical power for self-focusing, P_{c2}, by the following equation -

$$P_{c2} = 1.48 \times 10^{-9} c\lambda^2 / n_0 n_2 \quad \text{(SI units)}$$

providing that the input beam is gaussian, the depth of focus of the lens is
less than the sample length and that the response time of the material is
shorter than the pulse length (quasi-CW). n_2 is in turn related to the
molecular hyperpolarisability, γ, by -

$$\gamma = n_0^3 n_2 / L^4 N$$

where n_0 is the linear refractive index, L is the Lorentz field correction, taken
here to be $(n_0^2 + 2)/3$, and N is the molecular density.

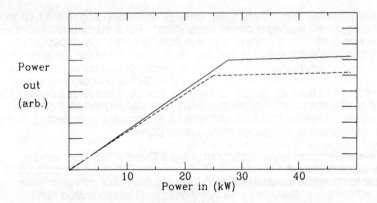

Figure 2 : Typical plots for nitrobenzene (dashed line) and
bis(trimethylsilyl)ferrocene (solid line) with 75mm focusing lens.

A simple power-limiting experiment using a single set of lenses does not unambiguously give n_2 since other phenomena can give rise to similar behaviour as discussed above. It is necessary to confirm the experiments with a number of different lenses. Figure 2 shows a typical result for bis(trimethylsilyl) ferrocene (abbreviated BTMSF in the tables) and nitrobenzene - which was used as a standard. Table 1 gives the results for all the materials under study and shows the effect of changing the lens focal length. The figures for the concentration of molecules are calculated by simple measurement of the volume of 1 gm of the molten material. Both hafnocene and zirconocene are dichlorides and decompose slowly under melt even in a helium atmosphere. These results are thus less accurate than for the more stable materials, where the experimental errors are about 20%.

Table 1: Results

Material	Concentration (molecs/m^3)	Lens (mm)	Critical Power (kW)
Nitrobenzene	5.9×10^{27}	75	25
BTMSF	1.9×10^{27}	50	26
BTMSF		75	28
BTMSF		100	20
BTMSF		150	25
Ferrocene	3×10^{27}	75	28
Ruthenocene	5×10^{27}	75	12
Ruthenocene		150	15
Hafnocene	4×10^{27}	75	30
Zirconocene	4×10^{27}	75	38

There are a number of mechanisms that can lead to the observed defocusing/limiting; these are positive n_2 (causing self-focusing), negative n_2 (causing defocusing), two photon absorption and material breakdown. The last three mechanisms are intensity dependent phenomena and thus are dependent on the inverse square of the focal length of the lens. The results on bis(trimethylsilyl) ferrocene are clearly independent of the focusing lens' focal length, a nine-fold variation being expected for intensity-dependent phenomena, a similar result is given for ruthenocene. This indicates that the mechanism is a simple non-linearity corresponding to the real part of $\chi^{(3)}$. Previous results give P_c for nitrobenzene as 25 kW and for CS_2 - a benchmark material - about 14 kW[4]. The resulting value of n_2 is in agreement with other published values for this pulse length[5,9].

Table 2 summarises the derived n_2's and c's for the different metallocenes studied. These results show that the molecular hyperpolarisability of bis(trimethylsilyl) ferrocene to be about 3x that of nitrobenzene and for the other metallocenes to be about 1-2x nitrobenzene. The values of n_0 are approximate for the melt phase and assumed, within experimental error, to be the same as the liquid derivative.

Table 2: Measured n_2 coefficients and calculated values of γ.

Material	n_2 (m^2/W)	n0	c $(V^{-2}m^5)$
Nitrobenzene	1.3×10^{-17}	1.56	1.8×10^{-45}
BTMSF	1.3×10^{-17}	1.55	5.5×10^{-45}
Ferrocene	1.1×10^{-17}	(1.55)	2.9×10^{-45}
Ruthenocene	2.2×10^{-17}	(1.55)	3.9×10^{-45}
Hafnocene	1.0×10^{-17}	(1.55)	2.4×10^{-45}
Zirconocene	0.8×10^{-17}	(1.55)	1.9×10^{-45}

n_2 in nitrobenzene and CS_2 arises from molecular re-orientation, which is related to $\Delta\alpha^2$ - the polarisability anisotropy. The electronic third order coefficient (γ) is much smaller[5]. Either mechanism could give rise to the observed non-linearity in the metallocenes. To distinguish between them it is normal to use pulses shorter than the rotational relaxation time. This can be estimated using the Debye formula -

$$\tau_r = 4\pi\eta a^3/kT$$

where η is the viscosity and a the radius of the molecular volume. This gives a value of about 500ps for BTMSF, longer for the melt. Further information on the mechanism can be obtained by comparing the measured n_2's obtained with linearly and circularly polarised light. Table 3 below gives the measured P_c's for nitrobenzene, BTMSF and Ruthenocene using a 75mm focal length lens.

Table 3: Linear vs Circular Polarisation.

Material	P_c(circ)	P_c(circ)/P_c(lin)
Nitrobenzene	58	2.1
BTMSF	18	0.7
Ruthenocene	7	0.6

The ratio shown in the table is equal to that of n_2(linear)$/n_2$(circular) since n_2 is inversely proportional to P_c. For molecular reorientation, the theory developed by Shen[10] predicts that this ratio should be 4, however experimental measurements on CS_2 have yielded values of about 2 [5,11]. Other, more complex, theories suggest different ratios but all agree that the n_2 for linearly polarised light should be larger than that for circularly polarised light if the mechanism is molecular reorientation. The nitrobenzene results are similar to existing results[5] but the BTMSF and ruthenocene results are anomalous. Measurements on solids where electronic and electrostrictive mechanisms dominate have found ratios of 1.1-1.3[12]. It is suggested here that the results for BTMSF and ruthenocene are more likely to arise from an electronic mechanism than a rotational reorientation. The similarity of the results on the melt phase measurements suggests a similar mechanism in all the metallocenes studied.

Since the relaxation time of any molecular reorientation is likely to be > 100ps in these systems it should be possible to resolve the mechanism using 20ps mode-locked laser pulses. We are now setting up a system to study ferrocene crystals and to measure liquids using shorter pulses for this purpose. The technique described above is best used on low absorption materials where the imaginary component of $\chi^{(3)}$ is small; for many other metallocenes this is not the case and degenerate four wave mixing on crystals at longer wavelengths is required.

4 CONCLUSIONS

Use of the molten phase has allowed measurements of the non-linear refractive index of certain metallocenes. The values measured are similar to those of CS_2 or nitrobenzene. The molecular origin of the observed non-linearites may be electronic although further experiments are now in progress to determine this. The metallocene family is particularly interesting as the wide range of different metal ions and ring substituents which can be used should allow systematic attempts to enhance the non-linearity.

ACKNOWLEDGEMENTS.

The Director of Research and Technology, British Telecom plc, for permission to publish this paper.

REFERENCES

1.   D.S. Chemla and J. Zyss (eds), 'Nonlinear Optical Properties of Organic Molecules and Crystals' Academic Press, London 1987.
2.   C.C. Frasier, M.A. Harvey, M.P. Cockerham, H.M. Hand, E.H. Chauchard and Chi H. Lee, J.Phys.Chem, 1986, 90, 5703.
3.   J.C. Calabrese and W. Tam, Chem.Phys.Lett, 1987, 133, 244.
4.   M.J. Soileau, J.B. Franck and T.C. Veatch, NBS Special Publ.,1980, 620, 385.
5.   M.J. Soileau, W.E. Williams and E.W. Van Stryland, IEEE J.Quant.Elect., 1983, QE-19, 731.
6.   W.E. Williams, M.J. Soileau and E.W. Van Stryland, Opt.Commun., 1984, 50, 256.
7.   G. Marr and T.M. White, J.Chem.Soc.(C), 1970, 1789.
8.   J.H. Marburger, Prog. Quant. Elect., 1975, 4, 35.
9.   M.J.Weber (ed), "Handbook of Laser Science and Technology Vol 3 Part 1" CRC Press, Florida, 1986.
10.   Y.R. Shen, Phys Lett. 1966, 20, 378.
11.   C.C. Wang, Phys Rev., 1966, 152, 149.
12.   A. Feldman, D. Horowitz and R.M. Waxler, NBS Spec. Publ., 1972, 372, 92.

Polymers

Overview: Non-linear Optical Organics and Devices

D.R. Ulrich

DIRECTORATE OF CHEMICAL AND ATMOSPHERIC SCIENCES, US AIR FORCE OFFICE OF SCIENTIFIC RESEARCH, BOLLING AIR FORCE BASE, WASHINGTON, DC 20332-6448, USA

ABSTRACT

Since 1982 the USAF has had a major basic research program in nonlinear optical organics with a major focus on polymers and some work on organometallics. Emphasis has been on both second and third order nonlinear effects. The Air Force laboratories as well as the Defense Advanced Research Projects Agency have been transitioning several of the basic research results to device application. Some of the current and projected directions for polymer development and device design are discussed.

INTRODUCTION

The use of organics for nonlinear optical processes has been gaining increased attention, particularly over the past two years. There has been a surge in the establishment of research programs in industry and academia in the U.S., Europe and Japan with the major focus on organic polymers. Some emphasis is also being placed on inorganic polymers and organometallics at a smaller level of effort.

Polymers are the subject of intense reserch because of the ability to tailor molecular structures which have inherently fast response times and large second and third order molecular susceptabilities. Polymers provide synthetic and processing options that are not available with the single crystal and multiple quantum well MQW classes of NLO materials, as well as excellent mechanical properties, environmental resistance and high laser damage thresholds.

In 1983 AFOSR initiated the first U.S. federally funded research program in nonlinear optical organics. In 1983 the Defense Science Office of the Defense Advanced Research Projects Agency and the US Army Night Vision Laboratory became coinvestors in the NLO organics research with AFOSR. Building on the work of AFOSR, in 1986 the Polymer Branch of the Air Force Wright Aeronautical Laboratory/Materials Laboratory, AFWAL/ML, in collaboration with the Frank J. Seiler Laboratory at the US Air Force Academy started an in-house and contract research program. By 1988 the Rome Air Development Center and the AFWAL/Avionics Laboratory were issuing requests-for-proposal (RFP). In April, 1988 the grantees, contractors and Air Force and other federal agency scientists and engineers met at the National Academy of Sciences to review the total program.

This paper reviews in part some of the results of the program and the current and projected directions for organic polymer and organometallic development and device design.

NONLINEAR OPTICAL APPLICATIONS FOR POLYMERS

The Air Force will have many uses for electrooptical and all-optical systems based on devices designed with materials having important optical properties. These include optical computing and optical storage, optical signal processing and optical sensor and vision protection against laser radiation.

Eighteen months ago there were four or five optical device groups in the U.S. which were either working on device architectures using NLO polymers or were doing some conceptualizing along these lines. They included the groups at Lockheed, University of Arizona, and the University of Southern California. These groups, the Air Force laboratories, DARPA and some industrial firms with an investment in telecommunications were surveyed in 1987 as to where NLO polymers would fit in.

The results are shown in Figure 1. Second order polymers were viewed as playing a major role in optical signal processing in such areas as spatial light modulators and neural nets. At the time optical communications were thought to be the domain of inorganic crystals, particularly lithium niobate or potassium dihydrogren phosphate. Now with the advances in polymers with properties commensurate with lithium niobate, this opinion is rapidly changing.

Where Will NLO Polymers Fit In?

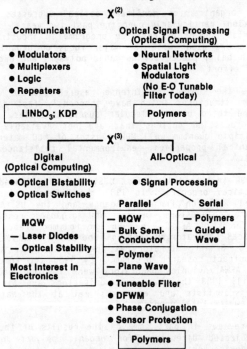

Figure   1   -   The role of polymers in nonlinear and electrooptical applications.

Third order polymers were going to play a major role in all-optical signal processing as well as tuneable filters, degenerate four wave mixing, phase conjugation and sensor protection. While it was accepted that polymers would find a role in all-optical signal processing, there was a majority position which predicted use in parallel processsing in polymer plane wave guides. A minority opinion stressed their role in serial processing in guided wave devices. All agreed that since the polymer area is embryonic considerable research has to be accomplished before the targets can be established.

At that time little role was seen in polymers in digital (optical) computing. While multiquantum well devices were thought to be the material of choice, it was stressed that this application is strongly dominated by electronics. However, with the promise of recent $\chi^{(3)}$ measurement of 10^{-8} to 10^{-9} esu in transparent polymers and proposed and theoretically calculated approaches to reach 10^{-7} or higher, this conclusion is being reconsidered.

ATTRIBUTES AND REQUIREMENTS OF POLYMERS FOR NONLINEAR OPTICS

Polymers are being investigated and developed as NLO materials for several reasons which are listed in Figure 2. The primary driver is the electronic origin of the nonlinear polarization. Organic polymers possess large, nonresonant optical susceptibilities whose origin lies in ultrafast lossless excitations of highly charged correlated pi-electron states. The optical behavior is nonresonant since the nonlinear optical polarizaiton is electronic with little or no lattice phonon contribution.

Why NLO Polymers?

- Subpicosecond Response Times

- Large, Nonresonant Nonlinearities

- Low DC Dielectric Constants

- Low Switching Energy

- Broadband

- Low Absorption

- Absence of Diffusion Problems

- Potential for Resonant Enhancement

- Ease of Processing and Synthesis Modification

- Room Temperature Operation

- Environmental Stability

- Mechanical and Structural Integrity

Figure   2   -   The advantages of NLO polymers.

Polymers offer time responses ranging over fifteen orders of magnitude, including the large nonresonant electronic nonlinearities (fsec-psec), thermal and motional nonlinearities (nsec-msec), configurational and orientational nonlinearities (μsec-sec), and photochemical nonlinearities. The contribution of each of these to nonlinear optical processes needs to be understood in order to design polymers with ultrafast response times. Measurement then becomes a critical issue.

Another major attribute of polymers is their low dc dielectric constants, being of the order of 3 as compared to 28 in the inorganic single crystal lithium niobate for example. In polymer waveguiding films the low dc dielectric constant means shorter time constants and small velocity mismatch. That is, the travelling wave device may be designed to achieve matching of optical and microwave velocities. The much longer dielectric constant in lithium niobate results in a loss of phase matching over shorter waveguide lengths. This results in higher drive voltages and power requirements, limiting frequencies accessible. There is some debate here with the inorganic crystal community since some advocates claim that comparison should be made with crystals such as KTP, Potassium Titanium Phosphate rather than lithium niobate (KTP has higher SHG efficiency, but less laser damage thresholds). However, the NLO polymer argument is supported by substantial device performance data whereas the inorganic crystal argument at this point rests more on opinion.

In second order polymers a large electrooptic effect r greater than 30 pm/V, the stability of electrically poled states, and low optical loss of 0.1 dB/cm are required. For second harmonic generation a $\chi^{(2)}$ of 10^{-7} esu, a low birefringence of 0.1, no two-photon absorption and transparency for doubling especially at 0.85μm is required.

While the switching time of conjugated polymers is several orders faster than hybrid lithium niobate and semiconductor devices, the power per bit is in the range of 1 watt, considerably higher than the other optical switching technologies. One focus of research is to reduce the switching element power requirement. According to the device design relationships, power requirements are inversely proportional to achieving a large intensity dependent index of refraction, n_2 .

This requirement is reflected in the optical performance comparison of polymers to gallium arsenide/gallium aluminum arsenide multiple quantum wells (MQWs). The figure of merit for relative comparison is the ratio of the energy required to induce switching, given by n_2, to the product of switching time and the absorption coefficient associated with switching. The switch on/switch off time, or recovery period for polymers is femtoseconds compared to nanoseconds for the semiconductors; the adsorption coefficient of polymers is about 1/10,000 that of gallium arsenide. For these reasons the nonresonant polymer figure of merit is high compared to a moderate value for the resonance enhanced MQW case.

While the resonant enhanced n_2 of the MQW is high compared to the moderate nonresonant n_2 for the polymers, nonresonance means that the polymers show this over broadband transparency from the ultraviolet to the near infrared. The high semiconductor n_2 in contrast is wavelength specific, being limited to wavelengths close to the bandgap. Another important attribute of polymers is that they show all the NLO processes, which MQWs show NLO third order, but no optical amplification or third harmonic generation.

In devices based on third order response, nonlinearities of $n_2 > 10^{-16}$ m^2/w and an absorption coefficient of less 0.1 cm^{-1} are required over the next few years depending on application. In the long term $n_2 \sim 10^{-14}$ m^2/w at 1.3 and 1.55μm, no vibrational overtone absorption and improved thermal conductivity will be required. While off-resonance nonlinearities of 10^{-10} to 10^{-11} esu were reported for mainchain and side-chain polymers last year, they still imply devices with large operating intensities.[2]

Ease of processing is another requirement, particularly for the deposition of optically clear films by nonvacuum techniques. Solubility is the key issue. The films must be able to transmit or guide light in addition to having large nonlinearities. Absorption and scattering losses need to be minimized, a loss of less than 0.1 dB/cm being required for wave guiding. Techniques for the fabrication of optically flat surfaces need to be refined.

NLO POLYMER CLASSES

The current classes of organic NLO polymers under investigation in the AFOSR program are listed in Figure 3. While inorganic polymers and organometallics are also under investigation, metal containing delocalized electron polymers for third order will be discussed.

Class	Examples		NLO Function
ISO Tropic	Glasses Alloys Composites		$\chi^{(2)}, \chi^{(3)}$
Bond-Alternation	Ladder Polymers PTL, PQL Polyacetylene Polythiophene		$\chi^{(3)}$
Liquid Crystalline Polymers (LCP)	Side Chain LCPs		$\chi^{(2)}$
Rigid Rod Aromatic Heterocyclics	PBT PBO BBL	LCPs	$\chi^{(3)}$
Polydiacetylenes			Mostly $\chi^{(3)}$ Some $\chi^{(2)}$

Figure   3   -   Status of nonlinear optical and electrooptical polymers.

STATUS OF SECOND ORDER POLYMERS AND DEVICE APPLICATION

Considerable advances have been made in the past year in isotropic polymers, which include glasses, alloys, and blends and composites for functions. Figure 4 shows the results of poled polymers where the active NLO unit is attached to the polymer backbone as a pendant side chain. Control of orientation and symmetry is achieved by poling in an external field at elevated temperatures resulting in second order susceptibilities larger than inorganic crystals.[3]

Second Order Polymers for Electrooptical Devices

● **Pendant Side Chain Structure**

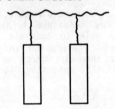

● **High Activity**	Polymer	LINbO₃
For SHG	$\chi^{(2)}$ = 120 pm/V	10 pm/V
For Electrooptics	r = 35 pm/V	30 pm/V
FOM = $\frac{r}{\varepsilon}$	10	1

● **Excellent Secondary Properties**

Spin Coatable for Thin Film Waveguides, 2-4 Micron

Low Dielectric Constant ($\varepsilon_{Polymer}$ = 3; ε_{LINbO_3} = 30)

Low Loss (<1 db/cm at 830 nm)

Melt Processable for Optics

Tg ~ 120°C

Figure   4   -   Comparison of poled second order polymers with the inorganic crystal lithium niobate.

The polymer had a second harmonic generation (SHG) after poling that is several times higher than lithium niobate. In addition, for first time in an NLO polymer the electrooptic coefficient is equal to that of an inorganic crystal. This is a major achievement since it was thought that an organic polymer with the same $\chi^{(2)}$ as an inorganic material would probably have a much smaller electrooptic coefficient r. The r in lithium niobate comes from lattice phonons and the contribution to r in organic polymers was essentially electronic. These polymers had a glass transiton temperature of 120°C. Accelerated life tests indicate that the second order activity should stay within 90% of the original value for five years. In April, 1988 it was reported that optically clear polymers with a loss less than 1dB/cm at 1.3 μm and $\chi^{(2)}$ larger than 50 pm/V had been achieved.[4] These are spin-coated from common solvents.

In another study on poled isotropic polymers by Marks, Carr and Wong, assemblies of appropriate nonlinear chromophores having noncentrosymmetry as well as high chemical stability and suitable processability are being constructed.[5] A computationally efficient SCF-LCAO MECI π-electron theoretical approach has been developed to aid in cheomophore design and to better understand molecular electronic/structure architectural features which give rise to high quadratic molecular optical nonlinearities (β). Selected high-β chromophores are then covalently linked via several synthetic procedures to robust, glassy film-forming chloromethylated or hydroxylated polystyrenes. By this procedure, it is possible to achieve very high chromophore densities in polymeric films with good optical transparency and stability characterisitics. Spin-coating of these polymers onto ITO-coated conductive glass, followed by drying and poling above Tg yields robust films with high SHG efficiencies. As an example, in Figure 5 films of poly(p-hydroxystyrene) functionalized with N-(nitro-phenyl)-L-prolinol exhibit d_{33} = 18 x 10^{-9} esu at 1065 nm (ca. 16 times the corresponding value for KDP).

These factors which effect the temporal stability of the SHG capacity of the poled polymers are being investigated. For example, the hydrogen bonding network is a factor in extending lifetime. Figure 6 shows there is short decay which is not affected by annealing, the reduced amplitude arising from the unremoved polar solvent. After annealing rests reported after 325 days show that the long decay component is significantly improved.

Enhancement of the NLO effect is expected in pendant side chain structures by making the NLO polar group the side chain mesogen. Until two years ago there had been considerable interest in liquid crystalline polymers for third order NLO effects, particularly in rigid chain structures.[6] It was postulated that the natural cooperative alignment of the liquid crystalline molecular structure in cooperation with the highly charge-correlated pi-electron state would lead to large nonresonant nonlinearities.

However, Garito and coworkers concluded in December, 1987 that χxxxx (-3w:w,w,w,) is much more sensitive to the length of the chain than the conformation. Using his recently developed many-electron theory of second and third order nonlinear optical susceptibilities, microscopic descriptions of χxxxx were compared for cis- and trans-polyenes.[7]

Several groups in the U.S. and Western Europe have revived interest in liquid crystalline polymers for second order effects using NLO mesogens as backbone or pendant side chain groups, or in combination. At this meeting Garito extended his work to show that both β_{ijk} and γ_{ijkl} of conjugated linear and cyclic chains orginate from electron correlation behavior during virtual pi-electron excitation processes in effectively reduced spatial dimensions.[8] Since the nonlinear electronic excitations are naturally confined to quantum length scales, they are highly correlated and obey well-defined symmetry rules. Garito's conclusion that these excitation processes result in β_{ijk} values 100x times larger than any previously reported values is expected to accelerate efforts dedicated to the synthesis of liquid crystalline polymers for second order effects.

Since 1985, work by Griffin has investigated the synthesis of liquid crystalline polymers where the pendant moiety in these side chain polymers is simultaneously the liquid crystalline (mesogenic) moiety and the nonlinear optical species having a pi-donor/pi-acceptor conjugated electronic structure. To this end polyester liquid crystalline side chain homo- and

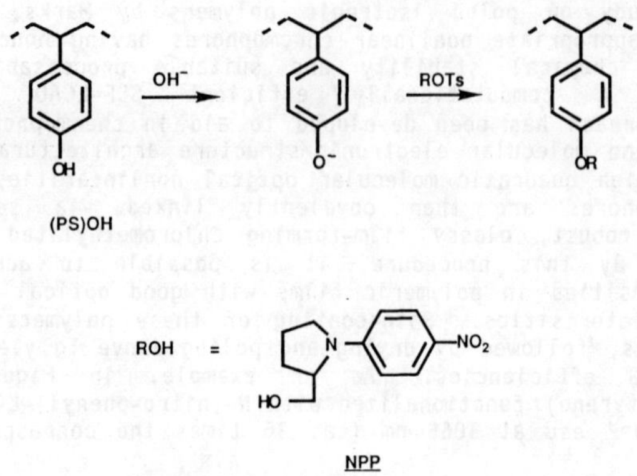

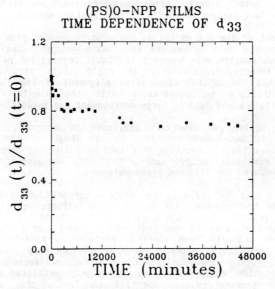

Figure   5   -   Functionalization of poly(p-hydroxystyrene) with NLO
                  chromophores.

(PS)O-NPP FILMS
TIME DEPENDENCE OF d_{33}

Figure   6   -   Temporal stability of poled poly(p-hydroxystyrene) with NLO
                  chromophores.

co-polymers have been synthesized. These materials can be made chiral by
empolying a chiral diol in the polymerization step, avoiding the formation of
a centrosymmetric system.[9]

Griffin and Prasad have fabricated an electrically poled Langmuir-Blodgett film from one such copolymer, (NBSBV)n-(C*SBV)m, and have examined its SHG properties. Eleectrooptic modulation in a monolayer was observed in the surface plasmon geometry, the measured $\chi^{(2)}$ being approximately 10^{-7}.[10] Griffin and Williams have made detailed dielectric measurements on one of these polymers and found it to be dual frequency addressable.[11] Polyamide side chain liquid crystalline polymers have been synthesized with improved retention of polar allignment.

The successful development of guest-host and side-chain polymer systems demonstrating large second order nonlinearities has made polymer thin films viable candiates for integrated optics. The state-of-the-art for second order devices is shown in Figure 7. The most fundamental electrooptic device is the modulator, a thin film waveguide electrooptic molulator employing one of three modulating systmes. These are the Mach-Zehnder interferometer, directional coupler, and rotation of the optical polarization. The Mach-Zehnder interferometer is the most common design. This high-frequency traveling wave device has been designed to provide a match in the optical and electron phase velocities.[12] Electrical modulating signals, operating at microwave frequencies, travel across electrodes at speeds which must be commensurate with the speed of light within the waveguide to achieve optimum performance.

Since the dielectric constant of the NLO polymers is on the order of 3, the traveling wave device may be designed to achieve close matching of optical and microwave velocities. The large dielectric constant of $LiNbO_3$ results in a rapid loss of phase matching over shorter waveguide lengths. This results in higher drive voltages and power requirements and limiting the frequencies accessible to 8 to 24 GHz. Because phase matching is possible for polymeric materials, the maximum frequency of single mode devices is limited only by electrode losses, being in the vicinity of 50 GHz using conventional electrodes.

State-of-Art $\chi^{(2)}$ Devices

● Traveling-Wave Electrode Mach-Zehnder Electrooptic Modulator

	$LiNbO_3$ Modulator	Polymer Modulator
Switching Voltage (V)	3 1/2–10 1/2	1.3 (0.7 With Higher $\chi^{(2)}$)
Power Requirement (W)	0.6–5	0.03
Maximum Frequency (GHz)	8–24	>50

— No Expected Velocity Mismatch Implies Higher Frequency Devices are Possible With Polymers

● Electrooptic Bragg Cell
— High Speed Radar Signal Processing
— 20 GHz as a Target

● Second Harmonic Generation
— High Efficiency Doubling of a Diode Laser

Figure 7 - Device state-of-the-art with second order polymers.

Lytel and coworkers recently reported the first measurement of an electrooptic effect in a poled-polymer channel waveguide above a GHz using the pendant side chain polymer shown in Figure 4.[13] Little or no dispersion in the electrooptic coefficient was observed. The losses in these polymers are below a dB/cm, which allows path lengths of several cm, and consequently, half-wave voltages approaching TTL levels.

This remarkable demonstration of performance was achieved by the concurrent development by Lytel of the selective poling procedure (SPP), in which active, poled channel wavelengths are fabricated in the aforementioned polymers in a single step. Only those regions of the material defined by the electrode pattern on the substrate are poled. Three prototype devices have been fabricated: a travelling-wave phase modulator, demonstrating 1.3 GHz modulation; a directional coupler; and a Y-branch interferometer.

Results of the research are also being adopted to meet a US Air Force requirement for a compact laser diode source in the 450 to 500 nanometer range for improved tactical optical data storage.[14] The high-efficiency frequency doubler is being developed also with one of the pendant side chain polymers of Figure 4. High speed radar signal processing with 20 GHz as a target is also being pursued with NLO polymeric materials.

In a later section the possibilities for second order polymers in implementing neural networks for optical computing are discussed.

STATUS OF THIRD ORDER POLYMER MATERIALS AND DEVICES

One of the goals of the current third order polymer research is to design multifunctional polymers having unique combinations of semiconductor, NLO and structural properties.

The objectives are to: (1) design polymers for maximum reliable nonlinear optical susceptibilities; (2) synthesize these electroactive polymers in a form exhibiting reasonable solubility in conventional solvents so that the polymers can be purified, characterized, and processed by convenient methods; and (3) explore the effects of charge transfer upon nonlinear optical activity.

Numerous theoretical calculations predict that optical nonlinearity in π electron systems increases with increasing π electron delocalization (or conjugation).[15] Electron delocalization depends upon orbital overlap between π-orbitals on adjacent atoms. Consideration of orbital overlap can be divided into two spatial factors. First, that overlap depends in an approximately exponential manner on the bond distance between atoms. Second, overlap depends upon the relative orientation of interacting orbitals with greater overlap observed for colinear π orbitals. From these considerations Dalton has pointed out that ladder polymers, with their planar configuration and uninterrupted π conjugation exhibit the molecular conformation necessary for optimum orbital overlap and hence maximum electron delocalization. These observations are supported by recent theoretical calculations. Also, ladder polymers exhibit symmetries appropriate for supporting polaron and bipolaron mid-gap state species. That is ladder polymers can be doped to charged lattices which will clearly exhibit different linear and nonlinear optical properties than the pristine polymers.

Because of the potential for optimum electron delocalization and because of the potential for supporting stable, charged lattices, ladder polymers were chosen as the most likely candidates to yield large nonlinear optical susceptibilities.[15]

Because of strong interchain Van der Waals interactions deriving from extensive electron delocalization and because of the rigid (low entropy) nature of ladder polymers, these materials have traditionally exhibited poor solubility which in turn has resulted in low polymer molecular weights and difficulty in processing these polymers. It is important to destabilize polymer-polymer interactions relative to polymer-solvent interactions. A logical approach to this objective is to introduce steric interactions associated with substituent groups which prevent tight packing of the ladder polymers. Thus, as a means of achieving improved polymer solubility and processibility, ladder polymers are derivatized by synthesizing such polymers from appropriately derivatized monomers.

Ladder polymers have traditionally been prepared by polycondensation.[15] The condensation reactions are required to complete each rung of the ladder and, in general, the two condensations occur at different rates (the condensations are thus referred to as being asymmetrical). To insure complete condensation and the fully-cyclized ladder structure yielding optimum electron delocalization, it is necessary to understand the polycondensation kinetics in detail. Moreover, the opportunity exists, by exploitation of the asymmetric polycondensation kinetics, to influence required polymer processing (e.g., fabrication of thin films) on the open-chain (or partially condensed) precursor polymer. Because of greater entropy and a conformation less favorable for polymer-polymer interaction, the precursor polymers are more soluble. Thus, an important route to processed ladder polymers, which have been purified and characterized by standard techniques, involves preparation and processing of precursor polymers and conversion to the final ladder polymer by thermal treatment.

Substituents can be used to fine tune the solubility of both precursor and ladder polymers in various solvents including both polar and nonpolar solvents. Solubility in water is influenced by terminating substituents with sulfonic acid or tetraalkylammonium groups; solubility in organic solvents can be influenced for alkyl, alkoxy, vinylamine, etc. groups.

Derivatization and precursor polymer synthetic methods represent the cornerstone of the approach.[15] This is very likely the most successful and systematic approach to the preparation of electroactive polymers.

One of the most straightforward schemes involves the preparation of dichloroquinone monomers derivatized with vinylamine groups employing a Mannich reaction.[15] Polycondensation of these monomers with aromatic amines can be carried out in simple solvents such as dimethylformamide (DMF) at ambient temperatures to yield highly processible open-chain precursor polymers. Optical quality thin films of the precursor polymer can be fabricated and converted to optical quality ladder polymer films by heating the films at approximately 250°C.

As synthesized polymers yield third order susceptibilities as high as 3×10^{-9} esu; these large values can be attributed to the combined effects of pi electron delocalization and charge transfer from the vinylamine substituent to the polymer backbone (Figure 8).[15] The experimentally observed dependence of nonlinear optical activity upon pi-orbital overlap is in agreement with recent calculations of Medrano and Goldfarb of Air Force Wright Aeronautical Laboratory/Material Laboratory for various lengths and conformations of rigid-rod macromolecules.[16] Consideration of trends observed to date suggest that nonlinear susceptibilities may be increased an additional one to two orders of magnitude by careful attention to optimizing electron delocalization. Third order susceptibilities in the range 10^{-8} to 10^{-11}

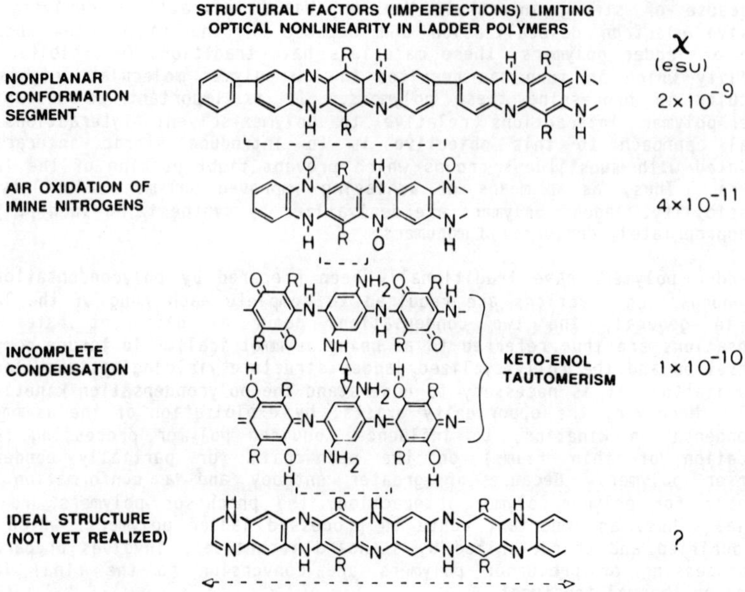

**STRUCTURAL FACTORS (IMPERFECTIONS) LIMITING
OPTICAL NONLINEARITY IN LADDER POLYMERS**

χ (esu)

NONPLANAR
CONFORMATION
SEGMENT

2×10^{-9}

AIR OXIDATION OF
IMINE NITROGENS

4×10^{-11}

INCOMPLETE
CONDENSATION

KETO-ENOL
TAUTOMERISM

1×10^{-10}

IDEAL STRUCTURE
(NOT YET REALIZED)

?

Figure   8   -   Structural design factors of NLO ladder polymers.

esu were observed for pristine polymers. At the time of observation, these were the largest values of third order susceptibility observed for any organic material. Recently, comparable values have been observed for related systems by a number of groups including Prasad[17] and Garito.[18]

The greatest variability in measured third order susceptibilities is observed for vinylamine-derivatized ladder polymers and can be attributed to interruption of pi electron conjugation associated with defects in the ladder structure. Defects have been identified as arising from air oxidation of imine nitrogens or incomplete condensation (Figure 8); air oxidation disrupts pi delocalization by introducing sp^3 centers. This work is unique in representing the first quantitative characterization of the perfection of the lattice for an electroactive polymer. The results suggest approaches for realizing the optimum optical nonlinearity predicted for ladder polymers.

Preliminary studies suggest that it is possible to enhance nonlinear optical susceptibility by more than an order of mangitude by chemical or electrochemical doping.[15] Moreover, in some cases, doping leads to improved transparency at visible light frequencies.

Dynamics of the dialkyl portions of the vinylanmine substituents appear to modulate coupling of the vinylamine pi-orbitals with the pi-system of the polymer backbone influencing the magnitudes of both the electronic and thermal components of the degenerate four wave mixing (DFWM) signal. Alicyclic substituents appear to be more sterically hindered leading to large electronic components of the third order susceptibility and to reduced thermal tails in the temporal response.

Very preliminary studies on oxidized and reduced (charged) lattices suggest that third order susceptibilities approaching 10^{-7} esu may be realized for doped organic materials.[15] Some support for this result is provided by preliminary studies of stacked organic metals.[19] However, the frequency dependence of the optical nonlinearity needs to be carefully defined and the nature of the modified polymer lattice characterized before doped materials can be considered for device application.

When ladder polymers are systematically cycled through their voltammetry, dramatic color changes are observed. Magnetic and optical measurements suggest that these changes can be associated with the generation of polaron and bipolaron states. For a number of systems, the optical phenomenon can be characterized as bleaching of the (visible) π-π^* interband transition together with the generation of (infrared) intraband transitions (Figure 9). Not only does this process produce a new optical lattice (which may or may not have greater optical nonlinearity but will most certainly exhibit different optical nonlinearity) but the phenonemon can immediately be applied to develop an optical switch (although the speed will be limited by the electrochemical response time).

Ladder polymers studied in this work exhibit unusually high optical damage thresholds (approximately 1 GW/cm^2) and show no saturation of the nonlinear response before damage.[15] Absence of saturability suggests nonresonant optical nonlinearity while linear absorption suggest some resonance absorption at measurement frequencies (e.g., 532 nm); the existence of both localized and delocalized electronic state and the existence of a variety of delocalized states (e.g., exciton, polaron, bipolaron, conduction band) may result in a decorrelation of linear and nonlinear effects.

The demonstrated concept of Dalton and Hellwarth for designing laser resistant polymers with $\chi^{(3)}$ of 10^{-8} to 10^{-9} esu and transparency in the visible has set the direction for a new US Army - US Air Force - Defense Advanced Research Projects Agency (DARPA) program for sensor and vision protection with importance to the NATO alliance. The envisioned way of protecting high gain optical military systems such as rangefinders, visual sensors, cameras, and the human eye against a rapidly tunable, pulsed, visible and near IR laser threat is predicated on demonstration of hybrid device concepts, potentially employing multiple NLO polymers. The device requirements dictate that the polymers should possess subnanosecond response times and millisecond recovery speeds, and should bave a broadband wavelength response (over visible and near IR). The most vigorous performance parameters are for eye protection, demanding responses in the picosecond/picowatt domain or smaller 100% transparency in the unradiated state, and should be able to dissipate 100% of the over-threshold energy incident upon the divice. An example is an optical limiter.

In another advance, organometallic polymer research by Garito, Dalton, Buckley and Prasad as well as Wegner has shown that very significant third order effects are obtained using phthalocyanine metal organic derivatives.[15,20-22] Garito has shown that planar structures with extended pi-orbitals will have high activity and speeds through saturable absorption (Figure 10).[23] Effective n_2 equal to AlAaAs mulitquantum structures are potentially achievable. Dalton has shown that lattice charge in addition to electron delocalization (interorbital spacing optimization) is another important factor to enhancing $\chi^{(3)}$ in metal containing delocalized electron polymer systems.[15]

A variety of organometallic macrocycle materials have been prepared and preliminary nonlinear optical measurements carried out. The observed behavior for phthalocyanines shows considerable variablitlity in optical nonlinearity with minor changes in chemical structure of the metallomacrocycle. These phenomena my be related to interstack dipoly coupling.

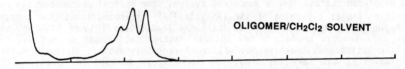

EFFECT OF CHEMICAL DOPING ON THE LINEAR OPTICAL
SPECTRUM OF A DELOCALIZED ELECTRON OLIGOMER:
INTRODUCTION OF VISUAL TRANSPARENCY AND
GENERATION OF A CHARGED (POLARON) LATTICE

INTERBAND π-π^* **TRANSITIONS**

OLIGOMER/CH$_2$Cl$_2$ SOLVENT

INTRABAND (POLARON) TRANSITIONS

OLIGOMER/SbCl$_5$/CH$_2$Cl$_2$ SOLVENT

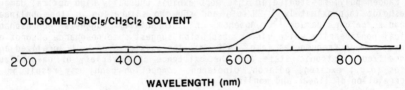

200   400   600   800

WAVELENGTH (nm)

Figure   9   -   Introduction of visible spectrum transparency in NLO doped ladder polymers.

● **Planar Structure With Extended π Orbitals**

M = Metal Atom

● **Shift In Focus From Conjugated Off-Resonance to Systems on Resonance With Narrow Molecular Extinction Coefficients**

● **High Activity Through Saturable Absorption**
   — **Effective N$_2$ Equals AlGaAs MQW Structures**

● **Excellent Secondary Properties**
   — **Spin Coatable For Thin Film Structure Applications**
       Optical Bistability
       Parallel Processing

Figure   10   -   Third order saturable absorption polymers.

Recent values of third order susceptibility to 10^{-7} esu have been realized in isolated cases. Dalton has shown that ultimately it will be important to extend consideration of oxidation/reduction (lattice charging effect) to the consideration of metal containing delocalized electron polymer systems as the variable redox states of metal should permit greater lattice change to be introduced and should permit realization of mixed valence centers.

Recent results by Singh, Prasad and Karasz, and Drury and Lusignea have developed high optical quality, highly oriented uniaxial poly(p-phenylene vinylene) (PPV).[24,25] Since the third order nonlinear optical susceptibility is a fourth rank tensor, a large orientational anisotropy has been observed. Device structures based on this anisotropy have been proposed using the concept of orientational bistability.[26] Femtosecond degenerate four wave mixing at 602 and 580 nm on a 10:1 stretch-oriented film shows a χ^3 value along the draw direction of 4×10^{-10} esu with a subpicosecond response.

The change in χ^3 value (actually the square root of the DFWM signal) as a function of film rotation with respect to the incident electric field vector yields the polar plot shown in Figure 11. The highest value of χ^3 is obtained when the electric vectors of all the four waves are parallel to the draw direction. The minimum value for χ^3 is for the orientation when all the electric vectors are perpendicular to the draw direction. The $\chi^{(3)}/\chi^{(3)}$ ratio is 37 which is a very high degree of orientational anisotropy." As shown by Xray, there is a high degree of polymer chain alignment along the draw direction.[27] This research then confirms that the largest component of χ^3, and therefore the microscopic nonlinearity tensor γ, is along the chain direction as is predicted from theoretical calculations of microscopic nonlinearity in pi-conjugated polymeric and oligomeric structures.

The PPV can be conveniently processed into various shapes using a water soluble precursor route. The oriented fibers have good mechanical strength such that even a two micron thick free-standing film can be prepared. Doping can produce high electrical conductivity along the draw direction, indicating a high effective pi-conjugation in this polymer--a criterion need for large χ^3. The film exhibits high optical damage threshold under picosecond and femisecond pulse illumination.

The state of the art for $\chi^{(3)}$ devices is shown in Figure 12. Internationally there has been far less progress in $\chi^{(3)}$ device development than in $\chi^{(2)}$ devices. The reason for this is that most of the research efforts in the United States, Europe and Japan have been in second order polymers.

While nonresonant third order nonlinearities of 10^{-10} to 10^{-11} esu had been reported one year ago, two orders of magnitude higher had to be achieved for practical device application. As reported in this section, polymers have now been achieved which demonstrate $\chi^{(3)}$ of 10^{-8} to 10^{-9} esu. While 10^{-10} esu was on the periphery for all-optical waveguide devices, these polymers could still find use in bistable optical switches and optically-controlled modulators and switches.

The new third order polymers will now make all-optical waveguide devices a realistic possibility. Third-order integrated optical devices will include distributed couplers, Mach-Zehnder interferometers and Bragg reflectors. The optical Mach-Zehnder modulator will utilize $\chi^{(3)}$ polymers which exhibit intensity dependent refractive indices.[29] In a typical Mach-Zehnder switch,

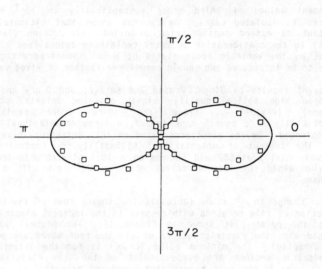

Figure   11   The polar plot (orientational anisotropy) of the square root
                      of the degenerate four wave mixing signal intensity
                      (proportional to $\chi^{(3)}$) foir the 10:1 stretch-oriented PPV
                      film. The squares are the observed data points; the solid
                      curve is the theoretical fit.

State-of-Art $\chi^{(3)}$ Devices

● **Far Less Progress Than $\chi^{(2)}$**

● **Large Payoff in Multiplexing and Demultiplexing**
  — High Speed
  — Handle Many Inputs
  — Very High $\chi^{(3)}$

● **Can Lead to:**
  — All Optical Interferometer and Optical Switch
  — Optical Bistability and Digital Optical Information
    Processing
  — Optically Induced Dynamic Grating and Real-Time
    Holography

● **Device Performance Depends on Material Properties in a
  Complicated Way**
  — Determined by Device Architecture
  — Parallel/Analog Processing of 2D Image
  — Phase Conjugate Optics and Image Processing
  — Self Focusing/Defocusing Applications

Figure   12   -   Device state-of-the-art with third order polymers.

the input beam is divided into two channels. When a high intensity modulating beam is introduced into a channel, the refractive index of the channel changes and it creates a phase difference of the input beam in the channel relative to the beam in the other channel. If the phase difference is equal to pi, the two cancel each when they are recombined. A fast optical Mach-Zehnder switch may be used for digital signal processing.

Optical bistability has been observed in a poly-4-BCMU polydiacetylene polymer quasi-wavelength interferometer by Singh and Prasad.[30] Biaxial NLO polymers as demonstrated with the rigid rod aromatic heterocyclic polymer, poly(p-phenylene-2, 6-benzobisthiazole), or PBT, and PPV offer additional device design options based on polarization bistability.[31]

Thin film devices such as etalons are now possible with the improved $\chi^{(3)}$ polymer.[32] A requirement defined for etalons at the June, 1988 NATO workshop on Nonlinear Optical Effects in Polymers over the next 3-5 years were polymers with $\chi^{(3)} \sim 10^{-7}$ esu and $\mathscr{L} \leq 0.1$ cm.

There will be large payoff in fast optical multiplex switching, based on high n_2 polymers, which combines optical signals from many channels converted to a single temporal signal for easy transmission.[33] As a high intensity pulse from a picosecond mode-locked laser transverses through a high n_2 medium, it sequentially opens the channels and thus creates a temporal signal. A similar design is used to demultiplex the transmitted signal to retrieve the individual signals. This can lead to digital optical information processing and other applications cited in Figure 12.

Device design with $\chi^{(3)}$ materials is often guided by the architecture or electrical parameters. Thus careful investigation of polymer properties concurrent with their operational performance in device prototypes should lead to parallel analog processing of 2-D images and other applications cited in Figure 12. For example the optical Kerr cell design can be used also for analog optical signal processing such as optical scanning.

NLO POLYMERS IN OPTICAL NEURAL NETWORKS

At AFOSR work is starting which will investigate the role of second order polymers in nonlinear optical neural nets and architectures. Neural networks are models of computation that are based on the way brains perform their computation. The human brain has somewhere on the order of 10^{11} to 10^{12} neurons, massive numbers of computational units. Each neuron is locally connected to 10^3 to 10^4 other neurons. They differ from serial networks, illustrated by digital computers, in interconnectivity because they are able to perform certain tasks very efficiently by parallel signal processing.

Optical implementation is particularly attractive.[34] In comparison to electronic devices, optical devices are inherently parallel and thus not limited by wire interconnections and cross-talk. Figure 13 compares the Figure of Merit for an optical neurocomputer with that of the brain and Very Large Scale Integration. As Malloy and Giles at AFOSR point out, 10^6 neurons is reasonable to achieve for an optical array, 10^{12} as in the human brain being very difficult.

There are several ways to implement optical interconnections, but most are limited by the dimensionality of interconnection medium. For the basic structure of an optical neural network (Figure 14), the interconnection cloud could be a two-dimensional spatial light modulator or a three-dimensional volume holographic element. The restrictions for the optical implications of

SIZE OF NEUROCOMPUTERS

(OR WHY OPTICS)

Figure of Merit for Neurocomputer Performance

$$F = F (N,I,S)$$

N = # of neurons
I = # of interconnections/neuron
S = speed of neuron update

$$F = (N I) \times (N I S) = N^2 I^2 S$$

System	N	$\underset{\text{(sec}^{-1})}{\underline{S}}$	I	NxI	F
Brain	10^{12}	10^3	10^3	10^{15}	10^{33}
Optical Neurocomputer	10^6	10^9	10^3	10^9	10^{27}
VLSI (fully connected)	10^3	10^9	10^3	10^6	10^{21}
VLSI (nearest neighbor)	10^6	10^9	O(1)	10^6	10^{21}

OPTICS OFFERS A LARGE # OF REAL (NOT SIMULATED) INTERCONNECTIONS (JUST LIKE BIOLOGY)

Figure   13   –   The role of optics in neurocomputers.[35]

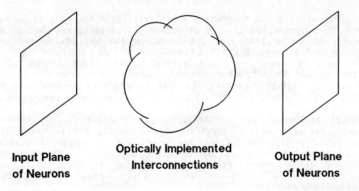

**Input Plane
of Neurons**

**Optically Implemented
Interconnections**

**Output Plane
of Neurons**

Figure   14   –   Basic structure of an optical neural network. The planes of
neuron could be connected to other planes not depicted.[36]

neural nets have yet to be defined. Architectures, dynamics and representation of information have to be expressed optically; models are starting to emerge. The architecture-imposed requirements on device design is an unknown which impacts the material properties requirement.

For the basic structure of an optical neural network (Figure 14), the interconnection cloud could be a two-dimensional volume holographic element.[34] In the near-term there is a need one- and two-dimensional arrays with 10^3 and 10^6 elements respectively with response times in the microsecond to nanosecond range. Multiple quantum well self-electrooptic effect devices (SEED) have a variety of desirable functions for optoelectronic neural net device arrays including wavelength compatibility with semiconductor sources and detectors, voltage levels compatible with electronic circuitry, transparent substrates, and the ratio of fast index change for switching to resultant thermal index change. (Malloy and Giles point out that in its simplest form the SEED switches off when a threshold intensity of light is reached).

Malloy and Giles see the ability to understand, control and tailor the nonlinear excitonic mechanisms in the III-V semiconductor multiple quantum wells as compared with the current state of understanding in polymers as the primary point of this section. Material requirements are still difficult to delineate because of the vague and equivocal device requirements. In the long term tailored polymeric materials are seen as viable for designing arrays; however, it remains to be determined how $\chi^{(2)}$ and $\chi^{(3)}$ polymers can mimic neuron slabs and arrays and how they can optically implement interconnections in optical, optoelectronic and hybrid optical/electronic architectures.

Malloy and Giles point out that if the optical device works on resonant transitions sensitive to temperature, many applications are ruled out.[34] The most recent approaches for $\chi^{(2)}$ and $\chi^{(3)}$ polymers in a nonresonant or near off-resonance mode offer some promise here provided that the calculated large low n_2 and absorptions can be achieved. In addition to being competitive with the aforementioned attributes for SEED optics, the electronic subpicosecond response times and other temporal responses of polymers offer considerable latitude for designing large element arrays working in the 10 nanosecond to 100 picosecond individual element range. Second order polymers in photo-addressable spatial light modulators (SLMs) operating at room temperature or third order polymers in optically bistable polymer etalons offer approaches for implementation of optical neural net designs.

SUMMARY

Some of the recent advances made in the synthesis and characterization of and polymers have been discussed. Poled isotropic polymers now perform as well as inorganic crystals. The chemistry which contributes to long term poling stability is starting to be understood. Considerable advances have been made in third order polymer design, synthesis and purity. Optical transparent ladder polymers with $\chi^{(3)}$ commonly measured at 10^{-9} esu and some samples with 10^{-8} esu being reported. The mechanisms for improving visible transparency through doping have been delineated in these polymers. They are now the subject of serious investigation for sensor and vision protection against pulsed laser threats. Mechanisms for synthesizing metal containing delocalized electron polymer systems with n_2 that can theoretically approach MQWs have been demonstrated through saturable absorption and lattice charging. A few isolated measurements of $\chi^{(3)} \sim 10^{-7}$ esu have recently been reported. High optical quality, stretch oriented PPV with a large anisotropy value in the $\chi^{(3)}$ now makes orientational bistability devices a reality.

Results of the $\chi^{(2)}$ polymer research are already being adopted to meet Air Force and DOD requirements for high-efficiency frequency doublers, travelling-wave phase modulators, optical shutters, waveguide multiplexers, and electro-optic Bragg cells. While $\chi^{(3)}$ polymer research is moving rapidly into sensor protection devices, $\chi^{(3)}$ polymer development is just now starting to approach device quality in terms of $\chi^{(3)}$ of 10^{-7} to 10^{-8} esu. Significant work still needs to be done to achieve optically transparent polymers with $\chi \leq 0.1$ cm^{-1}.

Both the second and third order polymer systems are long term candidates for optical implementation of neural net architectures. Material requirements are still difficult to delineate because of the vague and equivocal device requirements. Polymers are seen as viable for designing arrays because of their nonresonant mode promise.

ACKNOWLEDGMENTS

Special appreciation is expressed to Ms. Donna Proctor for her help in the preparation of this paper as well as to Dr. Donald L. Ball for his encouragement and support. We are indebted to the dedication, achievement and enthusiastic inspiration of all the investigators sponsored by AFOSR in this area, many of whom could not be listed here.

REFERENCES

1. NATO Workshop, Nonlinear Optics in Polymers, NICE-Sophia Antipolis, June, 1988.

2. D. R. Ulrich, "Nonlinear Optical Polymer Systems and Devices," MOL. CRYST. LIQ., Volume 160, pp. 1-32, (1988).

3. J. Riggs and J. Stamatoff, private communications; R. DeMaritino, et. al., NONLINEAR OPTICAL PROPERTIES OF POLYMERS, Materials Research Society Proceedings, Volum 109, A. J. Heeger, J. Orenstein, and D. R. Ulrich, Eds. Materials Research Society, Pittsburgh, Pennsylvania, 1988: pp. 65-77, J. Stamatoff, AFOSR Contract F49620-87-C-0115

4. J. Stamatoff, AFOSR Program Review, National Academy of Sciences, April 20-21, 1988; A. Buckley and J. Stamatoff, NATO Workshop, Nonlinear Opticas in Polymers, NICE-Sophia Antipolis, June 1988; J. Stamatoff, AFOSR Contract F49620-87-C-0115.

5. T. Marks, S. Carr and G. Wong, AFOSR Program Review, National Academy of Sciences, April 20-21, 1988; NATO Workshop, Nonlinear Optics in Polymers, NICE-Sophia Antipolis, June, 1988; T. Marks, AFOSR Grant AFOSR-86-0105.

6. D. R. Ulrich, SPIE OPTICAL ENGINEERING REPORTS, pp. 5A-7A, No. 43, July, 1987; D. R. Ulrich, POLYMER, Vol. 28, pp. 533-542, 1987, D. N. Rao, et. al., APPL. PHYS. LETT. Vol. 48, pg. 1187 (1986); A. F. Garito, et. al., MOLECULAR AND POLYMERIC OPTOELECTRONIC MATERIALS: FUNDAMENTALS AND APPLICATIONS, Proceedings SPIE, 482, G. Khanarian, 1986; pp. 2-11.

7. A. F. Garito, NONLINEAR OPTICAL PROPERTIES OF POLYMERS, Materials Research Society Proceedings, Volume 109, A. J. Heeger, J. Orenstein, and D. R. Ulrich, Eds., Materials Research Society, Pittsburgh, Pennsylvania, 1988; pp. 91-101; A. F. Garito, AFOSR Contract F49620-85-C-0105.

8. A. F. Garito, "Recent Developments in Nonlinear Optical Properties of Polymers," Abstract 2.2, International Conference on Organic Materials for Non-Linear Optics, Oxford, England, June 29-30, 1988; A. F. Garito, AFOSR Program Review, National Academy of Sciences, April 19-20, 1988, A. F. Garito, AFOSR Contract F49620-85-C-0105.

9. A. C. Griffin, AFSOR Program Review, National Academy of Sciences, April 20-21, 1988; A. C. Griffin, AFOSR Grant AFOSR-84-0249.

10. M. M. Carpenter, P. N. Prasad, and A. C. Griffin, "The Characterization of Langmuir-Blodgett Films of a Nonlinear Optical, Side Chain Liquid Crystalline Polymer," Thin Solid Films (in Press); P. N. Prasad, AFOSR Contract F49620-87-C-0042?; A. C. Griffin, AFOSR Grant AFOSR-84-0249.

11. A. C. Griffin and C. Williams, private communication; A. C. Griffin, AFOSR Grant AFOSR-84-0249; C. Williams, AFOSR Contract F49620-87-C-0111.

12. G. I. Stegeman, C. Seaton, and R. Zononi, NONLINEAR OPTICAL PROPERTIES OF POLYMERS, Materials Research Society Proceedings, Volume 109, A. J. Heeger, J. Orenstein, and D. R. Ulrich, Eds., Materials Research Society, Pittsburgh, Pennsylvania; pp. 53-64; R. Lytel, G. F. Lipscomb, and J. I. Thackara, ibid.: pp. 19-28; J. Stamatoff and J. Riggs, private communication; O. K. Kwon, F. R. W. Pease and M. R. Beasley, IEEE ELECTRON DEVICE LETTERS, USA, Vol EDL-8, No. 13, pp 582-585, December 1987

13. R. Lytel, NATO Workshop, Nonlinear Optics in Polymers, NICE-Sophia Antipolis, June, 1988; R. Lytel, "Advances in Organic Integrated Optic Devices," Abstract 3.5, International Conference on Organic Materials for Nonlinear Optics,: June 29-30, 1988 Oxford, England; R. Lytel, AFOSR Program Review, Natonal Academy of Sciences, April 20-21, 1988.

14. Recent Research Accomplishments of the Air force Office of Scientific Research, 1988, Bolling Air Force Base, Washington, D.C. 20332-6448, U.S.A. (in press).

15. L. R. Dalton and R. Hellwarth, AFOSR Contract F49620-87-C-0010; L. Dalton, AFOSR Contract F49620-85-C-0096; L. Dalton, NATO Workshop, Nonlinear Optics in Polymers, NICE-Sophia Antipolis, June, 1988; L. Dalton, AFOSR Program Review, National Academy of Sciences, April 20-21, 1988; L. Dalton, NONLINEAR OPTICAL PROPERTIES OF POLYMERS, Materials Research Society Proceedings, Volume 109. A. J. Heeger, J. Orenstein, and D. R. Ulrich, Eds., Materials Research Society, Pittsburgh, Pennsylvania, 1988: pp. 301-312; L. R. Dalton, NONLINEAR OPTICAL AND ELECTROACTIVE POLYMERS, P. N. Prasad and D. R. Ulrich, Eds., Planum Press, New York, 1988; pp. 243-272.

16. J. Medrano and I. Goldfarb, NATO Workshop, Nonlinear Optics in Polymers, NICE-Sophia Antipolis, June, 1988.

17. P. N. Prasad, private communication.

18. A. F. Garito, private communication.

19. P. G. Huggard, W. Blau, and D. Schweitzer, APPL. PHYS. LETT., Vol. 51, p. 2183 (1987).

20. A. F. Garito, AFOSR Contract F49620-85-C-0105.

21. P. N. Prasad, AFOSR Contract F49620-87-C-0097.

22. A. Buckley, AFOSR Contract F49620-86-C-0129.

23. A. F. Garito, AFOSR Contract F49620-85-C-0105.

24. B. P. Singh, P. N. Prasad, and F. E. Karasz, POLYMER (in press); F. E. Karasz, AFOSR Contract F49620-87-C-0027; P. N. Prosad, AFOSR Contract F49620-87-C-0097.

25. M. Druy and R. Lusignea, AFOSR Contract F49620-88-C-0065.

26. K. Otsuka, J. Yumoto, and J. J. Song, OPT, LETT., Volume 10, p. 508, 1985.

27. D. R. Gagnon, F. E. Karasz, E. L. Thomas, and R. W. Lenz, SYNTH. METAL., Volume 20, p. 85, 1987.

28. D. R. Ulrich "Nonlinear Optical Polymer Systems and Devices," MOL. CRYST. LIQ. CRYST., Volume 160, pp 1-32, (1988).

29. J. Riggs, private communicaton; G. Stegeman, private communication.

30. b. P. Singh and P. N. Prasad, J. OPT. SOC. AM. B5, 453 (1988).

31. R. Lytel, private communication; R. Lusignea and L. Domash, private communication; P. N. Prasad private communication; D. N. Rao, et. al., APPL. PHYS. LETT., Vol. 48, p. 1187 (1986); A. F. Garito, et. al., MOLECULAR AND POLYMERIC OPTOELECTRONIC MATERIALS: FUNDAMENTALS AND APPLICATIONS, Proceedings SPIE, 682, 1986: pp. 2-11.

32. R. Lytel, et. al. NONLINEAR OPTICAL AND ELECTROACTIVE POLYMERS, P. N. Prasad and D. R. Ulrich, Eds., Planum Press, New York (1988) pp. 415-426.

33. J. Riggs, private communications G. B. Kushner and J. A. Neff, NONLINEAR OPTICAL PROPERTIES OF POLYMERS, Materials Research Society Proceedings, Volume 109, A. J. Heeger, J. Orenstein, and D. R. Ulrich, Eds., Materials Research Society, Pittsburgh, Pennsylvania, 1988; pp. 3-17.

34. K. J. Malloy and C. L. Giles, NONLINEAR OPTICAL PROPERTIES OF POLYMERS, Materials Research Society Proceedings, Volume 109, A. J. Heeger, J. Orenstein, and D. R. Ulrich, Eds., Materials Research Society, Pittsburgh Pennsylvania, 1988; pp. 77-87.

35. K. J. Malloy and C. L. Giles, private communication.

36. D. Psaltis, special issue of APPLIED OPTICS on "Neural Networks," Vol. 26, 1987; N J. Malloy and C. L. Giles, private communication.

Studies of Ultrafast Third-order Non-linear Optical Processes in Polymer Films

P.N. Prasad

DEPARTMENT OF CHEMISTRY, STATE UNIVERSITY OF NEW YORK AT BUFFALO,
BUFFALO, NY 14214, USA

1 INTRODUCTION

There has been considerable interest during recent years in the
prospect of optical signal processing. The main attractive feature
is the gain in the speed of information processing by nonlinear
optical processes compared to that provided by electronics.
Therefore, a useful definition of an ultrafast process is one having
a rise time $\leq$ 1ps and a decay time $\leq$ 10ps. Organic polymeric
structures with extensive π-conjugation are ideally suited to meet
these specifications (1). These systems have a relatively large non-
resonant third-order susceptibility, $\chi^{(3)}$, which is derived from the
π-electron delocalization (1,2). Because of the inherent
distribution of conjugation length and the presence of defects it is
often very difficult to identify a truly non-resonant wavelength at
which no absorption occurs. Again, a practical definition of a non-
resonant wavelength used by our research group is one at which the
absorption is very small ($\alpha < 10 \text{cm}^{-1}$). Since non-resonant $\chi^{(3)}$
processes involve virtual transitions, they are by nature ultrafast.
In our laboratory, response time of non-resonant third-order
nonlinearity has been demonstrated to be in femtoseconds and limited
only by the optical pulse temporal resolution (3). However, we have
also found picosecond and subpicosecond response times for resonant
third-order optical nonlinearities in cases where extremely fast
excited state dynamics occurs (4).

As stated above the driving force behind the surge of interest
in this area has been the prospect of optical signal processing.
From this point of view, a important phenomenon is the intensity
dependent refractive index which provides a mechanism for light
control with light. The intensity dependent refractive index n_2 is
related to the third-order optical susceptibility $\chi^{(3)}$ by the
following relation:

$$n_2(\omega) = \frac{16\pi^2}{n_0^2 c} \chi^{(3)}(-\omega; \omega, -\omega, \omega) \tag{1}$$

In designing a device, important parameters are the magnitude, the response time and the sign of $\chi^{(3)}$. Other relevant factors are the anisotropy of $\chi^{(3)}$ (it is a fourth rank tensor), the dispersion of $\chi^{(3)}$, and the temperature dependence of $\chi^{(3)}$. Furthermore, near a resonance, $\chi^{(3)}$ might be a complex quantity. A variety of techniques have been used in our laboratory to investigate the behavior of $\chi^{(3)}$ as it is usually very difficult to obtain complete information on $\chi^{(3)}$ by a single technique. The technique we find particularly useful in obtaining the magnitude and response time of $\chi^{(3)}$ is degenerate four wave mixing (DFWM). In addition, the technique can conveniently be used to investigate the anisotropy of $\chi^{(3)}$ in oriented samples and to investigate its temperature dependence. One can also conveniently use this technique to obtain information on microscopic third-order nonlinearity, γ, for materials which are soluble. In such cases, the use of a number of concentrations permits one to obtain both the sign and the magnitude of averaged microscopic nonlinearity. However, the DFWM technique suffers from the complication that a variety of processes contribute to the signal and one has to use extreme care in the analysis of the data. This point has been discussed in detail in an earlier paper (3).

This paper will review some of the results of our comprehensive research program in the area of organic nonlinear optics. Results of some recent DFWM measurements of $\chi^{(3)}$ will be presented followed by a discussion of our recent approach involving theoretical and experimental studies of third-order optical nonlinearities in sequentially built and systematically derivatized structures. Such systematic studies are needed for developing our understanding of optical nonlinearities in organics. The relative merits of polymeric structures as nonlinear optical materials will be discussed. Finally, some future directions of research in relation to third-order optical nonlinearity will be presented.

2 MEASUREMENTS OF $\chi^{(3)}$

DFWM experiments have been conducted on a number of conjugated molecules and polymers in the form of films or solutions in various solvents. The experimental arrangement uses a cw mode-locked Nd-Yag laser; the fundamental output is compressed in a fiber optics pulse compressor, frequency doubled and then used to sync-pump a dye laser. The dye laser output is subsequently amplified in a three stage amplifier which is pumped by a 30Hz pulsed Nd-Yag laser. The output pulses from the amplifier are about ~350 femtoseconds wide with a pulse energy of ~0.5mJ. The beam passes through an appropriate attenuator to adjust and judiciously vary the power level. The

resulting beam is split into three portions which are incident at the
sample in a backward wave geometry described in detail in an earlier
paper (3). The time resolution of ~350fs is of considerable help in
determining the mechanism of nonlinearity. This point has also been
discussed in detail earlier (3). We have investigated a number of
polymers where the common names, structures and the observed $\chi^{(3)}$
values are listed below.

Of these materials, the polymer PPV appears to be the most
promising on both the combined nonlinear optical and material
properties. It has a relatively high non-resonant $\chi^{(3)}$ and a
response time in the subpicosecond range which is limited by our
optical pulse width (5). The film can be processed through a soluble
precursor route and stretch-oriented to produce good optical quality
free standing films of dimensions as thin as ~2μm; the oriented film
shows a considerable $\chi^{(3)}$ anisotropy (5).

Some of the general conclusions of our study of third order
optical nonlinearity in polymeric structures are now discussed. $\chi^{(3)}$
shows a strong dependence on the π-electron conjugation, the largest
component of the $\chi^{(3)}$ tensor being along the chain orientation
direction. The response time of the non-resonant $\chi^{(3)}$ in organic
structures is in femtoseconds and currently limited by the time
resolution of the optical pulse. In some cases, we have been able to
determine the sign of $\chi^{(3)}$. For small molecules or oligomers, the
solution study over a range of concentrations yields the sign. For
polymer films, we have used the intensity dependent phase shift
either in a waveguide or in a Fabry-Perot geometry. For the limited
number of cases we have investigated, we find $\chi^{(3)}$ to be positive
under non-resonant condition. In few cases we have investigated,
$\chi^{(3)}$ was found to be relatively insensitive to temperature, where
there were no structural changes.

The magnitudes, of $\chi^{(3)}$, although large for non-resonant
nonlinearity, are still not sufficient for optical switching using
diode lasers. The most important objective, therefore, is to enhance
this nonlinearity. One approach is to use resonance enhancement
without sacrificing the ultrafast speed. Resonant processes require
specific conditions to be met in order to be ultrafast. Since
radiative processes have fundamental limitations of speed, ultrafast
non-radiative relaxations are required. The polymeric structures
offer possibilities for mechanisms of ultrafast non-radiative
relaxation. First, low dimensionality of electronic structures in π-
conjugated linear polymers can enhance certain non-radiative
relaxations such as diffusion limited processes. Second, conjugated
structures which yield bond alternation permit very rapid
photoinduced conformational deformation. In trans-polyacetylene,
with a degenerate ground state, photoexcitation across the band gap
has been suggested to lead to rapid conformational deformation on a
subpicosecond time scale to produce soliton-antisoliton pairs with

TABLE 1 $\chi^{(3)}$ values of various polymers.

Systems	Structure	$\chi^{(3)}_{1111}$ (in esu) $\lambda = 602$ nm
LARC-TPI		$\sim 2 \times 10^{-12}$
PBT		$\sim 10^{-11}$
PBO		$\sim 10^{-11}$
PPV		$\sim 4 \times 10^{-10}$
poly PHENYL ACETYLENE		5×10^{-11}
poly-n-BCMU red		3×10^{-10}
yellow		2.5×10^{-11}

$R = (CH_2)_{n}O\text{-}C\text{-}N\text{-}CH_2COO(CH_2)_3CH_3$

states near the midgap (6). This shift of oscillator strength also provides a mechanism for large $\chi^{(3)}$. We have investigated a solution processable polyacetylene-polymethylmethacrylate graft co-polymer and have indeed observed a $\chi^{(3)}$ which scales to $\sim 10^{-10}$ esu for pure polyacetylene composition, with a subpicosecond response time (4). Polythiophene is another example where the photoexcitation produces polaronic distortion to create states in the band gap. In this case, we have observed $\chi^{(3)}$ close to 10^{-9} esu and again having subpicosecond response (4,7,8). The $\chi^{(3)}$ is sufficiently strong to observe the DFWM signal from a 20 layer Langmuir-Blodgett film (each layer ~30Å thick) of a soluble long alkyl chain substituted polythiophene, poly-3-dodecylthiophene. Another example of a very fast resonant $\chi^{(3)}$ behavior is observed for phthalocyanine films for which a value of ~10^{-8}esu is observed with the response time of ~ 1ps (7,9). The resonant nonlinearity is sufficiently strong to observe the DFWM signal from a single monolayer (~35Å thick) Langmuir-Blodgett monolayer film of metal free tetrakis cumylphenoxy phthalocyanine. The ultrafast speed in this case is suggested to be derived from very fast exciton dynamics (9).

We have also investigated the effect of vibrational resonance in CARS and CSRS experiments (10). Of course, the experiments do not involve degenerate four wave mixing since there are two input wavelengths with output being at a different wavelength. The resonance enhancement by the -C≡C- vibration in poly-4-BCMU is found to be over two orders of magnitude.

Our study of electronically resonantly enhanced $\chi^{(3)}$ with ultrafast speeds still yields a value of < 10^{-8} esu. In order for π-conjugated materials to compete with the inorganic resonant $\chi^{(3)}$ materials, the currently reported resonant $\chi^{(3)}$ values are not large enough. It is possible that by controlling the excited state dynamics one should be able to enhance $\chi^{(3)}$ appreciably. However, the best chance for organics, from device application point of view, is with the non-resonant processes. The question then arises "Can we achieve considerably enhanced non-resonant nonlinearities in organics or is there a fundamental limitation?" The history of third-order nonlinear optical effects in organic structures is rather recent, with considerable research being done only very recently. As a result, our understanding of structure-property relationships for third-order nonlinearity is in its infancy. To develop our understanding of structure-property relationship which could eventually lead to predictive capability, we have taken an approach of theoretical calculations of microscopic nonlinearities and experimental measurements on sequentially built and systematically derivatized structures. This approach is discussed in the next section.

3 THEORY AND MEASUREMENTS ON SEQUENTIALLY BUILT STRUCTURES

Since π-conjugated structures have been shown to exhibit large non-resonant $\chi^{(3)}$ derived from π-electron conjugation, our starting point for theoretical calculations of third order microscopic nonlinearity γ involves conjugated oligomeric structures. For the calculation of microscopic nonlinearity one can take two different approaches represented in Table 2 below. We have taken the derivative method approach at the ab initio-SCF level. The details of the method and results of our calculation are described elsewhere (11). We find that for calculation of optical nonlinearities one needs to describe adequately the tail portion of the wave functions. Consequently, inclusion of diffuse polarization functions in the basis function is extremely important. At the simplest level of the time-independent SCF approximation, one neglects electron correlation and calculates the static (frequency independent) hyperpolarizabilities β and γ. However, we feel that away from any excited state resonances, these calculations can provide useful information on trends in systematically varied structures. Therefore, this method can conveniently be used to derive chemical insight into the role of substituents. Also, the ab-initio approach allows one to optimize the geometry.

We have computed α and γ for various finite chain oligomers in the polyene (alternate single and double bonds), polyyne (alternate single and triple bonds) and cumulene (all double bonds) series. The second hyperpolarizability γ is found to be anisotropic with the longitudinal component γ_{zzzz} along the chain direction growing very rapidly as the chain length grows. We have fitted the increase of both α and γ for a given oligomer series by the power law equation

$$F = A + B(N-\delta)^C \qquad (2)$$

In the above equation $F = \alpha$ or γ; A is included to account for the end effect; δ is incorporated to adjust for an effective conjugation length. N is the number of repeat units and C describes the power law. Our ab-initio calculation yields C in the range 1.3 to 1.5 for α and 3.2 to 3.4 for γ. In contrast, a free electron model predicts a N^3 dependence for α and a N^5 dependence for γ (12). Semiempirical calculations predict C to be in the range of 5.3 to 4.5 for γ (13,14). Garito and co-workers have used the sum-over states approach on trans-polyene series and found a N^5 dependence for γ (15). Another important conclusion of our calculation is that sign of γ is positive. In derivatized structures $R_1-C\equiv C-R_2$ we find that when R_1 and/or R_2 are substituted by NO_2, the value of γ increases.

We have been able to test these predictions by our measurements on sequentially synthesized thiophene oligomers with N=1-6 and derivatized thiophene (16). The γ values for the thiophene oligomers were obtained by DFWM studies of their solutions at various concentrations in THF. This method yields an orientationally averaged value $\langle \gamma \rangle = \frac{1}{5}(\gamma_{zzzz} + \gamma_{xxxx} + \gamma_{yyyy} + 2\gamma_{yyzz} + 2\gamma_{xxzz} + 2\gamma_{xxyy})$. The orientationally averaged $\langle \alpha \rangle$ values were

TABLE 2   Microscopic Theory of Optical Nonlinearity

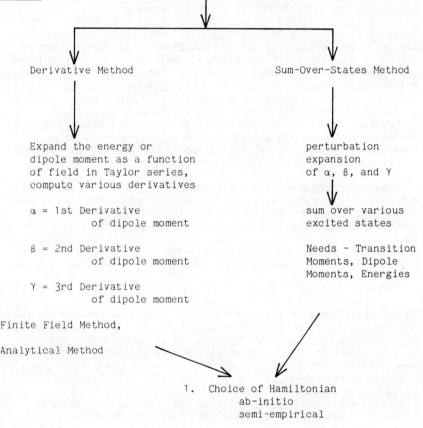

Derivative Method   Sum-Over-States Method

Expand the energy or   perturbation
dipole moment as a function   expansion
of field in Taylor series,   of α, β, and γ
compute various derivatives

α = 1st Derivative   sum over various
      of dipole moment   excited states

β = 2nd Derivative   Needs - Transition
      of dipole moment   Moments, Dipole
                                               Moments, Energies
γ = 3rd Derivative
      of dipole moment

Finite Field Method,

Analytical Method

1. Choice of Hamiltonian
        ab-initio
        semi-empirical

2. Choice of Basis Sets
        valence sets
        diffuse polarization functions

obtained from the refractive index measurements in THF solutions of
various concentrations. Our result shows that γ is positive for all
the oligomers as is predicted by our ab-initio calculations of the
various conjugated structures. The fit of the repeat unit dependence
of the experimental values using the power law equation (2) yields C
= 1.69 for <α> and C = 4.05 for <γ>. The <α> power law is in good
agreement with that predicted by our ab-initio calculation, but the
observed power law for <γ> does not agree as well with our ab-initio
calculation.

The value of $\langle\gamma\rangle$ for the 2-nitro substituted terthiophene is found to be larger than that for the unsubstituted terthiophene, again in agreement with the prediction of our ab-initio calculations.

4 CONCEPT OF EFFECTIVE CONJUGATION

Effective conjugation length can be defined as the chain length (repeat unit) over which the π-electrons are delocalized. The current microscopic theories as well as experimental investigation of optical nonlinearities show strong dependence of γ on π-conjugation (1). Therefore, a viable route to enhance γ would naturally be to increase the conjugation length by going to higher oligomers or even to the infinite chain limit of polymers. However, both semi-empirical and ab-initio calculations predict that as the number of repeat unit (N) increases, the value γ/N, which determines the bulk nonlinearity $\chi^{(3)}$, levels off (17,18). In other words, there is a theoretical limit of effective conjugation length for nonlinearity. One may then think that by making a finite chain oligomer of this size, one can reach the effective limiting nonlinearity. The question may arise: "Can we actually achieve the theoretical limit of effective conjugation in experimental studies?" Our experimental work indicates that it may be difficult to reach this theoretical limit in actual conjugated systems. There may be an experimental limit of effective conjugation imposed by either rotation along a single bond or the presence of a chemical defect. In systems like poly-p-phenylene or poly-thiophene, the effective conjugation is highly dependent on the angle between the two rings, the most effective conjugation being when the two rings are co-planar. One can optimize the geometry using ab-initio calculations, but such calculations are done for isolated molecules. The geometries in solution or solid phase may be quite different. Therefore, a difference in effective practical and theoretical limits of conjugation may result.

For poly-3-dodecylthiophene we have found that $\chi^{(3)}$ values are roughly the same for chemically and electrochemically prepared samples, even though the molecular weights of the two samples are ~ 10,000 (N≈10) and ~ 200,000 (N≈200) respectively (19). One would naturally conclude that the effective conjugation does not extend much beyond 10 repeat units. Even a more dramatic example is provided by poly-p-phenyl where our measurement shows that the γ value starts leveling off beyond N=3 (terphenyl), presumably because the effective conjugation does not extend beyond 3 or 4 rings (20). In case of polyacetylenes one can also envision the interruption of effective conjugation length by the presence of chemical defects such as sp^3 hybridized carbon centers.

5 POLYMERS VS OLIGOMERS

In view of the discussions of the above section, one may then ask
these questions: "Do we need polymers? Aren't oligomers of well
characterized structures more desirable?" In our view, it is too
early to tell. However, our feeling is that processable polymers
offer some distinct advantages over small size molecules and
oligomers. While sequentially built and systematically derivatized
oligomers are useful for improving our microscopic understanding,
certain distinct merits of polymeric structures make them suitable as
nonlinear optical materials. Some of these merits of polymeric
materials are as follows:

(i)   Linear polymers may possess appropriate structures for
       ultrafast processes. Examples are solitonic, polaronic and
       bipolaronic conformational deformations.

(ii)   Polymeric structures offer structural flexibility for grafting
       multifunctional groups by the use of derivatization of side
       chain and/or main chain.

(iii) Polymer processing can conveniently allow fabrication of
       various device structures such as fibers, planar waveguides,
       channels etc.

(iv)   Mechanical, thermal, and environmental stabilities as well as
       optical damage threshold of polymeric structures are much
       better than those of low molecular weight organic structures.

6 FUTURE DIRECTIONS

The area of nonlinear optical effects in organic molecules and
polymers is emerging as a frontier of science and technology.
Efforts are rapidly expanding in this area and, therefore, one can
look very optimistically towards the future. One area in which much
research is needed is our understanding of structure-property
relationship so that predictive capability for structures with large
$\chi^{(3)}$ can be developed. For this objective one needs to develop a
theoretical understanding of microscopic nonlinearities as well as of
bulk effects such as the roles of local field, excitons, polaritons,
intermolecular charge-transfer, etc. We need a great deal of
experimental input on sequentially built and systematically
derivatized structures. As is apparent from discussions of the
previous sections, we can not rely on the conjugation effect alone.
Other avenues of enhancing γ must be developed. For this objective,
studies of derivatized structures are of great significance. We have
observed a profound effect (reduction) on $\chi^{(3)}$ upon chemical doping
of polythiophenes and phthalocyanines (7-9). Therefore, the effect
of dopants is another area which needs to be explored.

Finally the optical quality of polymeric materials is another area which needs to be addressed. Organic films with large $\chi^{(3)}$ often tend to be unprocessable and extremely lossy due to both residual absorption and refractive index inhomogeneities. One needs to develop processing control and proper characterization of polymeric films to produce high optical quality and structurally controlled materials for guided-wave applications.

7 ACKNOWLEDGEMENTS

The author wishes to thank his group members Drs. J. Swiatkiewicz, B. P. Singh, R. Burzynski, J. Pfleger, M. Samoc, A. Samoc, J. Obhi, H. Nalwa, Miss P. Chopra, Mr. L. Carlacci, Mr. M. T. Zhao, Mr. Y. Pang, Mr. X. Huang, Mr. P. Logsdon, Mr. M. Casstevens, Mr. M. Carpenter and Mr. He whose works have contributed to this review. The author also thanks Dr. D. R. Ulrich of AFOSR for his encouragement. This work was supported by the Directorate of Chemical Sciences - the Air Force Office of Scientific Research and AFWAL/ML Polymer branch through contract number F4962087C002 and the Office of Innovative Science and Technology - the Defense Initiative organization through contract number F496208760097. Partial supportsfrom NSF solid state chemistry program grant number DMR8715688 and Health Instrument Device Institute at Buffalo are also acknowledged.

REFERENCES

1. "Nonlinear Optical and Electroactive Polymers" Eds. P. N. Prasad and D. R. Ulrich, Plenum Press (New York, 1988).
2. "Nonlinear Optical Properties of Polymers" Materials Research Society, Symposium Proceeding Vol. 109, Eds. A. J. Heeger, J. Orenstein and D. R. Ulrich, Materials Research Society (Pittsburgh, 1988).
3. P. N. Prasad in reference 1, p. 41.
4. P. N. Prasad, J. Swiatkiewicz and J. Pfleger, <u>Mol. Cryst. Liq. Cryst.</u> (in Press).
5. B. P. Singh, P. N. Prasad and F. E. Karasz, <u>Polymer</u> (in Press).
6. W. P. Su and J. R. Schrieffer, <u>Proc. Nat. Acad. Sci. USA</u> 77, 5626 (1980).
7. P. N. Prasad in reference 2, p. 271.
8. P. Logsdon, J. Pfleger and P. N. Prasad, to be published.
9. J. Pfleger, M. Casstevens and P. N. Prasad, to be published.
10. J. Swiatkiewicz, X. Mi, P. Chopra and P. N. Prasad, <u>J. Chem. Phys.</u> <u>87</u>, 1882 (1987).
11. P. Chopra, L. Carlacci, H. F. King and P. N. Prasad, <u>J. Phys. Chem.</u> (in Press).
12. M. P. Boggaard and B. J. Orr, <u>Int. Rev. Sci. Phys. Chem. Ser.</u> 2, 149.
13. O. Zamani-Khamiri and H. F. Hameka, <u>J. Chem. Phys.</u> 72, 5903 (1980).
14. C. P. de Melo and R. Silbey, <u>Chem. Phys. Lett.</u> 140, 537 (1987).

15.   A. F. Garito, A Topical Workshop on Organic and Polymeric
      Nonlinear Optical Materials, ACS Polymer Chemistry Division, May
      16-19, 1988, Virginia Beach.
16.   M. T. Zhao, B. P. Singh and P. N. Prasad, J. Chem. Phys. (in
      Press).
17.   D. N. Beratan, J. N. Onuchic and J. W. Perry, J. Phys. Chem. 91,
      2696 (1987).
18.   G. J. B. Hurst, M. Dupuis and E. Clementi, private
      communication.
19.   B. P. Singh, H. S. Nalwa and P. N. Prasad, to be published.
20.   M. T. Zhao, B. P. Singh and P. N. Prasad, to be published.

Perspectives for Optically Non-linear Polymers in Electro-optic Applications

G.R. Möhlmann

AKZO RESEARCH LABORATORIES ARNHEM, CORPORATE RESEARCH, APPLIED PHYSICS DEPT., POSTBUS 9300, 6800 SB ARNHEM, THE NETHERLANDS

1 ABSTRACT

Organic materials, incorporating conjugated π-electron charge-transfer molecules, show promising features for application as optically nonlinear materials. Especially polymers, containing the above mentioned molecules as active groups, may find use in optically nonlinear devices owing to their relative ease of processing. The optically nonlinear coefficients and corresponding effects can be large in organic materials. Sometimes, attenuation of light by absorption limits the applicability of organic molecules in guided wave structures; for free space applications these drawbacks may be less severe. The relation between the molecular properties of active side groups in polymers and the corresponding macroscopic nonlinear effects will be discussed, especially for second order applications (Pockels effect, frequency doubling, etc.).

2 ORIGIN OF OPTICAL NONLINEARITY

Optical nonlinearity in materials manifests itself by effects such as electro-optic effects, frequency doubling, frequency tripling and mixing, etc. Roughly, the origin of optical nonlinearity can be divided into two main categories:

- field induced orientational changes of anisotropically polarizable (molecular) building blocks. These molecular building blocks often are electrically neutral. Such orientational changes give rise to phenomena like the Kerr-effect, etc.,

- field induced displacements of the centres of positive
and negative charges (ions, electrons). Such induced
electric polarizations are responsible for the
Pockels-effect, frequency conversions, etc.

The nonlinear effects discussed in this article will
mainly depend on charge displacements by dc- and ac-
electric fields; optical fields can be considered as very
rapidly varying electric fields in this case.

Historically, the inorganic single crystals (KDP,
lithium niobate, etc.) received most attention for
nonlinear optical (nlo) applications. Currently, organic
materials are gaining interest as nlo-materials owing to
their attractive properties in comparison to the inorganic
single crystals.

In inorganic single crystals, the positive and
negative ions can be considered as the materials' building
blocks. However, ions have relatively large masses and are
bound to a particular region of the crystal; therefore,
ionic lattice deformations are relatively slow and
restricted to relatively small distances. Nevertheless,
considerable optically nonlinear effects have been
observed in inorganic materials (frequency doubling in
KDP, Pockels effect in lithium niobate, etc.).

For increasing frequencies of the polarizing
(electric) fields, the heavy ions in inorganic crystals
cannot follow the rapid field amplitude oscillations
anymore, resulting in small response amplitudes. Support
for this feature is the often much smaller coefficient for
frequency doubling, in comparison to the Pockels
coefficient in the same material, as found in most
inorganic crystals. Clearly, at optical frequencies, the
ions cannot follow the rapid field oscillations; the
electrons in the crystal are the dominant contributors to
the nonlinear phenomena at optical frequencies.

Organic materials, consisting of molecular building
blocks incorporating conjugated π-electron systems may
offer better (second order) nonlinear properties than the
inorganic materials. The charges to be displaced in such
organic systems are electrons with very small masses;
moreover, the conjugated π-electron system allows large
electron displacements. Combining these two features,
large polarizations are expected, being able to follow
very rapidly oscillating electric (optical) fields. In
addition, organic chemistry allows the synthesis of
molecules optimized for nonlinear applications.

In organic materials, large optical nonlinear effects have indeed been found. In the single crystal material 2-methyl-4-nitro-aniline (MNA), the coefficients for the linear electro-optic (Pockels) effect, and that for frequency doubling, are found to be about the same, suggesting that electronic displacements only are responsible for both effects. This feature allows electro-optic effects to be exploited at high frequencies, and also permits large frequency doubling effects.

An important constraint, however, exists with respect to the occurrence of even order nonlinear polarizations. No centrosymmetry is allowed in systems to be applied for second, fourth, etc., order nonlinear effects. For the uneven order polarizations, no such symmetry constraint is present; e.g. frequency tripling can take place in all materials, the magnitude of the effect being of course dependent on the degree of induced polarization.

3 ORGANIC OPTICALLY NONLINEAR MATERIALS

Several forms exist, in which organic materials are available for (second order) optically nonlinear applications. They are listed below and will be briefly discussed.

Single Crystals

A number of organic single crystals have been identified and synthesized, showing considerable optically nonlinear effects. Among them are: urea, MNA, POM, MAP, NPP[1]. Only those crystals are suited that show spontaneous noncentrosymmetry (n.c.s.); a small fraction of all crystals do, however, exhibit this feature. Structural tailoring of the molecular building blocks sometimes permits to realise n.c.s. in single crystals. Tailoring of organic molecules is an important feature in the development of materials, enabling to achieve optimised materials.

However, single crystals have the drawback that they are not easily processable with respect to optical waveguide formation. Also, the growth and shaping of crystals of sufficient optical quality is difficult; the realisation of large areas, or waveguiding structures in three dimensions, is virtually impossible. For frequency doubling and simple electro-optic architectures, for which a minimum amount of processing and shaping is required, single (bulk) crystals may find applications.

Langmuir-Blodgett Films

This technique enables one to achieve relatively
large areas of optically nonlinear material, by
transferring monolayers of suitable organic molecules from
the surface of a liquid onto a substrate. By taking the
proper substituted molecules, a high degree of polar
order (n.c.s.) can be achieved spontaneously in the
deposited layer.

However, it appeared to be difficult to deposit many
successive layers by the LB-technique, while maintaining
the necessary n.c.s. Currently, the polar order can not be
maintained over more than a few layers only. For many
monomode waveguides, guiding layers of the order of 1
micron thickness are required, corresponding to about 500
or more monolayers, if the LB-film is used as the
waveguide.

Polymeric Solid Solutions

In order to cover large areas with optically
nonlinear films of sufficient thickness for monomode
waveguiding, nonlinear polymers can be deposited via
spincoating or dipping techniques. Thicknesses ranging
from 0-20 microns can be obtained. One approach is to
dissolve a polymer host, together with optically nonlinear
guest molecules, in a suitable solvent, followed by
deposition onto a substrate. During the spinning or
dipping step, the solvent evaporates, leaving a smooth
film behind. The achieved thickness depends on the
concentration of the solution, on the spinning or dipping
speed, and on the type of substrate.

However, after the deposition process, the system is,
most probably, centrosymmetric (c.s.) By applying a strong
electric field (e.g. $1*10^{+8}$ V/m), the permanent dipoles of
the optically nonlinear molecules in the polymer matrix
are oriented by the field, thus breaking the symmetry and
yielding a n.c.s. system. Electric field poling of
initially isotropic systems yields optically nonlinear
materials. In order to enable the nlo-moieties to
rearrange themselves in the polymer host during the poling
step, the polymer has to be heated to above its glass
transition temperature (T_g), where the polymer exhibits a
strongly reduced visocsity. By successive cooling to below
the T_g, while keeping the poling field on, the obtained
polar order can be frozen-in. The temporarily reduced
viscosity of polymers, by heating them above T_g, is an
important property, allowing molecular reorientations.

Polymeric solid solutions have several disadvantages. The amount of optically nonlinear guest molecules that can be dissolved in a polymer host is restricted to about 10% (by weight) or less, only. Moreover, at elevated temperatures, orientational relaxation and segregation of the rather freely movable guest molecules can occur, causing the sample to become electro-optically inactive again, and highly scattering owing to crystallite growth.

Main Chain Polymers

In order to increase the concentration of active molecules in a polymer matrix, and decrease the possibility for segregation and of orientational relaxation after poling, it may be better to incorporate these molecules into the polymeric macromolecules themselves. If the active molecules are built into the backbone, so called main chain polymers are obtained. Disadvantage of this type of polymers is that large molecular fragments must change their orientations in order to achieve a n.c.s. system during the poling step. The reorientation of the backbone experiences a high viscosity.

Side Chain Polymers

By attaching optically active molecules as side chains to the polymer backbone, the concentration of these molecules in the polymeric system can also become rather high. Moreover, the side groups are again not free to migrate, preventing segregation and orientational relaxation after electric field poling. However, contrary to main chain polymers, only side groups need to be poled; this particular process experiences a lower resistance than orientating whole backbones.

Advantages of the side chain polymer approach are the large areas that can be covered and poled. The polymers are compatible with many types of substrate. The polymeric properties can be tailored by changing the structure of the backbone, the active group or the connecting spacer, etc. Further, low temperature deposition and shaping techniques are possible.

4 MAGNITUDE OF THE SECOND ORDER NONLINEAR COEFFICIENT

The magnitude of the macroscopic (bulk) optically nonlinear susceptibility $(\chi^{(2)})$ can be expressed as:

$$\chi^{(2)} = N * \beta * F * P \tag{1}$$

where N is the number of active molecules per unit volume, β the relevant component of the molecular nonlinear coefficient, F a (correcting) local field factor, and P the degree of polar ordering of the relevant active molecules. The different factors appearing in Equation 1 will be discussed below.

Number Density N

N is rather large in organic single crystals and LB-films. In solid solutions, N has to be kept rather low to prevent segregation; in main chain and side chain polymers N can be fairly high again, although not that high as in single crystals.

Local Field Factor F

The local field factor F takes the influence of the surroundings, with respect to the nonlinear effect, into account. For non polarizable matrices, the nonlinear coefficient approaches that corresponding to the gas phase (isolated molecules). In highly polarizable or strongly polar environments, the effective nonlinear coefficient will generally increase. The factor F lies in the range 3-20, depending on the matrix properties.

Molecular Nonlinear Coefficient

The magnitude of β depends on several molecular properties and can be approximated by:

$$\beta = SUM_l [\Delta\mu_{lk} * M^2_{lk} * L/2] \tag{2}$$

where $\Delta\mu_{lk}$ is the difference in magnitude between the dipole moments of a lower lying electronic state k and a higher lying electronic state l, respectively. M^2_{lk} is the square of the corresponding transition moment, and L is a factor taking the energy (U) of the applied photons and the transition energy (U_o) into account. The factor L depends on the nonlinear effect under study (e.g. frequency doubling, Pockels effect). In the case of the Pockels effect, the expression for L is:

$$L = [1/(U_o+U)^2 + 1/(U_o-U)^2 + 2/(U_o*(U_o+U)) +$$

$$+ 2/(U_o*(U_o-U))] \tag{3}$$

It can be derived from the above presented Equations 2 and 3 that to obtain large nonlinear coefficients, the molecules should show the following properties:

- Large change in dipole moment upon excitation; molecules exhibiting these effects are so called Charge Transfer (CT) molecules. Such molecules often contain a strong electron donating group (e.g. $-NH_2$) and a strong electron accepting group (e.g. $-NO_2$), connected via a conjugated π-electron system. A typical example of a CT-molecule is 4-dimethylamino-4' -nitrostilbene (DANS), as shown in Fig. 1.

- The transition moments of the corresponding CT-transitions must be large; therefore, CT-molecules exhibiting large molar extinctions for the relevant transitions, are favoured. The connection of the electron accepting and donating groups by a conjugated π-system permits a large transition probability (absorption coefficient).

- The term L can be optimized by decreasing the value for U_0; molecules with electronic transitions shifted towards the red part of the spectrum are favoured. Some factors in the term L contain the difference between U_0 and U. By bringing the energy of the applied photons in the experiment closer to that of the transition energy, this difference becomes smaller, making the term L larger.

The above mentioned parameters can be optimised, applying molecular engineering. Another practical way to increase the molecular nonlinear coefficient is to increase the length of the conjugated π-electron system. The separation between electron donor and acceptor will become enhanced and thus $\Delta\mu_{1k}$ after the electron charge-transfer has taken place. Also, the absorption coefficient increases upon lengthening the conjugation and, moreover, the energy of the electronic transitions decrease; all mentioned facts contribute to larger values of L.

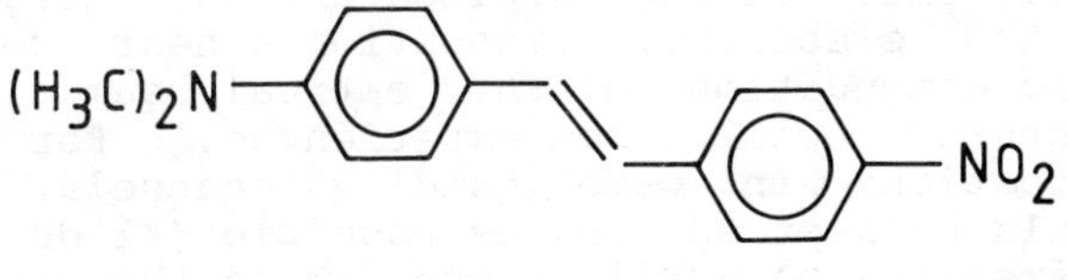

$(H_3C)_2N$ —⟨○⟩— \=—⟨○⟩— NO_2

Figure 1 Structure of the charge-transfer molecule 4-dimethylamino-4'-nitro-stilbene (DANS)

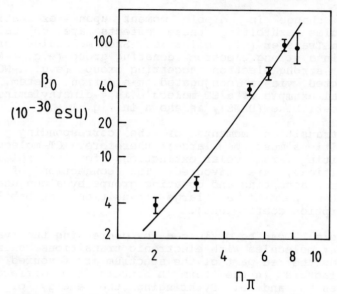

Figure 2 Measured β-values as a function of
the length of the conjugated π-electron
system for some p-methoxy-nitro-substituted
molecules.

A series of experiments, revealing the relation
between the molecular β and the length of the conjugated
π-system has been carried out[2]. The experiments consisted
of dc-SHG measurements. The results are shown in Fig. 2,
and it is seen that β increases for with conjugation.

There is a practical limit for bringing the
excitation energy of the nonlinear molecules down, for
increasing the corresponding absorption coefficient, or
for bringing the applied photon and transition energies
closer together. For many applications, the applied photon
energy is dictated by external conditions and is often not
freely selectable. If the applied photon energy is too
close to the electronic transition (near resonance),
unacceptable attenuation of the optical beam in guided
wave structures will be experienced. For optical
telecommunications in wave guiding channels, optical
losses should be kept as low as possible (<1 dB/cm). For
unguided (free space) applications, where the optical path
length through the nonlinear materials can be very short,
higher absorption coefficients may be allowed. Depending
on the particular application, a trade off exists between
the magnitude of the linear electro-optic coefficient and

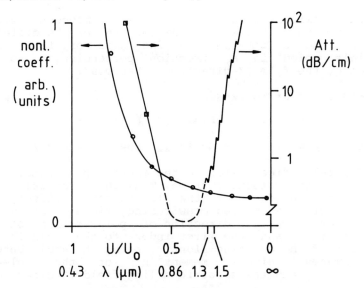

Figure 3 Relation between the linear electro-optic coefficient and optical attenuation as a function of the applied photon energy and wavelength in the case of DANS-polymers.

the corresponding optical loss due to absorption. In Fig. 3, these two curves are shown for the polymer incorporating DANS, as a function of the ratio of the applied photon and transition energies, and of the wavelength. Infrared (vibrational) absorptions are shown, too.

Polar Order Parameter P

This parameter is also important for the magnitude of a materials' macroscopic optical nonlinearity. In the case of isotropic systems, the net polar order is zero; $\chi^{(2)}$ is also zero in that case and no second order optical nonlinear effects occur. Via electric field poling of the active side groups (their permanent dipole moments become aligned parallel (but also anti-parallel) to the direction of the applied field) a net polar order can be induced, and a $\chi^{(2)}$ unequal to zero results.

If the active side chain molecules have structures similar to that of DANS, the direction of the permanent dipole moment almost coincides with that of the largest molecular β-value, and also with the long axis of the molecule. For an idealized one-dimensional active molecule (molecular axis is z-axis), only one β-component exists: β_{zzz}. For such molecules, the expression for the different tensor elements $\chi^{(2)}_{IJK}$, in the case of thin film materials poled in a direction perpendicular to the surface of the film (Z-direction), becomes[3]:

$$\chi^{(2)}_{ZZZ} = N*F*\beta_{zzz}*<\cos^3\theta> \tag{4}$$

$$\chi^{(2)}_{XXZ} = \chi^{(2)}_{XYZ} = \chi^{(2)}_{XZX} = \chi^{(2)}_{YZY} =$$

$$\chi^{(2)}_{ZXX} = \chi^{(2)}_{ZYY} = N*F*\beta_{zzz}*<\cos^3\theta * \sin^2\theta>/2 \tag{5}$$

Capital indices refer to the macroscopic frame, small indices to the molecular frame. The third capital index denotes, in the case of the electro-optic effect, the direction of the applied (switching) field; the second index refers to the direction of polarization of the incoming light, and the first index to the polarization direction of the outgoing light; θ being the angle between the permanent dipole moment μ_o of the molecule (z-direction) and the electric poling field (Z-direction).

The degree of polar order can be approximated by:

$$<\cos^3\theta> = \mu_o * E/ (b * k * T) \tag{6}$$

where E is the electric poling field strength, k the Boltzmann constant, and T the temperature in Kelvin; b is a factor depending on the state of the system and is equal to 5 for isotropic sytems, and equal to 1 for ideal liquid crystalline systems.

Liquid crystallinity (lc) seems to be important to achieve a large polar order and thus a large $\chi^{(2)}$ in the system. Owing to different statistics for lc and non-lc systems, respectively, polar order is enhanced by the presence of axial order (in lc-systems). However, it has been derived on the basis of a theoretical model by extending the Maier-Saupe theory, that at high poling field strengths, initial presence of lc may not be so important to achieve considerable polar order[4]. For poling field strengths of the order of $1*10^{+8}$ V/m, electric field induced axial order is created in initially isotropic systems, enhancing polar order.

The effect of electric field induced axial order (and thus induced lc character) can be visualised as follows. The active molecular side groups in the polymer show a strong anisotropy of their polarizability. Owing to the rod shape of the conjugated π-electron system, the polarizability along the (long) molecular z-axis is much larger than in a direction perpendicular to that. Anisotropically polarizable species tend to align themselves parallel or antiparallel to the direction of a dc- or slowly oscillating ac-electric field. At $1*10^{+8}$ V/m, the induced axial alignment is considerable, limiting the necessity to start with an initially (field free) lc system. This feature facilitates the design and synthesis of optically nonlinear side chain polymers very much. For low poling field strengths, however, initial lc character remains favourable to obtain considerable polar order.

The above mentioned model[4] shows that for a field strength of $1*10^{+8}$ V/m, a polar order of about 0.3 should be achieved for molecular systems with a dipole moment of 5.8 D, a polarization anisotropy equal to 24.7 $Å^3$, and with a T_g of 338 K. Polar order parameters in organic single crystals are often not larger than 0.2-0.3; so, polymeric materials are of the same order as crystals in this respect. A typical example of a side chain polymer, developed and patented by Akzo, resembling the above mentioned properties, is presented in Fig.4.

Knowing the estimated polar order parameter, one can make estimates of the achievable $\chi^{(2)}$-values in side chain polymers after poling. By inserting the necessary values into Equation 1, for polymers with side groups such as shown in Fig.4 (β_{zzz} is $67*10^{-30}$ cm⁵/esu for 1064 nm incident light; N is about $1.5*10^{+21}$ cm⁻³, F is taken equal to 3, and $<cos^3\theta> = 0.3$), one derives $\chi^{(2)}_{zzz} = 90*10^{-9}$ esu ($=38.3*10^{-12}$ mV); the often used corresponding value of the electro-optic coefficient r_{33} ($r = 2*\chi^{(2)}/n^4$; n is the refractive index), is equal to $11.4*10^{-12}$ m/V. The above mentioned r-value for the polymer, obtained via a simple theoretical model, is about an order of magnitude

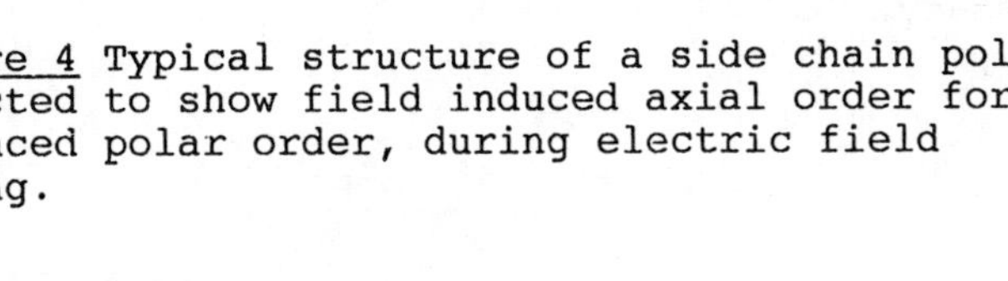

Figure 4 Typical structure of a side chain polymer, expected to show field induced axial order for enhanced polar order, during electric field poling.

lower than that corresponding to the crystal lithium
niobate $r_{33} = 32*10^{-12}$ mV).

5 EXPERIMENTAL NONLINEAR DATA OF POLYMERS

Measurements of the $\chi^{(2)}$-values of poled samples
containing nlo-Akzo-polymers are in agreement with the
above mentioned theoretical models.

In thin film (few microns thick) samples, containing
polymers with active groups as shown in Fig.4, equiped
with coplanar or planparallel electrode configurations,
the $\chi^{(2)}_{zzz}$-value has been measured at low poling field
strenghts (several MV/m). If the thus obtained
$\chi^{(2)}_{zzz}$-values are extrapolated to a poling field strength
of $1*10^{+8}$ V/m, the poled polymeric materials would exhibit
about 0.1 times the corresponding electro-optic effect of
lithium niobate, at the NeHe-wavelength of 632.8 nm[5].

Another Akzo side chain polymer, incorporating a
DANS-like moiety, and contained in a Fabry-Perot type
resonator, has been poled with a field strength of about
300 V/micron. The Fabry-Perot resonator consists of a
28-layer Bragg reflector (GaAs/AlAs on GaAs) on one side,
and of a gold mirror on the other side. Via measurement of
the shift of the central transmitted wavelength,
$\chi^{(2)}_{xxz}$-value could be measured, and as was found to be of
the same order of that of lithium niobate[6].

6 MAGNITUDE OF THE ELECTRO-OPTIC EFFECT IN POLYMERS

Applying the equation to calculate the change in
refractive index (dn) induced by an electric field through
the Pockels-effect:

$$dn = \chi^{(2)} * E / n \tag{7}$$

one derives (by inserting the theoretically estimate value
for the polymer of Fig. 4: $\chi^{(2)}_{zzz} = 38.3*10^{-12}$ m/V, a
switching field E = 3 V/μm, and n = 1.6) the value dn =
$7.2*10^{-5}$. In order to achieve a phase shift of π with this
material for light of 820 nm, an optical path length equal
to the ratio $820*10^{-9}/2*7.2*10^{-5} = 5.7*10^{-3}$ m (= 5.7 mm)
is necessary.

REFERENCES

1. D.S. Chemla and J. Zyss, 'Nonlinear Optical Properties of Organic Molecules and Crystals', Academic Press, Orlando, 1987, Vol. 1, Part II, Chapter II-1, p. 23.

2. R.A. Huijts, Proceedings NATO workshop 'Polymers for Nonlinear Optics', Sohpia Antipolis 19-24 June 1988, France, to be published.

3. D.J. Williams, <u>Angew. Chem.</u>, 1984, <u>96</u>, 637.

4. C.P.J.M. van der Vorst and S.J. Picken, SPIE Conference Proceedings, 1987, <u>866</u>, 99.

5. C.P.J.M. van der Vorst, private communication, 1988.

6. P. Lagasse, private communication, 1988.

The Synthesis and Third-order Optical Non-linearities of Soluble Polymers Derived from the Ring Opening Metathesis Copolymerization of Cyclooctatetraene and 1, 5-Cyclooctadiene

S.R. Marder and J.W. Perry

JET PROPULSION LABORATORY, CALIFORNIA INSTITUTE OF TECHNOLOGY, 4800 OAK GROVE DRIVE, PASADENA, CA 91109, USA

F.L. Klavetter and R.H. Grubbs

THE ARNOLD AND MABEL BECKMAN LABORATORIES OF CHEMICAL SYNTHESIS, DEPARTMENT OF CHEMISTRY, CALIFORNIA INSTITUTE OF TECHNOLOGY, PASADENA, CA 91125, USA

There is currently a substantial research effort devoted to the development of new organic materials with large cubic optical nonlinearities and to gaining insight into the fundamental factors that lead to large optical nonlinearities[1-3]. Recent experimental[4-6] and theoretical[7-9] studies indicate that extended conjugation leads to large cubic susceptibilities. It would be desirable to develop general methodologies for the synthesis of new materials that combine electronic benefits arising from extended conjugation and the processabilty advantages generally associated with polymers. Ring opening metathesis polymerization represents a promising new route to materials with large third-order optical nonlinearities since inherent in the mechanism of the reaction is the *preservation of the degree of unsaturation present in the monomer.* This is in contrast to many radical, cationic, anionic and Ziegler-Natta polymerizations in which an olefinic monomer will usually lead to a polymer with a fully saturated backbone.

We wish to report some preliminary results on the nonlinear optical properties of a new class of polymers synthesized by the ring opening metathesis polymerization of highly unsaturated monomers. Mixtures of cyclooctatetraene, **1**, and 1,5-cyclooctadiene, **2**, have been polymerized using the tungsten carbene complex **3** as the catalyst.[10] The solution polymerization of monomers **1** and **2** in tetrahydrofuran leads to an essentially random copolymer.[11] The monomers contain exclusively "cis" double bonds and no attempt has been made to isomerize the double bonds in the resulting polymer. The general constitution of the copolymers is shown in Figure 1 (an all-trans geometrical configuration is shown for simplicity).

The polymerizations of mixtures of **1** and **2**, where the mole fraction of **1** was varied from 0.10-0.50 were typically performed as described in the following procedure. The monomer mixture (100 μl, ~0.85 mmol) in 400 μl of a tetrahydrofuran solution containing 8.0 mg (0.01 mmol) of **3** were allowed to react, for 6 hours. Then 1200 μl of a tetrahydrofuran solution containing 36 mg (0.42 mmol) of $(CH_3)_3CC(O)H$, which destroys the catalyst,[10] and 24 ul (0.17 mmol) of mesitylene, an internal standard for gas chromatographic studies (*vide infra*) was added.

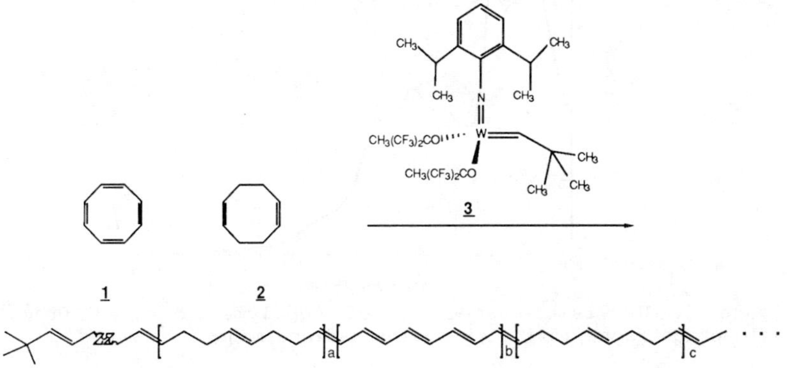

Figure 1. The synthesis of cyclooctatetraene/ 1,5-cyclooctadiene copolymers.

We chose these reaction conditions to ensure that the resulting polymer was soluble in tetrahydrofuran and to prevent any further metathesis reactions from occurring after the desired reaction time. If the reaction is run for a longer time or at higher monomer concentrations, the solutions of polymers containing higher concentrations (>40%) of cyclooctatetraene form a gel. It should be noted however that the conditions we describe have not been optimized for yielding the polymer with the highest $\chi^{(3)}$. After quenching, examination of the reaction mixture by gas chromatography allowed us to determine the number of moles of cyclooctatetraene and 1,5-cyclooctadiene incorporated into the polymer. These results shed light into mechanistic aspects of the polymerization which will be discussed elsewhere.[12]

The optical spectra of the materials (an example of which is shown in Figure 2) demonstrate the essential features of the polymers studied. On the basis of previous studies of polyenes[13] and conjugated polymers[14] we assign the bands at 316, 332, and 344 nm to segments of the polymer with five conjugated double bonds, bands at 384, 404, and 432 to segments of the polymer with nine conjugated double bonds and bands at 432, 464 and 496 to segments of the polymer with thirteen conjugated double bonds. The segments with 5,9, and 13 double bonds arise from incorporation of 1, 2, or 3 consecutive cyclooctatetraene monomers respectively into the polymer. In the materials that we have examined there are few (<2% by weight) segments with greater than 13 double bonds. We can say qualitatively, there is an *increase in both the percentage of olefinic carbons and the average conjugation length in the polymer as the percentage of cyclooctatetraene in the monomer mixture increases* as was seen in the previously reported copolymer films.[11]

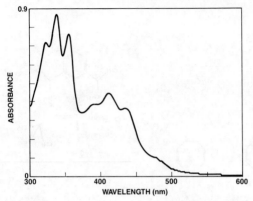

Figure 2. UV-visible spectrum of copolymer derived from 50% cyclooctatetraene, 50% 1,5-cyclooctadiene monomer mixture.

Nonlinear optical properties of the polymer mixtures were studied as a function of the composition. The third-order susceptibilities of the copolymer solutions were determined using wedged cell third harmonic generation (THG) techniques.[15,16] The 1.907 μm Raman shifted (H$_2$ gas) output from a Q-switched Nd:YAG laser was used as the fundamental for THG measurements. A two-channel system was used to provide shot-to-shot normalization of the polymer solutions signals to a THG signal from a red color filter. The THG signals (636 nm) were isolated with filters and monochronomators and were detected with photomultiplier tubes. A thick window wedged THG cell[16] was used for the polymer solutions. The fused silica cell windows were 1" in diameter and 0.5" thick. The wedge angle between the windows was ~0.5°. A 20 cm focal length lens was used for focussing into the wedge cell. Typically, 200-300 μJ pulse energies were used for the measurements. Wedge THG interference fringes were observed by translating the cell normal to the laser beam. For the polymer solutions studied, simple fringing behavior was observed as shown in Figure 3.

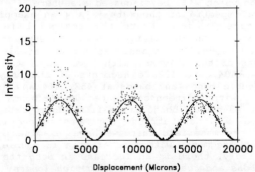

Figure 3. THG fringes of a solution of copolymer derived from 50% cyclooctatetraene, 50% 1,5-cyclooctadiene monomer mixture.

The THG pattern from the thick window cell is given by:[16]

$$I(3\omega) = A \sin^2(\Delta\psi/2) \tag{1}$$

where $\quad A = 4|s^2|(1-a\rho)^2,$

and $\qquad \Delta\psi = \pi x/l_c \tan(\theta/2),$

with $\qquad a = (\eta^1{}_\omega + \eta^1{}_{3\omega})/(\eta^1{}_{3\omega} + \eta^g{}_{3\omega}),$

and $\qquad \rho = (\chi^{(3)}/\Delta\varepsilon)_l/(\chi^{(3)}/\Delta\varepsilon)_g.$

$|s^2|$ is a scaling constant, l_c is the coherence length, x is the displacement of the cell as it is translated, θ is the wedge angle, η is the index of refraction for the liquid (l) or the window (g), and $\Delta\varepsilon = \eta^2{}_\omega - \eta^2{}_{3\omega}$.

The THG fringes were fit to the expression above to obtain the amplitude A and l_c for a solution. Measurements were also made using toluene as a reference liquid in the same cell for a relative $\chi^{(3)}$ determination. $\chi^{(3)}$ for a given solution was calculated using a simplified expression:

$$\chi^{(3)} = 1/l_c \, [(1-\sqrt{R}) \, (\chi^{(3)} l_c)_g + \sqrt{R} \, (\chi^{(3)} l_c)_r] \tag{2}$$

where $R = A/A_r$ and the subscript r refers to the reference liquid. The $\chi^{(3)}$ of a solution is given by :

$$\chi^{(3)}{}_{soln.} = \sum_i N_i L(3\omega) L^3(\omega) <\gamma_i> \tag{3}$$

where N_i is the number density , L is the local-field factor, $<\gamma>$ is the ensemble average hyperpolarizability, and the sum is over all i components. For an ideal solution we can write:

$$\chi^{(3)}{}_{soln.} = \sum_i x_i \chi^{(3)}{}_i \tag{4}$$

where x is the mole fraction and $\chi^{(3)}$ is the susceptibility of the species i in the local field of the solution. For a two component solution the polymer susceptibility is given by:

$$\chi^{(3)}{}_{polymer} = (\chi^{(3)}{}_{soln.} - \chi^{(3)}{}_{solv.})/x_{polymer} + \chi^{(3)}{}_{solv.} \tag{5}$$

It is emphasized that the $\chi^{(3)}$ calculated in this way corresponds to the neat polymer with the local field factor of the solution.

Table 1 summarizes the results of the THG studies along with results on the polymer composition determined from gas chromatographic analysis. The gas chromatographic results show that the mole fraction of the cyclooctatetraene in the polymer differs from the mole fraction in the monomer mixture. There is less cyclooctatetraene incorporated into the polymer than one would expect based on the the monomer composition indicating that 1,5-cyclooctadiene reacts faster than cyclooctatetraene. The $\chi^{(3)}$ values listed in Table 1 are based on the mole fraction of polymer

in the solution. Table 1 also includes $\chi^{(3)}$ of the copolymers
normalized to a constant amount of cyclooctatetraene.

Table 1. Summary of Composition and Third-order Optical
Nonlinearities of Cyclooctatetraene, and 1,5- Cyclooctadiene
Copolymers.

mole fraction of 1 in monomer mixture	mole fraction of 1 in polymer	mmol 1 in polymer	$\chi^{(3)}$ polymer[a,b]	$\chi^{(3)}$ polymer/ mmol 1 in polymer
0.10	0.08	0.064	50	790
0.20	0.15	0.108	90	850
0.40	0.27	0.159	210	1310
0.50	0.32	0.166	240	1460

a: $\chi^{(3)}$ reported in esu X 10^{-14}
b: We have independently measured the $\chi^{(3)}$ of cyclooctatetraene,
1,5-cyclooctadiene and $(CH_3)_3CC(O)H$ and have determined that
correcting for their contributions only alters the solution $\chi^{(3)}$
values by 1-2%.

Figure 4, which shows a plot of the $\chi^{(3)}$ of the copolymers
verses mmol of cyclooctatetraene incorporated into the polymer,
demonstrates that as the amount of cyclooctatetraene in the
polymer increases (with concomitant increase in the average
conjugation length) $\chi^{(3)}$ increases nonlinearly. This is in
qualitative agreement with theoretical calculations of the
conjugation length dependence of $\chi^{(3)}$ and with earlier experimental
results on finite polyenes. However since neither the cis/trans
compositions nor the detailed distribution of conjugation lengths
in the copolymers have yet been determined, the results are not
directly comparable to theory. Work is in progress to afford such
a comparison.

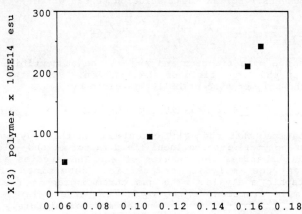

mmol cyclooctatetraene in polymer

Figure 4. Plot of $\chi^{(3)}$ of the copolymers verses mmol of
cyclooctatetraene incorporated into the polymer.

The magnitude of $\chi^{(3)}$ of the copolymer with 32% cyclooctatetraene is ~2.4x10^{-12} esu. By comparison a solution measurement on β-carotene (with 11 conjugated double bonds) gave a value of $\chi^{(3)}$ = 9.1x10^{-11} esu, which is in agreement with recent measurements.[17] Measurements on neat polyacetylene have given a value of 1.3x10^{-9} esu at 1.9 μm.[6] While the $\chi^{(3)}$ of the copolymer is modest, this work demonstrates that the ring opening metathesis synthetic methodology can be used to produce materials with substantial nonlinearities and is flexible enough to allow tailoring of materials properties. Since functionalized cyclooctatetraene derivatives can be polymerized in a similar way,[11] electronic/functional group effects can be examined. It should be noted that transparent uniform films of the soluble polymers can readily be prepared by spin coating. Processability is an important issue for applications of such materials in nonlinear optical devices such as nonlinear waveguides[18]. Additionally scattering losses in the polymer films must be minimized. By using the ring opening metathesis polymerization methodology it may be possible to develop soluble, processable materials with low scattering losses and high nonlinearities. Such routes are currently under investigation.

Acknowledgments:

The research described in this paper was performed in part by the Jet Propulsion Laboratory, California Institute of Technology as part of its Innovative Space Technology Center. SRM thanks the National Research Council and the National Aeronautics and Space Administration for a National Research Council Resident Research Associateship at the Jet Propulsion Laboratory. FLK thanks the National Science Foundation for a predoctoral fellowship. RHG wishes to acknowledge support from the National Science Foundation. JWP acknowledges support from the Strategic Defense Initiative Organization.

References and Footnotes:

1. 'Nonlinear Optical Properties of Organic Molecules and Crystals,' Volumes 1 and 2, D. S. Chemla and J. Zyss, ed., Academic Press, Orlando, 1987 .

2. 'Nonlinear Optical Properties of Organic and Polymeric Materials,' D.J. Williams, ed., ACS Symp. Ser.233, American Chemical Society, Washington 1983.

3. 'Molecular and Polymeric Optoelectronic Materials: Fundamentals and Applications,' G. Khanarian, ed., Proc. SPIE, 682, SPIE-The International Society for Optical Engineering, Bellingham, 1987.

4. C. Sauteret, J.-P. Hermann, R. Frey, F. Pradere, J. Ducuing, R. H. Baughman and R. R. Chance, *Phys. Rev. Lett.*, 1976, *36*, 956.

5. G.M., Carter, Y.J. Chen and S.K. Tripathy, *Appl. Phys. Lett.*, 1983, *43*, 891.

6. F Kajzar, S. Etemad, J. Messier and G.L. Baker, Synth. Met.,
    1987, 17, 563.
7. G.P. Agrawal, C. Cojan and C. Flytzanis, Phys. Rev. B: Solid
    State, 1978, 17, 776.
8. D.N. Beratan, J.N. Onuchic and J.W. Perry, J. Phys. Chem.,
    1987, 91, 2696.
9. A.F. Garito, J.R. Heflin, K.Y. Wong and O.Zamani-Khamiri in
    'Nonlinear Optical Properties of Polymers,' Materials
    Research Society Symposium Proceedings, 109, A. J. Heeger,
    J. Orenstein and D.R. Ulrich, ed., Materials Research
    Society, Pittsburgh, 1988, p. 91.
10. R.R. Schrock, J. Feldman, L.F. Cannizzo and R.H. Grubbs,
    Macromolecules, 1987, 20, 1169.
11. F.L. Klavetter and R.H. Grubbs, J. Amer. Chem. Soc.,
    accepted for publication.
12. R.H. Grubbs, F.L. Klavetter and S.R. Marder, to be
    published.
13. F.Bohlmann and H.-J. Mannhardt, Chem. Ber., 1956, 89, 1307.
14. A. Winston and P. Wichacheewa, Marcomolecules, 1973, 6, 200.
15. G.R. Meredith, B. Buchalter and C. Hanzlik, J. Chem. Phys.,
    1983, 78, 1543.
16. F. Kajzar and J. Messier, J. Opt. Soc. Am. B, 1987, 4, 1040.
17. S.H. Stevenson, D.S. Donald and G.R. Meredith, in 'Nonlinear
    Optical Properties of Polymers,' Materials Research Society
    Symposium Proceedings, 109, A. J. Heeger, J. Orenstein and
    D.R. Ulrich, ed., Materials Research Society, Pittsburgh,
    1988, p. 103.
18. G.I. Stegeman and C.T. Seaton, J.Appl. Phys., 1985, 58, R57.

Side Chain Liquid Crystalline Polymers for Non-linear Optics

A.C. Griffin and A.M. Bhatti

DEPARTMENTS OF CHEMISTRY AND POLYMER SCIENCE, UNIVERSITY OF SOUTHERN MISSISSIPPI, HATTIESBURG, MS 39406, USA

INTRODUCTION

Side chain liquid crystalline polymers are attractive materials for nonlinear optics (nlo); in particular, second harmonic generation. The primary synthetic approach to side chain liquid crystalline polymers has traditionally been free radical polymerization of a vinyl monomer - acrylate or methacrylate (1). This type of monomer is desirable as it is relatively easily prepared and conditions for free radical polymerization of acrylates and methacrylates have been well documented. There are, however, special difficulties associated with free radical polymerization of an interesting class of nlo materials - nitroaromatics. The nitroaromatic group is a retarder of free radical polymerization intercepting propagating radicals by reaction on the aromatic ring or on the nitro group itself (2). Stilbenes containing a nitroaromatic structure are desirable for nlo, but the stilbene unit itself is also potentially reactive with a free radical center (3).

There have been reports of use of vinyl nitroaromatic monomers for synthesis of nlo polymers by free radical polymerization. Le Barny et al. (4) have used a variety of nitroaromatic acrylates as comonomers and De Martino et al. (5) have polymerized, via the methacrylate, a nitrobiphenyl species. Griffin et al. (6) reported nitro-aromatic imine and stilbene acrylates as comonomers and in addition found evidence of crosslinking using polymeri-zation conditions involving multiple charges of initiator.

Although another synthetic route to nitroaromatic side chain liquid crystalline polymers has been described

(6) - polyesters, not involving a free radical polymerization, it was considered of interest to further
investigate free radical polymerization of nitroaromatic
stilbenes. An advantage of homopolymerization of such
species would be the density of nitroaromatic pendant
groups along the polymer chain. Since neither homopolymerization of relevant nitroaromatic acrylates nor of
nitrostilbenes containing methacrylates has, to our
knowledge, been reported, we undertook the homopolymerization of the monomers $\underline{1}$ to $\underline{4}$ shown below.

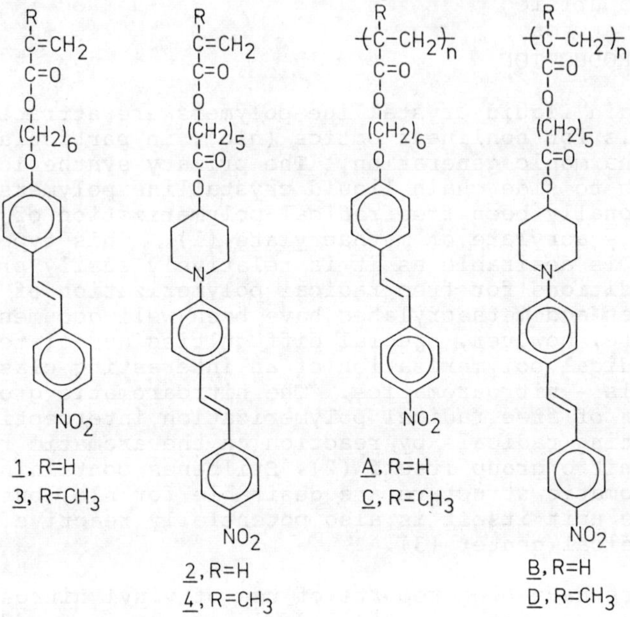

$\underline{1}$, R=H
$\underline{3}$, R=CH$_3$

$\underline{2}$, R=H
$\underline{4}$, R=CH$_3$

$\underline{A}$, R=H
$\underline{C}$, R=CH$_3$

$\underline{B}$, R=H
$\underline{D}$, R=CH$_3$

EXPERIMENTAL

Synthesis of monomers $\underline{1}$ and $\underline{2}$ has been described
earlier (6). Monomers $\underline{3}$ and $\underline{4}$ were synthesized by
analogous routes, by substituting acrylic compounds with
corresponding methacrylic derivatives.

Polymerizations

Table 1 summarizes polymerizations under different
conditions. Experimental details of four general
approaches are as follows.

Without Solvent/Single Charge of Solid Initiator. A
known mass of a monomer was intimately mixed with an
appropriate mole fraction of the solid initiator using a
mortar and pestle. The mixture was taken in a small
round-bottom flask and swept with prepurified nitrogen.
Then the flask was heated in an oil bath at appropriately
set temperature (above the monomer melting point) and the
melt was stirred for 5-10 minutes and then taken out.

Without Solvent/Single Charge of Liquid Initiator. A
known mass of a monomer was taken in a small 3-neck round-
bottom flask, equipped with a rubber septum, and swept
with prepurified nitrogen. Then it was heated in an oil
bath at a preset temperature. When melted, a suitable
mole fraction of the liquid initiator was injected through
the septum and the melt was stirred for 5 minutes to 1
hour. Then the flask was taken out.

In Toluene/Single Charge of AIBN. A known mass of a
monomer was taken in a small 3-neck round-bottom flask,
equipped with a rubber septum, and containing a small
volume of dry toluene. Then the mixture was swept with
prepurified nitrogen and heated in an oil bath at a
preset temperature. After allowing a few minutes to
equilibrate, an appropriate mole fraction of AIBN,
dissolved in a small volume of toluene, was injected
through the septum. Stirring of the reaction mixture was
done for 30 minutes.

In Toluene/Multiple Addition of AIBN. The monomer
was taken in a round-bottom flask and melted in·an oil
bath at a preset temperature. Stirring was started and
then appropriate installments of AIBN were added at 2-
minute intervals. For example, in Table 1, 2 mol%x10
means 2 mol% additions at 2-minute intervals 10 times.
Heating and stirring was continued for a further 5-10
minutes.

Polymers were separated by repeatedly dissolving in
chloroform and precipitating with ether until monomer-
free. For runs where toluene was used as solvent, it was
evaporated before dissolving the product in chloroform.
Absence of monomer was detected by TLC on silica gel or
alumina plates/50% ethyl acetate + 50% hexane. Polymers
were dried under high vacuum.

Methacrylate polymers precipitated out of toluene, as
soon as formed. Polymer D had poor solubility in
chloroform. It was purified by repeatedly washing with
warm 50% chloroform + 50% ethanol. In those instances

where polymers formed colloidal solutions in chloroform +
ether, centrifugation at 8000-9000 rpm for 10 minutes
proved successful for their isolation.

All polymers were investigated for crosslinking and
liquid crystallinity by polarized light optical microscopy. Crosslinking was inferred by appearance of gel
formation upon cooling from the isotropic liquid phase.
In cases of only slight crosslinking, the appearance of a
viscous, weakly birefringent fluid phase was often seen on
cooling from the isotropic liquid. Polymers having more
extensive crosslinking cooled directly from the isotropic
liquid to a non-birefringent glass (visually transparent).
All polymers with no crosslinking or minimal crosslinking
were liquid crystalline.

RESULTS AND DISCUSSION

Chemical structures of the monomers and polymers (shown in
Introduction) contain a nitroaromatic stilbene as part of
the pendant group.

Collected in Table 1 are the various experimental
conditions employed for polymerization. Not shown in the
Table are results from a variety of polymerization runs
using monomer 1 and 1-2 wt% AIBN in benzene and toluene at
temperatures in the 70-85°C range. These runs employed
degassed samples in a sealed ampoule under an inert
atmosphere. In each case virtually no polymer formation
was seen, the monomer being recovered.

As can be seen from the Table, acrylates can be
polymerized neat with AIBN or other, higher temperature
initiators; but in low yield and usually with some
crosslinking. Using a solution (toluene) polymerization,
both a single charge of initiator and multiple charges
were employed. The multiple charge method was carried out
in air. Reasonable yields were obtained for these runs,
but crosslinking occurred - albeit less so for the
methacrylate (Run 11). A complicating feature of these
multiple charge runs is the possibility of inhibition and
crosslinking due to oxygen in air. As is described in the
comment on Run 1, the solution NMR (H-1 and C-13) of the
crosslinked polymer is virtually identical to that of the
monomer in the spectral region of the pendant group.
Therefore, crosslinking must involve a relatively small
number of interchain bonds.

It was found for a single charge of initiator that
methacrylate was superior to acrylate (Runs 9, 10, 12,

Table 1 Comparative Polymerization Data on Acrylate and Methacrylate Monomers

Run No. (Monomer)	Initiator (Concn.)	Solvent	Bath Temp.	Rxn. Time (Atmos.)	Comments
1 (1)	AIBN (2 mol%x10)	Toluene	99-100°C	30 min	34% yield (crosslinked but H-1 & C-13 NMR satisfactory)
2 (1)	AIBN (2 mol%)	None	120°C	10 min (under N_2)	Very low yield (low crosslinking)
3 (1)	AIBN (6 mol%)	None	130°C	5 min (under N_2)	3% yield (more crosslinking than in Run No. 2)
4 (1)	t-Bu$_2$O$_2$ (0.9 mol%)	None	130°C	20 min (under N_2)	No polymer
5 (1)	t-Bu$_2$O$_2$ (2.7 mol%)	None	130°C	1 hr (under N_2)	Very low polymer formation
6 (1)	t-BuOOCOPh (4 mol%)	None	130°C	1 hr (under N_2)	2% yield (almost no crosslinking)
7 (2)	AIBN (10 mol%)	None	150°C	10 min (under N_2)	7% yield (low crosslinking)
8 (2)	AIBN (1 mol%x20)	Toluene	98-99°C	40 min	30% yield (more crosslinking than in Run No. 7)
9 (1)	AIBN (10 mol%)	Toluene	99-100°C	30 min (under N_2)	21% yield (less crosslinking than in Run No. 1)
10 (3)	AIBN (10 mol%)	Toluene	99-100°C	30 min (under N_2)	43% yield (No observable crosslinking)
11 (3)	AIBN (2 mol%x10)	Toluene	99-100°C	30 min	43% yield (low crosslinking)
12 (2)	AIBN (10 mol%)	Toluene	99-100°C	30 min (under N_2)	20% yield (less crosslinking than in Run No. 8)
13 (4)	AIBN (10 mol%)	Toluene	99-100°C	30 min (under N_2)	40% yield (No observable crosslinking)

and 13) both in terms of yield and extent of crosslinking. Methacrylates converted nicely to polymer with no observable crosslinking. This difference in extent of crosslinking is attributed to the greater stability of the propagating radical from methacrylate monomer (tertiary) and its concomitant greater selectivity. This increased selectivity and increased steric bulk at the radical center is thought to minimize reactions leading to crosslinking-attack of aromatic ring carbons or addition to stilbene C=C carbons.

In summary, it was found that polymerization of nitroaromatic stilbene methacrylates in solution under nitrogen using a single charge of initiator can provide an acceptable route to high nlo density side chain liquid crystalline polymers. Analogous polymerization of acrylates proceeds in lower yields with crosslinking.

ACKNOWLEDGEMENT

This research was supported by the U.S. Air Force Office of Scientific Research through contract number F49620-88-C-0068.

REFERENCES

1. H. Finkelmann in 'Polymer Liquid Crystals', A. Ciferri, W. R. Krigbaum and R. B. Meyer, eds., Academic Press, New York, 1982, p. 35.
2. G. Odian,'Principles of Polymerization', 2nd Ed., Wiley & Sons, New York, 1981, p. 248.
3. G. F. D'Alelio, 'Fundamental Principles of Polymerization', Wiley & Sons, New York, 1952, p. 49.
4. P. Le Barny, G. Ravaux, J. C. Dubois, J. P. Parneix, R. Njeumo, C. Legrand and A. M. Levelut, in 'Molecular and Polymeric Optoelectronic Materials: Fundamentals and Applications', G. Khanarian, ed., Proc. SPIE, 1987, 682, 56.
5. R. DeMartino, D. Haas, G. Khanarian, T. Leslie, H. T. Man, J. Riggs, M. Sansone, J. Stamatoff, C. Teng and H. Yoon, in 'Nonlinear Optical Properties of Polymers', A. J. Heeger, J. Orenstein and D. R. Ulrich, eds., Materials Research Society, Symp. Proceedings, 1988, 109, 65.
6. A.C. Griffin, A. M. Bhatti and R. S. L. Hung, in 'Nonlinear Optical and Electroactive Polymers', P. N. Prasad and D. R. Ulrich, eds., Plenum Publishing Corp., NY, NY, 1988, p. 375.
7. A. C. Griffin, A. M. Bhatti and R. S. L. Hung, Mol. Cryst. Liq. Cryst., 1988, 155, 129.

Free Carrier Contribution to Picosecond Phase Conjugation in a Polydiacetylene Gel

J.M. Nunzi and F. Charra

CENTRE D'ETUDES NUCLÉAIRES DE SACLAY, 91191 GIF-SUR-YVETTE CEDEX, FRANCE

Polydiacetylenes (PDA) are known to possess very large optical nonlinearities. However, nonlinearities may have very different origins. Thus, force anharmonicity which is the main process leading to frequency up conversion cannot fully explain degenerate effects. Those require special treatments and are strongly time scale dependent.

In a recent experiment[1], we saw that 4-BCMU gels exhibit a strong two photon absorption (TPA) at 1.06 μm. This gave rise to phase conjugation (PC) whose efficiency was proportional to the fifth power of laser intensity at pump powers greater than 1 GW cm^{-2}. The most part of the two photon excitations decayed thermally during the first few picosecond[2], giving rise later to periodic acoustic oscillations.

Moreover part of the two photon excitations at 1.06 μm in PDA is known to decay into free electron hole pairs. They give rise to photoconductivity[3] as well as internal polarizations[4]. Our goal here is to see how such free electron hole pairs contribute to picosecond PC at 1.06 μm.

I - FIFTH ORDER NONLINEARITY : Two photon excitations in PDA gels contribute to fifth order PC via an effective $\chi^{(5)}(\omega \; ; \omega,\omega,-\omega,\omega,-\omega)$ susceptibility. To understand its symmetry properties, consider an isolated rod like chain with unit vector $\hat{i}$ in the laboratory reference frame $(\hat{x}, \hat{y}, \hat{z})$. As a consequence of π conjugated backbone one dimensionality, it only

301

possesses one relevant fifth order molecular hyperpolarisability component δ, namely δ_{iiiiii}. We assume
here that each part of the polymer chain have such a
symmetry, the remaining being only a few percents.
Averaging over all $\hat{i}$ space orientations and neglecting local field corrections, we get $\chi^{(5)}$ of an isotropic amorphous PDA gel: $\chi^{(5)}_{xxxxxx} = N \delta_{iiiiii}/7$, where N is the number of rod like molecules per unit
volume and:

$$\chi^{(5)}_{yxxxxy} = \chi^{(5)}_{yxxxyx} = \chi^{(5)}_{xxxxyy} = \chi^{(5)}_{xxxxxx}/5 \qquad (1)$$

These are the only components relevant to our experiments and those having an odd number of x or y are
zero.

Our δ hyperpolarizability is a cascade of
TPA, described by the imaginary part of the third order hyperpolarizability γ_{iiii}, and of the linear exci-
ted state induced polarizability change α_{ii} ; $\delta \propto \gamma.\alpha$.

Such a system exhibits polarization memory as
long as the induced excited state survives and keeps
the anisotropy of the absorbing segments. This manifests upon reading the interference grating of two
cross polarized beams. Light intensity is constant
along its interference direction and anisotropic molecules which record light polarization at each point
exhibit birefringence upon reading.

II - **EXPERIMENTS** : The samples are 4-BCMU red gels[1]
with concentrations between 5 and 10% in N,N-dimethy-
laniline (DMA). They do not exhibit observable linear
absorption nor diffusion nor birefringence at 1.06 μm,
our working wavelength.

We use the same experimental set up as in
ref. 1, with an improved noise rejection. Erasure of a
thermal grating[5] shows that our laser pulses are coherent with a coherence time equal to the 33 ps pulse
width. In order to simplify intensity dependences, the
forward pump is made ten times brighter than the backward pump. So that the main term in fifth order PC is
the one involving three times the forward pump.

We first check PC in the pure solvent DMA. It
is 1/4 that of CS_2 in the same configuration. We then
check PC in gels at intensities higher than 1 GW cm^{-2}.
The signal is always I^5 dependent and (1) is verified
for the effective $\chi^{(5)}$. Signal to noise ratio exceeds
10^3 in all studied configurations while conjugate

reflectivity is 10^{-3} to 10^{-4}. Since zero time signal is at least 4 times that of CS_2, pure solvent has little effect.

III - RESULTS

a) Background signal : Fig. 1 depicts the usual signal time dependence on delaying the bacward pump pulse cross polarized to the two others. The whole signal follows a $\chi_{yxxxyx}^{(5)}$ process where the strong pump I_1 is $\hat{x}$ polarized. The maximum at 3.3 ns is the first acoustic oscillation of the large spaced grating (Λ = 10 μm), speed of sound in DMA is v = 1550 ms^{-1}. Time evolution fits a $\sin^4$ ($\pi tv/\Lambda$) law[5] which is the hydrodynamic impulse response of a spatially periodic thermal excitation and produces a negative index modulation due to dilatation of the medium.

Fig. 2a is an expansion of Fig. 1 in the time range 100 to 800 ps after excitation. We see a quasiconstant background signal which is 5% zero time signal and always one order of magnitude above noise. Its lifetime is greater than 1 ns and smaller than 1 s (the effect desappears between two pulses) and it cannot be explained by acoustic effects.

In order to precise the sign of the index modulation, we make this background interfere with the acoustic response of a small spaced grating (Λ = 0.35 μm) with known negative index modulation. Such an interference is possible since our pulses are coherent[6]. Thus we put a 10% dielectric mirror in contact to the cell, opposite to the probe beam path. The probe beam reflection interferes with the forward pump, adding a small spaced grating to the large spaced one. This small spaced grating is read out by the backward pump and interferes with the regular conjugate pulse after mirror reflection. Signal modulation is 100 time less than for the large spaced grating, and has the same magnitude as the background. The interference pattern is reported on Fig. 2b with two least squares fits to check the background phase. The best fit is obtained with two components in phase. Thus, as for the acoustic response, the background index change is negative.

b) Polarization memory : Anisotropy of the background signal appears on polarization gratings study, when the probe beam is cross polarized to the two parallel

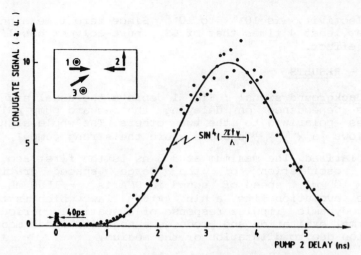

Figure 1 - *Signal evolution on delaying pump pulse 2 in the 5% red gel at 1 GW/cm² pump power. Response at t = 0 has a nearly gaussian shape with 40ps half width. Solid line is a least squares fit of the acoustic signal. Inset depicts polarisation of incoming beams. 1 and 3 income at 6°.*

pump beams. The fast response has the same halfwidth as for intensity gratings and there is no visible acoustic oscillations since this is an isotropic solvent effect. The delayed read pump beam has been referenced with a photodiode to account for travel errors.

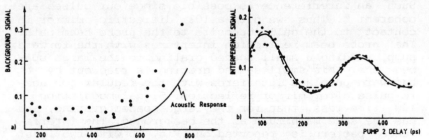

Figure 2 - *Background response of an intensity grating. (a): expansion of fig. 1 before appearance of acoustic oscillations (solid line). (b): signal obtained by interference of the background effect with a small spaced grating created by reflection of probe beam 3. Dashed line: fit with two components in quadrature. Solid line: fit with two components in phase.*

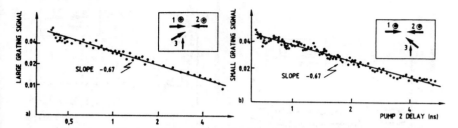

Figure 3 - Logarithmic plot of the signal evolution of
polarisation memory, after initial fast response.
Insets depict polarizations and angles of incoming
beams. (a): large spacing grating, (b): small spacing
grating. Both fit the same temporal law (solid lines)
at this time scale.

Fig. 3a and b are log-log plots of the tempo-
ral response of a large (Λ = 10 μm) and of a small
(Λ = 0.35 μm) spaced polarization grating, respectively. These two polarization memories have the same
magnitude as the background intensity effect and fit a
$t^{-\alpha}$ decay law with α = 0.67 $\pm$ 0.05 for both large and
small gratings.

c) Static electric field effects : The long lived
intensity and polarization gratings are sensitive to a
strong (3.10[5] V cm[-1]) DC electric field. The 10% polymer gel is contained between two transparent electrodes, perpendicular to the pumps direction, making a
150 μm path cell. A constant 5000 V tension is chopped
into 10 μs pulses and applied to the sample in synchronism with one laser shot over two, early enough to
avoid stray fields during measurement.

DC electric field induces a general 10 to 30%
rise of the high laser intensity conjugate signal. It
induces the same rise in the TPA coefficient measured
looking at non linear transmission[1]. Quadratic DC Kerr
effect measured in a Kerr gate produces here a polarization rotation less than 10[-2] rd.

As concerns background effects, we observe an
almost homothetical 100 to 300% enhancement of both
intensity (Fig. 4a) and polarization memories (Fig.
4b). Since their lifetime does not apparently change,
only the number of long lived excited species increases under DC field.

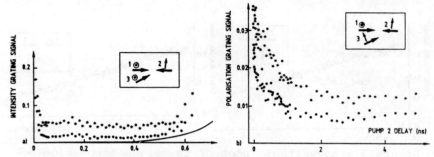

Figure 4 - *Effect of a* 3.10^5 V/cm *continuous electric field on the background signal for a 10% red gel. (a): background intensity grating, (b): polarization memory (see insets). Full points: without field; Stars: with the applied field.*

In order to vary DC field strength applied to the excited chains, we rotate the cell untill its normal is at 45° to the strong pump beam polarization. The number of long lived species, which is proportional to the square root of the background signal, increases almost quadratically with the electric field strength, which is defined as the projection of DC field direction on strong pump polarization.

IV - DISCUSSION : Polarization memory reveals that the background signal is due to excited species bound to the polymer. The simplest description is given by one dimensional free electron hole pairs. Dimensionality comes from the verification of (1). Negative polarizability and DC electric field sensitivity infer carrier behaviour.

DC electric field acts in reducing initial geminate recombination. The square dependence of this effect upon electric field strength has been discussed by Bässler[3]. The overall signal and TPA increase are explained by symmetry breaking of the two photon forbidden excitonic state upon electric field.

The number N of free electron hole pairs involved into the background response is roughly estimated from a simple Drude model[1]. With account of the factor 5 given by anisotrpy (eq. 1), we find an order of magnitude N $\approx$ 10^{16} cm^{-3}.

The decay of the intensity memory cannot be

analysed since it is hindered by the acoustic oscillations. As concerns polarization memory, small and large spaced gratings follow the same decay law. Thus, it
is not governed by migration through the grating.
Moreover, the decay is neither governed by orientation
lost upon migration through polymer fibrils since such
a decay is exponential[7] and the persistence length of
a red polymer rod (ca. 1 μm)[8] is greater than our
small grating interfringe.

Photoexcitation power law decays are often
encountered in semiconducting polymers at these time
scales[9]. They are mainly due to one dimensional trap
limited diffusion[10]. The deep traps acting in that
experiment as centers where either orientation or
polarizability of the excitations is lost.

In conclusion, between 30 ps and 5 ns, two
photon excitations of a PDA gel behave like electron
hole pairs bound to the polymer chains.

REFERENCES

[1] J.M. Nunzi and D. Grec, J. Appl. Phys., 62, 2198
    (87).
[2] F. Charra, Rapport CEA-DEIN/INT/LPEM/ 87-42,
    (87).
[3] H. Bässler, in "Polydiacetylenes", D. Bloor and
    R.R. Chance, eds., Nijhoff, Boston, 135 (85).
[4] P.A. Chollet, F. Kajzar, J. Messier, in ref. 3,
    p. 317.
[5] J.M. Nunzi and D. Ricard, Appl. Phys. B, 35, 209
    (84).
[6] F. Hache, P. Roussignol, D. Ricard and C.
    Flytzanis, Opt. Com., 64, 200 (87).
[7] Z. Vardeny, J. Strait, D. Moses, T.C. Chung and
    A.J. Heeger, Phys. Rev. Lett., 49 1657 (82).
[8] K.C. Lim, A. Kapitulnik, R. Zacher, S. Casalnuo-
    vo, F. Wudl and A.J. Heeger, in ref. 3, p. 257
    (85).
[9] D.L. Wiedman and D.B. Fitchen, Synth. Metals, 17,
    355 (87).
[10] B. Movaghar, B. Pohlmann and D. Wurtz, in ref. 3,
    p 177 (85).

Second Harmonic Generation by Composite Materials

P.D. Calvert, N. Azoz, and B.D. Moyle

SCHOOL OF CHEMISTRY AND MOLECULAR SCIENCES, UNIVERSITY OF SUSSEX,
BRIGHTON BN I 9QJ, UK

1 INTRODUCTION

The need for optical switching and detection devices to be
used in association with optical communications and computing has led to interest in non-linear optical properties
of materials which could be used as thin layers in integrated optics. Inorganic oxides, such as lithium niobate,
can be grown as suitable films but many polymers and molecular materials show very promising properties and may be
used if they can be developed in a suitable format.

Materials under investigation for second harmonic generation include many organic compounds which might be
formed as crystalline films(1) or as aligned fibres in
glass capillaries(2). Also under study are polymers with
dyes attached to the chains, or as solutes, where the necessary molecular alignment can be produced by an applied
electric field (3).

We have been investigating composite films of organic
crystals in polymers, where the crystals are formed during
solidification of the polymer (4). For such composites to
be of use in optoelectronics, it will be necessary to limit
light scattering and to control the crystallite morphology
to optimise the properties. Based on analogies with bio-
logical materials, we believe that the polymer matrix can
be used to control the size, orientation and packing of the
crystals to produce composites with novel optical and electrical properties. The sensitivity of second harmonic generation by powders to particle morphology (5), make this a
good basis for the study of composite structure.

Much of our recent work has focussed on the 3-nitro-
aniline crystallized within polymethylmethacrylate. We
will discuss the phase diagram of this system as it relates
to crystal growth, and compare light scattering studies
with second harmonic generation. Having obtained control
of crystallite size, one would also like to control
orientation and methods of doing this will be discussed.

2 PHASE BEHAVIOUR

The morphology of the composite will depend on whether the
polymer is crystalline or amorphous, and the extent of
solubility of the solute in the polymer, as well as on the
solidification conditions. Above their melting point, the
nitroanilines are completely soluble in polymethylmethacrylate and polystyrene. On cooling any given composition
will crystallize out when the equilibrium solubility is exceeded. At lower concentrations of solute, the dissolution
temperature becomes close to the glass transition temperature so that rapid cooling may not lead to crystallization
but to a supersaturated glassy solution.

Figure 1 shows the measured melting curve for 3-nitro-
aniline in polymethylmethacrylate and the glass transition
temperature as estimated from recrystallization behaviour.

3 SCATTERING AND SHG FROM ISOTHERMAL CRYSTALLIZATION

The apparatus for SHG and scattering measurements has been
decribed before (4). A photomultiplier was used to measure
frequency-doubled light scattered from a film sample exposed to a pulsed Nd-YAG beam at $1.06\mu m$.

Figure 2 shows the dependence of SHG on nitroaniline
content for films of 3-nitroaniline and 2-methyl-4-nitro-
aniline in PMMA formed by casting from solution and drying.
It can be seen that, for 3-nitroaniline the intensity of
second harmonic is not very dependent on concentration once
a critical value corresponding to the stable supersaturation is exceeded. For 2-methyl-4-nitroaniline the intensity builds up much more slowly. This is apparently due to
a morphological difference in that 3-nitroaniline forms
very elongated crystals which largely scatter light forwards, while 2-methyl-4-nitroaniline gives more nearly
equiaxed crystals which scatter over a much wider range,
figure 3. In both cases the SHG intensity is also very
variable.

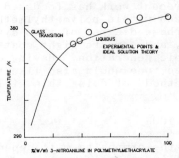

Figure 1:
Phase behaviour of polymethylmethacrylate/3-nitroaniline,
showing polymer glass transition and nitroaniline melting.
The melting corresponds closely to ideal solution theory.

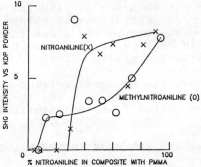

Figure 2:
SHG intensity in forward direction for various compositions
of 3-nitroaniline and 2-methyl-4-nitroaniline in PMMA

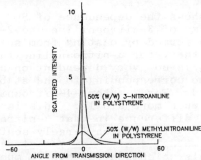

Figure 3:
Scattered intensity (as photomultiplier output voltage)
versus angle for 50wt% composites of 3-nitroaniline in PMMA
and 2-methyl-4-nitroaniline in PMMA.

A strategy to maximize the SHG output of these materials is to first modify the morphology, in order to minimize the scattering, and then to look for morphological factors which affect SHG without affecting scattering. A first step is to reduce the diameter of the 3-nitroaniline crystals as far as possible below the wavelength of light. In addition the crystals should be oriented in a single direction. Needle-like crystals growing at high volume fractions will naturally tend to be locally parallel but will give rise to strong scattering if the orientation direction varies through the sample.

To form fine crystals under controlled conditions, we melted composites containing in the region of 30% 3-nitroaniline and then crystallized them by annealing at various temperatures in the region of 20-50°C. In these circumstances, the crystals are growing into highly viscous polymer melt. Scattering studies from these samples give a particle size in the range of 0.1-0.2μm with relatively little variation. Depsite the small change in particle size, there are large variations in the forward SHG intensity and the forward scattering intensity, as shown in figure 4.

4 ORIENTED SAMPLES

To resolve better the importance of orientation for SHG, we have used several methods to produce parallel arrays of crystals of 3-nitroaniline. To exploit the high orientability of crystalline polymers, we explored ways of incorporating 3-nitroaniline into polyolefins. Simple blending does not work because the polymer is immiscible with 3-nitroaniline and 2-methyl-4-nitroaniline in the liquid state. The nitroanilines can be incorporated by swelling polyethylene with a hot solution of nitroaniline in xylene. On drying, much of the nitroaniline is trapped within the polymer. The swelling temperature is quite critical. Successful samples were prepared by swelling a linear low density polyethylene at 88°C in a 30 %(w/v) solution of nitroaniline in p-xylene. The concentration of nitroaniline, incorporated in the polyethylene, increased with swelling time and temperature.

Drawing (elongation) of this composite of polyethylene and 3-nitroaniline did not orient the nitroaniline but only the polymer, as seen from Laue x-ray diffraction. Presumably the 3-nitroaniline crystals broke rather than rotating during drawing. A highly oriented composite was made by

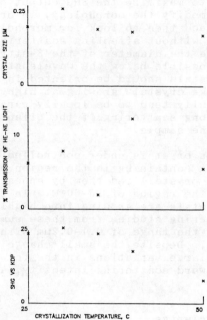

Figure 4:
Effect of crystallization temperature on crystal size, forward light transmission and forward SHG intensity for a composite of 27wt% 3-nitroaniline in PMMA.

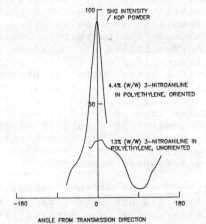

Figure 5:
SHG intensity versus angle for oriented and unoriented composites of 3-nitroaniline in polyethylene.

drawing the polymer first and then swelling it, while
clamped at constant length, with nitroaniline in xylene.

Light scattering gave a particle size of 0.13μm in unoriented polymer. As can be seen from figure 5, the SHG
from the oriented polymer is much more concentrated within
the forward direction than it is in unoriented composite.
In both cases the actual levels of SHG measured are very
high, up to 100 times that of KDP powder. This is despite
the fact that the polymer only contains 4% of nitroaniline.

Oriented samples were also prepared by crystallization
in a temperature gradient stage, which consists of a heated
block and cold block separated by a narrow gap. The
sample, a film on a glass slide, is slowly pulled from the
hot side to the cold side. This produced highly transparent composites with high levels of forward SHG output.
These samples are compared with other methods of preparation in table 1. It can be seen that high SHG outputs are
produced by crystallization within a crystalline polymer
matrix, which would be expected to impose a very small
crystal diameter, and by crystallization within a
temperature gradient.

The SHG level has not been normalised between different types of samples. We have found that it increases linearly with sample thickness, in a series of solution cast
films, but it must also depend on crystal content and scattering. The coherence lengths for 3-nitroaniline are of
the order of microns, and dependent on the propagation direction. Our crystals are usually less than one micron.

5 APPLICATIONS

Optical materials are generally selected for their low levels of scattering and it thus seems intrinsically unlikely
that composite materials could be used in optics. On the
other hand, it does seem possible that suitable composites
could have electro-optical properties comparable with single crystals, and it is much easier to align crystals in a
film than to align individual molecules. Moreover biolog-
ical optics are generally based on highly complex multi-
phase structures. Hence, it is not clear that composites
are necessarily to be avoided. The potential for these
materials will really be seen only in the context of a device which is designed to use them.

The polymer plays a largely passive role in these materials and a polycrystalline film might work equally as

well. However, the polymer does add toughness and durabil-
ity and allows a much higher level of morphological control.

6 CONCLUSIONS

We have shown that composites of second harmonic generating
crystals in polymer films can be grown with a controlled
microstructure. The forward intensity of second harmonic
is very sensitive to the extent of scattering by the
structure. Other factors such as crystal size, film thickness and orientation are also undoubtedly important. In
future work we plan to resolve these effects by forming
closely controlled structures.

TABLE 1
Morphology and Forward Second Harmonic Intensity

Sample	%(w/w)	Thickness μm	Forward Scatt.%	Crystal Size μm	SHG /KDP
NA Powder in oil	80	200	19.6	40	60
NA Powder in oil	80	400	1	40	40
Soln. cast NA/PMMA	50			2-10	8
Soln. cast MNA/PMMA	50			0.5	3
NA/PMMA quenched	27	110	7	0.3	26
NA/PMMA quenched	40	70	7	0.2	2
NA/PMMA oriented	70	50	90	40	140
NA/PE unoriented	13	450	0.9	0.1	30
NA/PE oriented	4.4	60	16		95

REFERENCES

1. J.Zyss, J. Mol. Electron., 1986, 1 25.

2. B.K.Nayar, ACS Symposia, 1983, 233, 153.

3. J.B.Stamatoff, A.Buckley, G.Calundann, E.W.Choe,
    R.DeMartino, G.Khanarian, T.Leslie, G.Nelson, D.Stuetz,
    C.C.Teng, H.N.Yoon, SPIEJ, 1986, 682 85.

4. B.D.Moyle, R.E.Ellul and P.D.Calvert, J.Materials Sci.
    Lett. 1987, 6, 167.

5. S.K.Kurtz, T.T.Perry, J.Appl. Phys., 1968, 39, 3798.

Monolayers and Langmuir–Blodgett Films

Non-linear Optical Effects in Langmuir–Blodgett Films

I.R. Peterson

JOHANNES-GUTENBERG-UNIVERSITÄT, INSTITUT FÜR PHYSIKALISCHE CHEMIE,
JACOB-WELDER-WEG, 15, D6500 MAINZ, FRG

1 INTRODUCTION

Small, smooth departures from linear response have
been studied in electronics for a long time and are
usually considered as a nuisance to be minimised, as
they introduce distortion into analog signals and
chaotic behaviour into high frequency amplifiers and
oscillators. Since these effects vary as the square
or cube of the signal level they are most troublesome
at high powers. Such effects can also occur in optics,
but are usually only of importance at the extremely
high instantaneous intensities provided by pulsed
lasers. One such phenomenon is self focussing, which
can cause serious damage to glass lenses.

These nonlinear effects show a positive side, too.
In electronics they have been used to control, convert
and even amplify signals at frequencies too high for
more conventional approaches. In work with pulsed
lasers, they have proved to be a very versatile way of
controlling the wavelength, envelope and duration of
the pulse[1].

The ability to control pulses of light is also
desirable in telecommunications, where the growing
cost-effectiveness of optical fibres is leading to the
increased transmission of data in the form of modulated
infra-red beams. The peak power levels in this case are
much lower than from a pulsed laser, so that the nonlinear response is proportionately quite small, but
here the manifest benefits of being able to switch the

fibre-optic signal directly without conversion to
electrical form has led to the development of new and
much more responsive materials[2]. Devices based
on them are now available commercially in small
quantities for evaluation. More speculatively, there
is also the possibility of using nonlinear effects for
signal processing. This is claimed to have advantages
of parallelism[3], speed[4] and speed-power product[5].

Each different application of a nonlinear effect
requires a different combination of material
parameters, and often the optimum material is organic.
For example, liquid crystals are of interest because of
their huge odd-order effects which can be used for
optical switching or data storage, but only in
situations which can tolerate their extremely slow
response. The present paper is concerned with another
class of organic material, the internal charge transfer
(ICT) dyes, which display extremely fast, all-
electronic effects. The second-order (square law)
response of materials based on chromophoric moieties
of this type are being seriously considered for
second harmonic generation and electro-optic
modulation applications[6,7].

To produce a good material in this category, it is
essential to find chromophoric moieties with large
molecular second-order coefficients, and to pack
them into the material as densely as possible using
sidegroups or auxiliary molecules. These considerations, however, do not guarantee a large second-order
response, because depending on their relative
orientation, it is possible for the contributions to
the bulk second-order effect of two chromophores to
cancel rather than reinforce. Hence it is also
necessary to impose an overall orientation during
material fabrication. There is inevitably a large
element of chance in finding a good material, because,
in spite of increasingly sophisticated
attempts[8,9], the theory of molecular packing is
not sufficiently developed to enable the prediction
of material performance.

However, it is clear that if the original chromophore
has only a weak second-order response, then the
resulting material will be poor. In view of the effort
involved in derivatisation, fabrication and
characterisation, it is of interest to have a quick and
reliable method of chromophore evaluation. This is at

the moment typically carried out by electric field induced second harmonic generation (EFISH) in which the molecules are dissolved in an organic solvent, oriented using a powerful electric field, and irradiated with laser pulses. From the intensity of light produced at half the laser wavelength it is possible to deduce the product $\mu\beta$ of the chromophore dipole moment and the appropriate second-order hyperpolarisability . Although the EFISH technique sounds simple, it suffers from experimental difficulties ranging from the conductivity of the solvent , the hazards associated with the use of high voltage, the difficulties in obtaining dimensionally precise cells, and the uncertainties about solvent–chromophore interactions, not to mention the fact that the chromophore dipole moment must be determined by a separate method before the required value of β is obtained.

2 CHROMOPHORE CHARACTERISATION USING MONOLAYERS

A new method, monolayer second harmonic generation (MSHG), has recently been reported. It avoids many of the above problems and directly yields a value of β. For MSHG, the chromophore is derivatised with a straight alkyl chain, typically docosyl, and a dilute solution of the material is spread on a water surface. Once the solvent has evaporated, the dye molecules are left on the surface as an oriented monolayer. In this technique the chromophores are oriented by the anisotropic forces at the water–air interface, which are many orders of magnitude greater than the effects of electric fields even at the breakdown threshold, so that the resulting spread of molecular tilt elevations is small[10,11].

The monolayer can either be examined in situ on the water surface[12], or more conveniently, transferred to a glass or quartz slide[13] while preserving much of the imposed orientation. Basically, the monolayer is illuminated with a high power laser, and the resulting second-harmonic light is detected. The increased order of the chromophores compared to EFISH offsets the reduced quantity of material, so that it is possible to use photomultipliers rather than more sensitive techniques such as photon counting. It has even proved possible to detect the second harmonic produced by a monolayer of carboxylic acid[14], which, while capable of internal charge transfer from the α-methylene donor to the carbonyl acceptor, is not normally considered to be a nonlinear optical material at all.

Because of the extreme thinness of the monolayer, the absorption of both the incident fundamental and the generated harmonic is very small, even at wavelengths where the molecules have electronic resonances. This means that it is possible with MSHG to measure the complete hyperpolarisability dispersion curve, something which is not possible with EFISH or crystal measurements.

Like EFISH, the technique has a number of special features. Just as dilution measurements are used in EFISH to check for aggregation effects, it is possible to measure monolayer SHG at different surface densities of the active molecules, either alone or mixed with specified inactive material, using the well-known techniques for handling monolayers[15]. Just as in EFISH, it is necessary to make more than one measurement, from which it is possible to deduce other parameters which are not of primary interest. In MSHG it is necessary to measure the response for two incident polarisations in order to dissentangle the effects of chromophore hyperpolarisability β and tilt elevation ψ. This is directly comparable to the Maker fringe measurements necessary in EFISH. Unlike EFISH, the molecular environment of dense-packed chromophores in MSHG more nearly resembles what it would be in an efficient nonlinear optical material.

Hence the new technique has a number of advantages over the existing one. Considering the large experimental error typically encountered in nonlinear optical measurements, and the differences in β expected due to different molecular environments, the agreement between the two techniques for several chromophores is nevertheless good. Table 1 compares the values measured for a number of different chromophores.

Compared to EFISH, the new technique does require extra synthetic effort in attaching the long aliphatic chain to the chromophore. However this may be achieved using a carbon-carbon, ether, amide, peptide or ester linkage, and is no problem for any competent synthetic chemist. Moreover, once having derivatised the chromophore to allow monolayer formation, it is also possible to measure its linear electro-optic coefficient, which is not possible with electric field aligned solution techniques. The linear electro-optic effect, also

Chromophore name and structure	β in units of nm^4/V [Ref]		
	EFISH	MSHG	MPE
Alkyl N-stilbazene RHN—⟨O⟩—⟨O⟩N+—Me		0.44 [16]	0.40 [16]
Dialkyl N-stilbazene RR'N—⟨O⟩—⟨O⟩N+—Me		0.84 [17]	
Dimethylamino- nitrostilbene Me$_2$N—⟨O⟩—⟨O⟩—NO$_2$	0.19 [18]		
Alkyl Gedye merocyanine R-N+⟨O⟩—⟨O⟩—O$^-$	0.42 [19]	1.13 [20]	
Alkyl hemicyanine R-N+⟨O⟩—⟨O⟩—NMe$_2$		0.11 [21]	0.28 [21]
N-alkyl-2-amino 5-nitropyridine RHN—⟨N⟩—NO$_2$		$\gtrsim$ 0.02 [22]	

Table 1. Values of molecular hyperpolarisability β measured using Electric Field Induced Second Harmonic, Monolayer Second Harmonic and Monolayer Pockels techniques. The chosen unit is related to other systems as follows: $1 \ nm^4/V = 8.85 \times 10^{-48} \ C^3 J^{-2} m^3 = 2.39 \times 10^{-27}$ esu

called the Pockels effect, is a consequence of
the Stark effect in which the energy levels of a
molecule are shifted by an electric field, and is of
direct technological interest for optical switches
and modulators. Although often considered to be a
linear effect as the dielectric constant and
electronic energy levels vary linearly with the DC
component of the electric field, when the optical
field is taken into consideration the induced
polarisation is seen to vary to second order in the
total field. In fact the theoretical formalism for both
SHG and the Pockels effect is identical, and the latter
can also be characterised by a value of the molecular
hyperpolarisability β.

The Pockels hyperpolarisability β of a range of ICT
chromophores has been measured by a variant of the
attenuated total reflection (ATR) technique[23]
in which the monolayer is transferred from the water
surface to a metal-coated glass prism. At certain
angles of incidence in the range corresponding to total
internal reflection, light falling on the metal
from the prism couples into propagating modes at the
metal-monolayer interface and is absorbed or
scattered, thus attenuating the reflection. When the
monolayer properties are modulated by applying a
sinusoidal electric field, the resulting cyclic
variation of intensity is measured using a phase-
sensitive detector. The technique is easily sensitive
to the effect in a dye monolayer and, moreover, gives
the sign of β which is not readily accessible using SHG
or solution-based techniques. Table 1 gives some
measured values. In all cases the Pockels hyperpolarisability is quite close to the SHG value
indicating that the effect is electronic in nature,
and that resonance effects are small.

3 LB FILMS COMPARED TO OTHER NLO MATERIALS

It has just been demonstrated that the monolayer
technique is capable of fabricating a densely-packed
oriented array of almost any chromophore. The monolayer
can be deposited on a wide range of optical substrates.
Having deposited one monolayer, it is possible then to
deposit a second, a third, and so on, thus building up
a film of 'bulk' material. This is the Langmuir-Blodgett
technique[15,24]. It has been proposed that multilayer
films produced in this way could themselves be used as
the nonlinear optical material in future devices[25,26].

In order to evaluate this proposal it is necessary to compare its advantages and disadvantages with those of alternative techniques. There are basically two competitive ways of forming bulk materials containing oriented ICT chromophores: crystal growth is one, and field poling of spin-coated chromophore-loaded polymers is the other. Both of the latter techniques have been developed to a much greater degree for this application than LB films, and it is possible to say with confidence where they may most logically be applied.

Crystals are well understood theoretically and are equilibrium systems, so that barring temperature-induced phase transitions, there should be no problems of material instability. The ideal crystal has no optical inhomogeneities which might give rise to scattering. Moreover it is possible to find crystals which are phase matchable for harmonic generation. Hence they are ideally suited for harmonic generation from high-powered lasers in an optical bench setup.

Application of nonlinear effects in tele-communications involves quite low powers, so that it is necessary to confine the light to a very small cross-section over interaction lengths of the order of millimetres in a one-dimensional waveguide. In addition, a single, isolated, device makes sense only in a limited number of applications, and it is far more desirable to have a technology in which a large number of devices are integrated on the one substrate. If an organic is to be the active medium, it must be fabricated on a structural support as an overlayer in which one-dimensional waveguides can be etched. Attempts have been made to grow large, thin overlayer crystals for this purpose, and the best results have been achieved by sandwiching the crystal between two glass plates. However even in this case there is a trade-off between layer thickness and crystalline domain size, and the removal of the top plate to allow patterning is difficult.

The poled loaded polymer film, on the other hand, is a natural candidate for overlayer techniques and hence telecommunications. Large area substrates can be spin-coated and poled rapidly and inexpensively, and there is no problem in achieving the micrometre-range thicknesses required for waveguiding. The problem is rather that the material is not phase matchable in

bulk, and the thickness of spin-coated films cannot be
controlled to sufficient accuracy for phase matching
in a waveguide, so that only their Pockels electro-
optic effect is usable.

These materials are not in a state of equilibrium,
as the anisotropic orientation of the chromophores is
achieved by applying a high external field at
temperatures above the polymer glass transition, and
then removing the field after cooling. Experimentally
the film electro-optic coefficient is found to decay
with time, and this problem is in fact one of the major
topics being addressed in the search for practical
polymer/chromophore combinations.

There are of course electro-optic modulators
already available commercially, based on lithium
niobate, one of the perovskite ferroelectrics, and
poled loaded polymers will be required to have some
competitive advantage. It is commonly believed that
ICT organics have already demonstrated much higher
electro-optic figures of merit than $LiNbO_3$[27]. This
is unfortunately not true. The present best loaded
polymer materials[28] have a $\chi^{(2)}$ of approximately
20 pm/V compared with the coefficient for lithium
niobate[29] of ~600 pm/V. Moreover the transparency
window of the ICT organics is much narrower than that
of lithium niobate.

There are, however, other advantages of poled loaded
polymers. By no means the least is their low fabrication
cost, but there is also a technical argument, valid for
all ICT organics. The perovskite ferroelectrics achieve
high values of $\chi^{(2)}$ because of the large changes of
nuclear positions with electric field. This also
inherently leads to a very large low-frequency
dielectric constant ε (~37 for $LiNbO_3$[30], compared to
the optical frequency value of 2-3). This means that in
a modulator, the control signal and optical wave
propagate at different speeds, placing an upper bound
on the modulation rate. Organics display a much lower
dispersion of dielectric constant, so that modulation
rates of hundreds of GHz should be possible. Moreover
the main applications are forseen at diode laser
wavelengths in the near infrared, so that the
absorption of the organics in the visible is not a
drawback.

LB films fit into this context with a distinctly different set of characteristics. Because of the painstaking, monolayer-by-monolayer assembly procedure, even micrometre-thick films will be quite expensive to fabricate, and there is absolutely no prospect of producing the millimetre-thick films required for optical-bench-type harmonic generation. For integrated electro-optic applications, however, LB films are a distinct possibility and should have the advantage over poled loaded polymers that the chromophores are much better ordered, so that shorter interaction lengths are possible.

The high correlation of chromophore orientation which has already been discussed applies strictly only within each monolayer. In many cases, LB materials deposit as Y-type films, that is to say, successive monolayers are deposited with oppositely-oriented molecules, so that their contributions to the second-order bulk hyperpolarisability cancel exactly. There are, however, two approaches to fabricating multilayers in which this does not happen. The first involves finding materials which deposit preferentially on either the downstroke or upstroke of the deposition cycle. The resulting films are called X- or Z-type, respectively. The second involves using pairs of materials, with monolayers of the first deposited only on the downstroke and of the other only on the upstroke, giving rise to an alternated Y-type or superlattice film. Active films of Z-type[31] and alternated Y-type[32] have been demonstrated.

The most outstanding advantage of LB films is the precision to which their thickness may be controlled. This suits them particularly for low-power, all-optical nonlinear effects, where very precise phase matching between pairs of propagating waveguide modes is essential. Whereas the electro-optic effect in ICT organics is exceeded by that of many perovskite ferro-electrics, their competitive position is much stronger for all-optical nonlinearities.

In all-optical waveguiding applications the nonlinear coupling between sets of phase-matched modes is critical. In uniaxial materials these must have have different symmetries, so that the coupling coefficient is composed of contributions of differing sign from different parts of the guide. With superlattice-type LB films it is possible to control

the orientation of the local hyperpolarisability
tensor so that the contributions from all points
reinforce.

All-optical applications of nonlinear effects
include parametric amplification and oscillation,
signal up- and down-conversion, and second-harmonic
generation. A particularly exciting possibility is
that of simulating third-order effects of the type
needed for all-optical digital computing by cascading
second- order nonlinearities. Cascaded effects were
first discussed by Meredith[33] in the context of EFISH.
When efficient harmonic conversion is achieved, these
could be several orders of magnitude larger than the
largest intrinsic third-order effects demonstrated in
organic polymers[34].

It can be seen from this discussion that no one ICT
organic-based technology is likely to become dominant,
but that each will find its niche in an overall picture.
Crystals will most logically be applied to harmonic
conversion of the output from high-powered lasers;
poled loaded polymers to integrated electro-optic switches
and modulators for telecommunications; while LB films, the
least developed of the three, could conceivably be most
suited for integrated all-optical signal processing.

4 EXPERIMENTAL STUDIES OF LB FILMS

The first question to be asked about a nonlinear optical
material produced by a specific fabrication technology
is how large the desired response is, and whether it
is stable with time over a reasonable temperature
range. This question is being answered for LB films by
a number of recent studies.

Figure 1 shows the variation with number of layers
of the measured maximal $\chi^{(2)}$ coefficient for several
different material combinations deposited in Y-type
superlattice form[32,35,36]. In this, and some other
reported studies[32,37] the variation of SHG intensity
with layer thickness does not follow the theoretically
expected square law dependence, indicating that the
bulk hyperpolarisability falls off as more layers are
deposited, and that the monolayer values are not always
a good guide to the performance to be expected in device
applications. Although there has been no definitive
study of the reasons for this, it is probable that the
ordering of the chromophores in the thick films becomes
worse. However it is apparent from the curve obtained

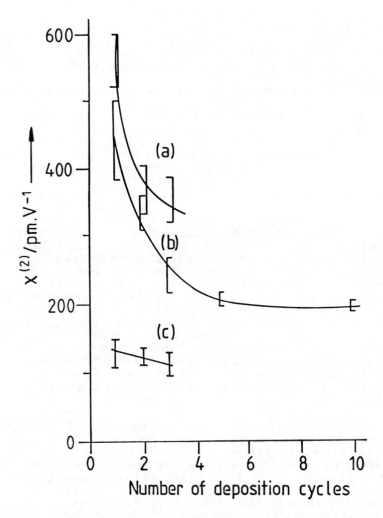

Figure 1. The variation with layer thickness of SHG
hyperpolarisability for three different
superlattice film molecular combinations
(Upstroke: Downstroke):
(a) 22-Gedye: 22-Tricosenoic acid[35]
(b) 22-Hemicyanine: 4HANS[32]
(c) 22-Hemicyanine: 22-Tricosenoic acid[37]

for the hemicyanine–4HANS combination in figure 1 ,
that the decrease of order does not necessarily persist

indefinitely, and that it is possible to achieve values of $\chi^{(2)}$ in thick films which are an order of magnitude greater than for poled loaded polymer films based on the same chromophores. Moreover some material combinations show the ideal dependence of SHG on thickness.

Because of problems foreseen with the temperature and mechanical stability of low molecular weight materials, there is a lot of interest in polymers for LB films. Some years ago, the preferred approach was to deposit films of small molecules which could then be polymerised. Various vinyl[39,40] and acrylate[41] derivatives were investigated, but the most popular monomers were the diacetylenic fatty acids[26,42]. Interest in this method waned when it became apparent that the film developed cracks and variations of thickness on polymerisation[43]. Recent effort in this direction has involved preformed polymers which appear to satisfy all requirements for mechanical stability and film quality at the expense of rather slow deposition speeds. However, like the films of low molecular weight material, there is a fall-off of bulk hyperpolarisability in thicker films[37], and problems of long-term decay of the nonlinear response. Moreover, it is not certain that the temperature range for polymer film stability will be significantly higher than for films of small molecules. Although films of many conventional LB materials are unstable at temperatures only slightly above ambient[44,45] some of the new nonlinear film compounds have melting points above 200°C.

One of the theoretical advantages of LB films, which has already been noted, is their suitability for use as waveguides with extremely well-defined propagation parameters. However, all the SHG measurements reported so far involve light in unbound modes propagating essentially across rather than along the film. The reason for this is the rather high values of attenuation of waveguide modes[46], caused by scattering rather than absorption. One possible extrinsic mechanism is microbial contamination[47], while long-range orientational order has been found to be the dominant intrinsic cause in films of fatty acids[48]. In both cases there exists the possibility of reducing attenuation through suitable measures.

The Pockels effect has been measured in multilayer films, too. The results of one study[49] are shown in

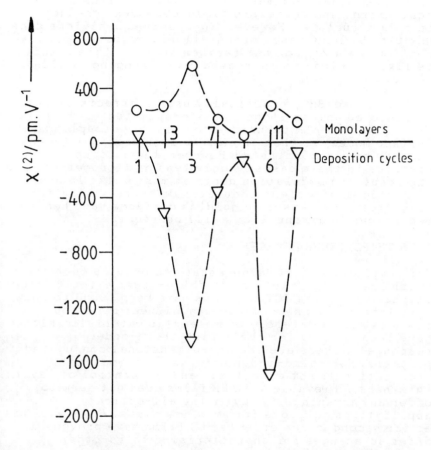

Figure 2. Variation of effective bulk Pockels effect
$\chi^{(2)}$ for a superlattice film of 22-hemicyanine
(upstroke) and 4HANS (downstroke) as a
function of layer thickness:O, real part;
∇, imaginary part.

figure 2 as a function of the number of active layers.
It can be seen that the apparent bulk hyperpolarisability is far from constant, and that it is
necessary to assume an imaginary part of $\chi^{(2)}$ much
larger than the real part in many cases. While the
imaginary part of $\chi^{(2)}$ is what is normally measured

for the Stark effect, it is clear that here the
anomalous behaviour must be due to variations in
scattering, and therefore field-dependent changes
in film structure. However the observed behaviour does
not fit the long-range orientational order model
mentioned above, so that further study of this effect
is likely to lead to increased understanding of film
structure.

Just as for SHG, practically usable effects cannot
be achieved unless light can be propagated in the film.
Of course, the ATR technique does involve coupling into
surface modes, so that in a sense such propagation has
already been demonstrated.However,the guiding
distance in these cases is only tens of micrometres, and
the resulting modulation depth at most 0.1%. In order
to design an efficient modulator or crossover switch,
significant increases in guiding distance, and hence
reductions in attenuation, will be required.

5 THIRD-ORDER EFFECTS

For completeness, it is necessary to mention the work
which has been carried out on third-order effects in LB
films. Just as the ICT organics have been found to show
very interesting second-order nonlinear behaviour, so
there is also a category of molecular materials which
have been studied for their fast third-order
response[50]. These are the one-dimensional conjugated
polymers. For third-order effects, chromophore
orientation is not essential, and the control of layer
thickness achievable with LB films does not seem to
offer any advantage in any of the all-optical
applications suitable for μm-thick films. However, just
as for second-order effects, LB films have proved to
offer advantages for characterising third-order
materials. They have been used to measure the response
of the red and blue forms of polydiacetylene for both
third-harmonic generation (THG)[51,52] and optical Kerr
effect[53]. Because of the very precise control of layer
thickness, LB films have been used to measure the phase
of the THG produced by a range of materials[54].

6 CONCLUSION

The LB technique has already proved to be a very useful
adjunct in basic scientific studies of nonlinear
optical effects in organic materials: as a quick
characterisation technique for the electro-optic, SHG

and THG coefficients of new chromophores; and for
measuring response in the presence of strong
absorption.

There is also a strong possibility that the second-
order response of LB films may find an area of practical
application. Demonstrated electro-optic performance
is comparable to that of the perovskite ferro-
electrics, while their all-optical nonlinearity can be
an order of magnitude greater than that of competing
materials. In order to achieve this goal, much more
research is necessary to understand and reduce the
levels of scattering. Encouragingly, the studies which
have been carried out in this area have not demonstrated
any fundamental obstacle.

7 BIBLIOGRAPHY

1.   P.Günter, in 'Proc. Europ. Conf. on Optical Systems
      and Applications, Utrecht 1980',SPIE, vol *236*, p8.
2.   R.A.Becker, R.H.Rediker and T.A.Lind, *Appl. Phys.
      Lett.*,1985,*46*,809.
3.   S.D.Smith,*Nature,*1985,*316*,319.
4.   A.Lattes, H.A.Haus, F.J.Leonberger and E.P Ippen,
      *IEEE Jnl. Quantum Electron,*1983,*QE-19*,1718.
5.   H.Peyghambarian and H.M.Gibbs,*Optical Engineering,*
      1985,*24*,68.
6.   D.J.Williams, '*Nonlinear Optical Properties of
      Organic and Polymeric Materials*',ACS Symposium
      Series No 233, Washington DC,1983.
7.   J.Zyss and I.Ledoux, *L'Echo des Recherches,*1987,
      127,19.
8.   J.Zyss and J.L.Oudar,*Phys. Rev.A*, 1982,*26*,2028.
9.   M.Hurst and R.W.Munn,*J. Mol. Electron.*,1986,*2*,35;
      ibid,1986,*2*,43.
10.   M.Vandevyver,*Thin Solid Films,*1988,*158*,in press.
11.   I.R.Peterson, G.J.Russell, J.D.Earls and
      I.R.Girling,*Thin Solid Films,*1988,*161*,in press.
12.   T.Rasing, G.Berkovic, Y.R.Shen, S.G.Grubb and
      M.W.Kim, *Chem. Phys. Lett.*, 1986,*130*,1.
13.   I.R.Girling, N.A.cade, P.V.Kolinsky and
      C.M.Montgomery, *Electron. Lett.*,1985,*21*,169.
14.   Z.Chen, W.Chen, J.Zheng, W.Wang and Z.Zhang,*Opt.
      Commun.*,1985,*54*,305.
15.   G.L.Gaines, '*Insoluble Monolayers at Liquid-Gas
      Interfaces*',Wiley,New York,1966.
16.   G.H.Cross, I.R.Peterson, I.R.Girling, N.A.Cade,
      M.J.Goodwin, N.Carr, R.S.Sethi, R.Marsden,
      G.W.Gray,D.Lacey, A.M.McRoberts, R.M.Scrowston
      and K.J.Toyne, *Thin Solid Films,*1988,*156*,39.

17. D.Lupo, W.Prass, U.Scheunemann, A.Laschewsky,
     H.Ringsdorf and I.Ledoux, *J. Opt. Soc. Am. B*,1988,
     5,300.
18. D.J.Williams,*Angew. Chem. Int. Ed. Engl.*,1984,*23*,
     690.
19. A.Dulic and C.Flytzanis,*Optics Commun.*,1978,*25*,
     402.
20. I.R.Girling, P.V.Kolinsky, N.A.Cade, J.D.Earls
     and I.R.Peterson,*Optics Commun.*,1985,*55*,289.
21. I.R.Girling, N.A.Cade, P.V.Kolinsky, J.D.Earls,
     G.H.Cross and I.R.Peterson,*Thin Solid Films*,
     1985,*135*,173.
22. G.Decher, B.Tieke, C.Bosshard and P.Günter,in
     press.
23. G.H.Cross, I.R.Girling, I.R.Peterson, N.A.Cade
     and J.D.Earls,*J. Opt. Soc. Am. B*,1987,*4*,962.
24. e.g.,Proc. 2nd Intl Conf. on Langmuir-Blodgett
     Films, Schenectady NY USA July 1985,*Thin Solid
     Films*,1985,*132-4*.
25. O.A.Aktsipetrov, N.N.Akhmediev, E.D.Mishina and
     V.R.Novak, *Pis'ma Zh. Eksp. Teor. Fiz.*,1983,*370*,
     175 (Russian);*JETP Lett.*,1983,*37*,207(Engl).
26. F.Grunfeld and C.W.Pitt, *Electron. Lett.*,1982,
     99,249.
27. A.F.Garito and K.D.Singer, *Laser Focus*,1982,*80*,
     59.
28. G.Khanarian, *ACS Workshop on Organic and
     Polymeric Nonlinear Optical Materials, Virginia
     Beach VA May 16-19* ,1988.
29. *'CRCHandbook of Laser Science & Technology.Vol
     III.Optical Materials.Part I. Nonlinear Optical
     Properties/Radiation Damage'*,Ed. M.J.Weber,
     CRC Press Inc, Boca Raton FA, 1986.
30. *'Electrooptic and Photorefractive Materials'*,Ed.
     P.Günter, Springer-Verlag, Berlin, 1979.
31. V.R.Novak and I.V.Myakov, *Sov. Tech. Phys. Lett.*,
     1985,*11*,159.
32. D.B.Neal, M.C.Petty, G.G.Roberts, M.M.Ahmad,
     W.J.Feast, I.R.Girling, N.A.Cade, P.V.Kolinsky
     and I.R.Peterson, *Electron. Lett.*,1986,*22*,460
33. G.R.Meredith and B.Buchhalter, *J. Chem. Phys.*,
     1983,*78*,1938.
34. C.Sauteret, J.-P.Hermann, R.Frey, F.Pradère,
     J.Ducuing, R.H.Baughman and R.R.Chance,*Phys. Rev.
     Lett.*,1976,*36*,956.
35. I.R.Girling, P.V.Kolinsky, N.A.Cade, J.D.Earls
     and I.R.Peterson, *Opt. Comm.*,1985,*55*,289.

36. I.R.Girling, N.A.Cade, P.V.Kolinsky, J.D.Earls, G.H.Cross and I.R.Peterson, *Thin Solid Films*, 1985, *134*, 135.
37. R.H.Tredgold, M.C.J.Young, R.Jones, P.Hodge, P.Kolinsky and R.J.Jones, in press.
38. R.C.Hall, G.A.Lindsay, B.Anderson, S.T.Kowel, B.G.Higgins and P.Stroeve, *Proc. Materials Research Soc. Dec 1987 Boston MA*, in press.
39. S.A.Letts, T.Fort and J.B.Lando, *J. Coll. Interface Sci*, 1976, *56*, 64.
40. A.Barraud, C.Rosilio and A.Ruaudel-Teixier, *J. Coll. Interface Sci.*, 1977, *62*, 509.
41. G.Fariss, J.B.Lando and S.E.Rickert, *J. Mater. Sci.*, 1983, *18*, 3323.
42. B.Tieke and D.Bloor, *Makromol. Chem.*, 1979, *180*, 2275.
43. G.Lieser, B.Tieke and G.Wegner, *Thin Solid Films*, 1980, *68*, 77.
44. R.H.Tredgold, M.C.J.Young, R.Jones, P.Hodge, P.Kolinsky, R.J.Jones, *J. Chim. Physique*, in press.
45. L.A.Laxhuber, B.Rothenhäusler, G.Schneider and H.Möhwald, *Appl. Phys. A*, 1986, *39*, 173.
46. W.L.Barnes and J.R.Sambles, *J. Phys. D*, 1987, *20*, 1125.
47. R.H.Tredgold, M.C.J.Young, P.Hodge and E.Khoshdel, *Thin Solid Films*, 1987, *151*, 441.
48. I.R.Peterson, J.D.Earls, I.R.Girling and W.L.Barnes, *J. Phys. D*, 1988, *21*, 773.
49. G.H.Cross, I.R.Peterson and I.R.Girling, *Proc. SPIE Conf. San Diego August 1987*, in press.
50. *'Nonlinear Optical Properties of Organic Molecules and Crystals'*, Vol 2, Eds. D.S.Chemla and J.Zyss, Academic, Orlando FA, 1987.
51. F.Kajzar and J.Messier, *Thin Solid Films*, 1985, *132*, 11.
52. F.Kajzar, J.Messier, J.Zyss and I.Ledoux, *Optics Commun.*, 1983, *45*, 133.
53. G.M.Carter, Y.J.Chen and S.K.Tripathy, *Appl. Phys. Lett.*, 1983, *43*, 891.
54. F.Kajzar, I.R.Girling and I.R.Peterson, *Thin Solid Films*, 1988, *158*, in press.

Non-linear Optical Studies of Organic Monolayers

Y.R. Shen

DEPARTMENT OF PHYSICS, UNIVERSITY OF CALIFORNIA CENTER FOR ADVANCED
MATERIALS, LAWRENCE BERKELEY LABORATORY, BERKELEY, CA 94720, USA

Second-order nonlinear optical effects are forbidden in a medium with inversion symmetry, but are necessarily allowed at a surface where the inversion symmetry is broken. They are often sufficiently strong so that a submonolayer perturbation of the surface can be readily detected. They can therefore be used as effective tools to study monolayers adsorbed at various interfaces.[1] We discuss here a number of recent experiments in which optical second harmonic generation (SHG) and sum-frequency generation (SFG) are employed to probe and characterize organic monolayers.

We first consider the application of SHG to organic molecular monolayers to measure the nonlinearity of some organic molecules.[2] As schematically shown in Fig. 1, insoluble amphiphilic molecules can be spread on water in a monolayer form, and their surface density can be controlled by varying the overall surface area. The second harmonic (SH) signal generated from the monolayer is then monitored and compared with that generated from a quartz crystal. From the results, the nonlinear susceptibility $\chi^{(2)}$ for the monolayer relative to that of quartz can be obtained and the nonlinear polarizability $\alpha^{(2)}$ for the molecules can be deduced in the following way.

If the interaction between molecules in the monolayer can be neglected, $\overset{\leftrightarrow}{\chi}^{(2)}$ and $N_s\overset{\leftrightarrow}{\alpha}^{(2)}$ are simply related by a coordinate transformation from the molecular (ξ,η,ζ) system to the lab (x,y,z) system.[3]

$$\chi^{(2)}_{ijk} = N_s\langle G^{\xi\eta\zeta}_{ijk}\rangle\alpha^{(2)}_{\xi\eta\zeta} , \qquad (1)$$

where N_s is the surface density of molecules, $G^{\xi\eta\zeta}_{ijk}$ is the

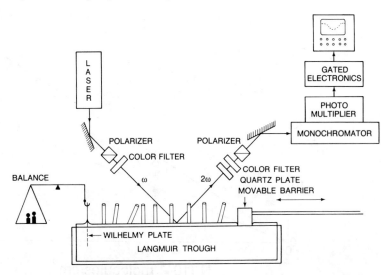

Figure 1   Experimental arrangement for studies of organic
monolayers spread on water in a langmuir trough by
simultaneous   measurements of surface tension and second
harmonic generation versus surface area.

transformation matrix, and the angular brackets denote an
average over the molecular orientational distribution. In
many cases, $\overleftrightarrow{\alpha}^{(2)}$ is dominated by a single element $\alpha_{\xi\xi\xi}^{(2)}$
along the molecular axis $\hat{\xi}$, and the molecules have
isotropic distribution in the azimuthal plane. One can
then easily show that the nonvanishing elements of $\overleftrightarrow{\chi}^{(2)}$ are

$$\chi_{zzz}^{(2)} = N_s \langle \cos^3\theta \rangle \alpha_{\xi\xi\xi}^{(2)}$$

$$\chi_{zxx}^{(2)} = \chi_{xzx}^{(2)} = \chi_{zyy}^{(2)} = \chi_{yzy}^{(2)} = \frac{1}{2} N_s \langle \sin^2\theta \cos\theta \rangle \alpha_{\xi\xi\xi}^{(2)} , \qquad (2)$$

where θ is the angle between $\hat{\xi}$ and the surface normal $\hat{z}$.
With a further assumption of a δ-function for the
orientational distribution, measurements of $\chi_{zzz}^{(2)}$ and $\chi_{zxx}^{(2)}$
allow us to determine $\alpha_{\xi\xi\xi}^{(2)}$ and θ separately.[3]

The validity of Eq. (1) means that the local-field
correction due to molecule-molecule interaction can be
neglected. The analogous relation in the linear case is
$\chi_{ij}^{(1)} = N_s \langle G_{ij}^{\xi\eta} \rangle \alpha_{\xi\eta}^{(1)}$ which would imply that the dielectric
constant $\varepsilon(\omega)$ for the monolayer is close to 1. In our
experiment, Eq. (1) can be verified by varying N_s. An
example is shown in Fig. 2,[2] where the square root of the
SH signal, which is proportional to $\chi^{(2)}$, is plotted

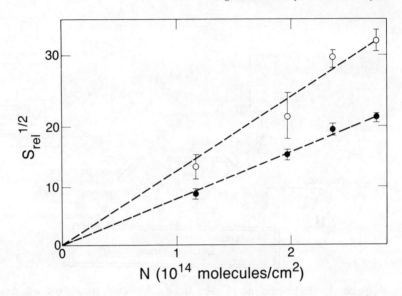

<u>Figure 2</u> Square root of the relative SHG intensity $S_{rel}^{1/2}$ as a function of surface density for 5CT spread on water. The input laser field was polarized at 45° to the plane of incidence, for which both the s-polarized (open circles) and p-polarized (filled circles) SHG outputs were measured.

against the surface density of 4"-n-pentyl-4-cyano-p-terphenyl (5CT). The linear relation between $\overset{\leftrightarrow}{\chi}^{(2)}$ and N_S is clearly satisfied. With other molecules of higher surface packing densities, deviation from the linear relation has been observed. The local field correction factors must then be included in the calculation.[4]

    One can also test experimentally the assumption that $\alpha_{\xi\xi\xi}^{(2)}$ dominates in $\alpha^{(2)}$. If this is not the case, we should find $\chi_{zzx}^{(2)} \neq \chi_{xzx}^{(2)}$ and $\chi_{zyy}^{(2)} \neq \chi_{yzy}^{(2)}$ even in the low N_S limit. On the other assumption of a δ-function distribution for θ, a crude estimate indicates that it is a good approximation if the spread of θ is less than 10° or if θ is close to 0° or 90°. Finally, $\overset{\leftrightarrow}{\chi}^{(2)}$ for adsorbed molecules is actually compost of three parts:

$$\overset{\leftrightarrow}{\chi}^{(2)} = \overset{\leftrightarrow}{\chi}_{mol}^{(2)} + \overset{\leftrightarrow}{\chi}_{water}^{(2)} + \overset{\leftrightarrow}{\chi}_{int}^{(2)} , \tag{3}$$

where $\overset{\leftrightarrow}{\chi}_{mol}^{(2)}$, $\overset{\leftrightarrow}{\chi}_{water}^{(2)}$, and $\overset{\leftrightarrow}{\chi}_{int}^{(2)}$ are nonlinear susceptibilities due to the molecular monolayer isolated from water, the bare water surface, and the interaction between the molecules and water, respectively. For organic molecules with relatively high nonlinearity, $\chi^{(2)}_{\text{water}}$ can be neglected. If the molecular properties are not very much affected by the adsorption on water, then $\chi^{(2)}_{\text{int}}$ may also be negligible compared with $\chi^{(2)}_{\text{mol}}$, leaving $\chi^{(2)} = \chi^{(2)}_{\text{mol}}$.

We have used the technique to measure $\alpha^{(2)}_{\xi\xi\xi}$ for a number of structurally related molecules (phenyl derivates).[2] The results are what one would normally expect, namely, $\alpha^{(2)}$ increases with the conjugation length and decreases with interruption of electron delocalization by twist between phenyl rings or by replacing phenyl ring by pyrimidine; $\alpha^{(2)}$ also increases with increase of charge transfer.

The above technique used to measure $\alpha^{(2)}$ has the disadvantages that not all molecules can be spread on water and that molecules may have their properties drastically changed when adsorbed on water. The latter is more likely to happen for molecules with large charge transfer between strong electron donor and acceptor end groups. Hemicyanine is an example.[5] The strong charge-transfer band of

$\text{R}-\overset{+}{\text{N}}\underset{}{\diagdown}\hspace{-0.3em}\bigcirc\hspace{-0.3em}\diagup\hspace{-0.3em}\diagdown\hspace{-0.3em}\bigcirc\hspace{-0.3em}\diagup\hspace{-0.3em}-\text{N}\overset{R'}{\underset{R'}{\diagdown}}$ responsible for the large second-order

nonlinearity of the molecule is effectively suppressed by protonation. Adsorption of the molecules on water with only a small amount of acid ($\sim 0.1\%$ H_2SO_4 by volume) could reduce $\alpha^{(2)}$ of the molecules by more than two orders of magnitude. Special precaution is needed to prevent such changes from happening.

Information deduced from SHG about the orientation of molecular adsorbates is also useful for fundamental understanding of the properties of an organic monolayer. For example, from the SHG measurements, it was found that the liquid expanded-liquid condensed transition of a Langmuir monolayer on water is associated with a sudden change in the molecular orientation.[6] Since SHG is applicable to any interface accessible by light, it can also be used to study how the orientation of surfactant molecules at liquid/liquid interfaces depends on the enironment.[7] This is relevant, for example, to the understanding of micellar structure. Consider the recent experiment on sodium 1-dodecylnaphthalene-4-sulfonate (SDNS) ($C_{12}H_{25}-C_{10}H_6SO_3Na$) at $C_{10}H_{22}/H_2O$ and $CC\ell_4/H_2O$ interfaces. The results yield a polar tilt angle θ of $21°$ and $38°$ for SDNA at the two interfaces, respectively, in comparison to $\theta = 13°$ at the air/water interface. Here, the nonlinearity is dominated by $\alpha^{(2)}_{\xi\xi\xi}$ arising from the naphthalene chromophore and the ξ-axis is along the charge-transfer direction from the SO_3 group to the

hydrocarbon chain. The larger θ is qualitatively
correlated with the weaker tail-tail interaction of the
hydrophobic part of the molecules in the nonaqueous
solution due to dielectric screening.

Reactions of molecules in a surface monolayer will
change the properties of the monolayer. They can be
monitored by SHG. Photo-polymerization of a monolayer is a
good example.[8] The problem is meaningful because two-
dimensional polymerization is expected to be very different
from the three dimensional one. Moreover, it is now
possible to vary the surface density and orientation of the
monomers and study their effects on polymerization. Figure
3 describes how the SHG from a monolayer of octadecyl
methacrylate (ODMA) on water changes as it is polymerized
by UV irradiation. The π electrons associated with the C=C
double bond in the monomers contribute significantly to the
nonlinearity of the monomers. Upon polymerization, the
linkage between monomers changes the double bond to a
single bond, and hence reduces the nonlinearity. This
explains the decrease of the SH signal with the irradiation
time in Fig. 3. The data can be used to test theoretical
models for surface polymerization, as shown in Fig. 3.

In some cases, SHG may be too weak to probe a polymer
monolayer, but the third-order nonlinearity of the polymer
is sufficiently strong so that third harmonic generation
(THG) can be used.[9] For example, a monolayer of poly-4-
BCMU (a processable polydiacetylene), yields a TH signal at
0.36 μm that shows an increase in correlation with the
conformational transition from a less conjugated yellow
form to a more conjugated red form. The surface density at
which the transition occurs agrees with the value deduced
from the observed transition in the bulk, suggesting that
in this case the surface and bulk properties are very much
alike.

We can also use SHG to study biologically interesting
molecules in monolayer form or embedded in membranes.[10]
Retinal chromophores are responsible for vision. These
molecules have structures exhibiting a long conjugation
length with an appreciable charge transfer. (The all-trans
retinal, ATR, illustrated in Fig. 4, is a representative
example.) One therefore expects a large $\alpha^{(2)}$ for such
molecules. Measurements of SHG at 0.266 μm from a
monolayer of retinal molecules on water indeed show a value
of $\alpha^{(2)}$ larger than 10^{-28} esu. Using the two-band model
that relates $\alpha^{(2)}$ with Δμ, the difference between the
dipole moments of the ground and the excited states, we can
then deduce Δμ. The latter quantity is important for

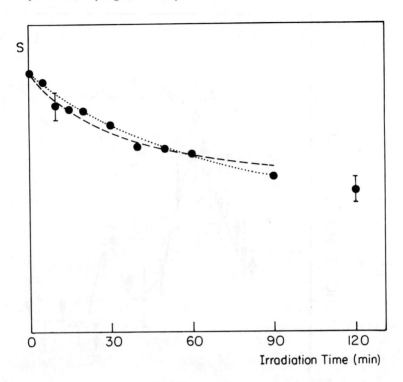

Figure 3 Relative SHG intensity from a monolayer of ODMA on water versus irradiation time for UV polymerization of the monolayer. The dotted and dashed curves are derived from theoretical models assuming first-order and second-order kinetics, respectively.

modeling the photochemistry of vision.

The large value of $\alpha^{(2)}$ would allow the study of retinal chromophores by SHG even when they are embedded in a membrane with a low surface density (but still in polar arrangement). However, it is possible that $\alpha^{(2)}$ may change with environment. A recent SHG experiment on a monolayer of protonated retinylidene n-butylamine Schiff base (PNRB) on water and on PNRB embedded in the purple membrane of Halobacterium halobium yielded $\alpha^{(2)}$ for the two cases separately. The results led to nearly the same $\Delta\mu$ for both cases, indicating that $\Delta\mu$ of the chromophore is insensitive to the environment.

By varying the frequency of the input laser, spectral

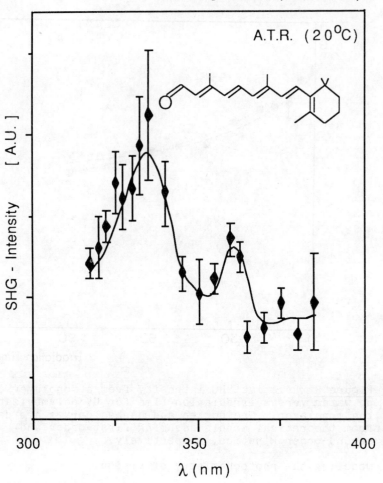

Figure 4 SHG spectrum of a monolayer of all-trans retinal
(ATR) at an air/water interface.

information about a retinal monolayer can also be obtained
from the SHG measurements.[10] Figure 4 depicts the SH
spectrum of a monolayer of ATR on water. Two resonant
peaks are seen, one at 335 nm and the other at 360 nm.
They can be assigned to one-photon $S_0 \rightarrow B_u$ and two-photon
$S_0 \rightarrow A_g$ transitions, respectively, both being blue-shifted
from the corresponding ones for ATR in solution.

The monolayer sensitivity of SHG requires the output

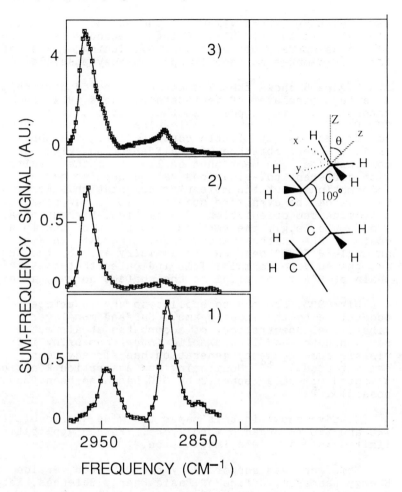

Figure 5 SFG spectra of a full monolayer of PDA with various (vis,IR) polarization combinations: 1) s-visible, p-IR; 2) p-visible, s-IR; 3) p-visible, p-IR. Inset: coordinate axes chosen for the terminal methyl group.

being in the visible so that high-gain photodetectors can be used. For spectroscopic studies, this limits the applications to probing of electronic transitions in molecules. An extension of SHG to sum-frequency generation (SFG), however, removes the limitation. In SFG for surface vibrational spectroscopy, two laser beams are simultaneously incident on the sample, one in the infrared

and the other in the visible. The output at $\omega_1 + \omega_2$ is
then also in the visible. Being a second-order process,
SFG is as surface-specific as SHG. Tuning of the infrared
input then makes surface IR spectroscopy possible.[11]

Figure 5 shows the SF spectra of the CH stretch modes
of a full monolayer of pentadecanoic acid (PDA) on
water.[12] All three peaks in the spectra originate from the
CH_3 terminal group of the molecules. Because of symmetry,
the CH_2 groups give little contribution. The three spectra
in Fig. 5 were obtained with different polarization
combinations. We mentioned earlier that the results of SHG
with different polarizations can be used to deduce
information about the molecular orientation. Here, with
SFG, we can be even more specific: it is now possible to
determine the orientation of a particular group of atoms in
a molecule, e.g., the terminal CH_3 group of the long alkane
chain of PDA in the present case. The data in Fig. 5 yield
an angle $\theta \approx 35°$ between the symmetry axis of the CH_3 group
and the surface normal. This indicates that the alkane
chain of PDA is parallel to the surface normal, as expected

Like SHG, SFG can be applied to any interface
accessible by the input beams. Surface monolayer
vibrational spectroscopy of adsorbates at air/metal, air/
semiconductor and liquid/solid interfaces using infrared-
visible sum-frequency generation has already been
demonstrated.[13-15] Monitoring the adsorption kinetics of
selected molecular species by SFG has also been found
possible.[15]

We have shown in this paper the wide applicability of
SHG and SFG to organic monolayers. Futher applications are
limited only by one's imagination.

This work was supported by the Director, Office of
Energy Research, Office of Basic Energy Sciences, Materials
Sciences Division of the U.S. Department of Energy under
Contract No. DE-ACO3-76SF00098.

REFERENCES

1.   See, for example, Y. R. Shen. <u>Ann. Rev. Mater. Sci.</u>,
      1986, <u>16</u>, 69.
2.   G. Berkovic, Th. Rasing, and Y. R. Shen, <u>J. Opt. Soc.
      Amer.</u>, 1987, B4, 945.
3.   T. F. Heinz, H. W. K. Tom, and Y. R. Shen, <u>Phys. Rev.
      A</u>, 1983, <u>28</u>, 1883.
4.   P. Ye and Y. R. Shen, <u>Phys. Rev. B</u>, 1983, <u>28</u>, 4288.

5. G. Marowsky, L. F. Chi, D. Mobius, R. Steinhoff, Y. R. Shen, D. Dorsch, and B. Riger, Chem. Phys. Lett. (to be published).
6. Th. Rasing, Y. R. Shen, M. W. Kim, and S. Grubb, Phys. Rev. Lett., 1985, 55, 2903.
7. S. G. Grubb, M. W. Kim, Th. Rasing, and Y. R. Shen, Langmuir, 1988, 4, 452.
8. G. Berkovic, Th. Rasing, and Y. R. Shen, J. Chem. Phys., 1986, 85, 7374.
9. G. Berkovic, R. Superfine, P. Guyot-Sionnest, Y. R. Shen, and P. Prasad, J. Opt. Soc. Amer. B, 1988, 5, 668.
10. J. Huang, A. Lewis, and Th. Rasing, J. Phys. Chem. (to be published); Th. Rasing, J. Huang, A. Lewis, T. Stehlin, and Y. R. Shen, (to be published).
11. X. D. Zhu, H. Suhr, and Y. R. Shen, Phys. Rev. B, 1987, 35, 3047; J. H. Hunt, P. Guyot-Sionnest, and Y. R. Shen, Chem. Phys. Lett., 1987, 133, 189.
12. P. Guyot-Sionnest, J. H. Hunt, Y. R. Shen, Phys. Rev. Lett., 1987, 59, 1597.
13. A. L. Harris, C. E. D. Chidsey, N. J. Levinos, and D. N. Loiacono, Chem. Phys. Lett., 1987, 141, 350.
14. R. Superfine, P. Guyot-Sionnest, J. H. Hunt, C. T. Kao, and Y. R. Shen, Surf. Sci. Lett. (in press).
15. P. Guyot-Sionnest, R. Superfine, J. H. Hunt, and Y. R. Shen, Chem. Phys. Lett., 1988, 144, 1.

Preparation and Characterization of Ultrathin Organic Films with Defined Thickness and Orientation

C. Bubeck*, T. Arndt, T. Sauer, G. Duda, and G. Wegner

MAX-PLANCK-INSTITUT FÜR POLYMERFORSCHUNG, POSTFACH 3148, D-6500 MAINZ, FRG

1. INTRODUCTION

There is strong interest in thin organic films with highly controlled thickness and molecular order in view of numerous potential applications. The Langmuir-Blodgett (LB) technique can be utilized to achieve the thickness control (1,2). The control of the molecular orientation and its characterization is crucial with respect to two aspects: the orientation perpendicular to the layer plane is important to obtain noncentrosymmetric structures whereas a control of the lateral orientation might be of considerable value to obtain an anisotropic refractive index for phase matching of guided waves. The intention of this contribution is to present a short survey on the recent progress in the formation of LB films of rigid rod polymers and to show how linear optical spectroscopy can be applied to determine the molecular orientation in ultrathin films.

2. BUILDING PRINCIPLES OF LANGMUIR-BLODGETT FILMS OF VARIOUS COMPOUNDS

A survey of different kinds of LB films is shown in Fig. 1. The classical LB films are built up with amphiphiles of low molecular weight such as fatty acids or their metal salts. The multilayers are stabilized by Van-der-Waals and ionic interactions in the multilayer. However, these films suffer from low mechanical and thermal stability. Furthermore their microcrystalline composition leads to considerable light scattering. By solid state polymerization of the multilayers (3,4) or by the use of preformed polymers (5,6) films with improved

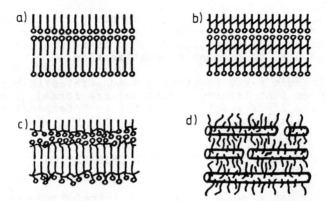

Figure 1 Comparison of the structure of Langmuir-Blodgett layers of different compounds (roughly simplified; view parallel to the layer plane):
a) classical amphiphiles
b) reactive amphiphile after solid state polymerization
c) prepolymerized amphiphilic polymer
d) Rigid rod polymer substituted with flexible side chains.

stability can be obtained. As Fig. 1b and 1c show, the molecular design of the monomers and polymers still relies on the classical building principles for LB films: to use a hydrophilic head group coupled to hydrophobic alkyl chains.

Recently it was shown that rigid rod polymers with flexible side groups can be used to built up multilayers. These polymers form liquid crystalline phases. Presently two examples exist: phthalocyaninato -polysiloxane (7) and polyglutamate(8). Both systems behave rather similar with respect to film formation and general structure. The following characteristic properties were found:

• Good transfer properties; the transfer ratio is 100% and stable up to high numbers of transferred layers.

• The rigid rods are oriented preferentially parallel to the dipping direction of the substrate, presumably due to shear forces during the transfer process.

• The anisotropy of the films can be determined with optical absorption spectroscopy. The dichroitic ratio is in the order of 2 - 5.

- The molecular order of the side groups is low (8). This originates from the substitution pattern of the n-alkyl side groups, that is deliberately designed as to prevent crystallization.

- Small angle x-ray scattering demonstrates in both examples that layered structures are formed. The measured layer spacing indicates an interpenetration of the alkyl chains.

3. DETERMINATION OF MOLECULAR ORIENTATION IN MONOLAYERS

Grazing incidence reflection (GIR) infrared spectroscopy (9) is becoming a common tool to investigate the structure of LB films (10,11). In a GIR set-up, the electric field is perpendicularly oriented to the surface. Information on orientation is derived from the fact that only transition moments with a component in that direction will be detected. Recently we presented a novel method for the determination of tilt angles in LB films (12). It is based on the analysis of the intensities of three modes with linearly independent transition moments as shown in Fig 2. A comparison and evaluation of the GIR spectrum of the thin film with the spectrum of an isotropic bulk sample gives the molecular orientation. The method has been verified with the model compound cadmium arachidate. The tilt angle found between

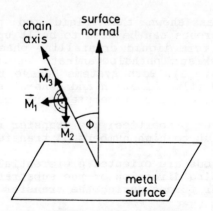

<u>Figure</u> <u>2</u> Geometry of the polarizations of the three IR modes used for the evaluation of the molecular tilt angle ϕ.

the chain axis and the surface normal is 15º to 17º. This value is in agreement with a recently reported value derived from small angle X-ray scattering of the same compound (13). The method can be applied to other systems such as dye molecules in LB layers. The principle of the method was derived for a case where the three vibrational modes constitute a rectangular coordinate system (12). If this is not the case, the three modes must at least be linearly independent and the formalism has to be replaced by a similar relationship according to geometry.

ACKNOWLEDGEMENT

This work was supported by the BMFT under the project number 03M 4008 E/9

REFERENCES

1. G.L. Gaines, "Insoluble Monolayers at Liquid-Gas Interfaces", Interscience Publ., New York 1966
2. H. Kuhn, D. Möbius and H. Bücher in: "Physical Methods of Organic Chemistry", ed. by A. Weissberger and P. Rossiter, Wiley Interscience 1972, Vol 1, part 3b, P. 577
3. B.Tieke, Adv.Polym.Sci., 1985, 71, 79
4. C. Bubeck, Thin Solid Films, 1987, 159, in press
5. R.H. Tredgold, Thin Solid Films, 1987, 152, 223
6. S.J. Mumby, J.D. Swalen and J. F. Rabolt, Macromol., 1986, 19, 1054
7. E. Orthmann and G. Wegner, Angew.Chem.Int.Ed.Engl., 1986, 25, 1105
8 G. Duda, A.J. Schouten, T. Arndt, G. Lieser, G.F. Schmidt C. Bubeck and G. Wegner, Thin Solid Films, 1988, in press
9. R.G. Greenler, J.Chem.Phys., 1966, 44, 310
10. J.D. Swalen and J. F. Rabolt in "Fourier Transform Infrared Spectroscopy", Vol.4, J.R. Ferraro and L.J. Basile (eds.), Acad. Press, New York, 1985, Chap. 7
11. T. Arndt and C. Bubeck, Thin Solid Films, 1988, in press
12. T. Arndt, C. Bubeck and G. Wegner, submitted for publication in Langmuir
13. P. Fromherz, U. Oelschlägel and W. Wilke, Thin Solid Films, 1988, in press

Effect of Mixing on Second Harmonic Generation and Chromophore Orientation in Langmuir–Blodgett Films

J. Bauer, P. Jeckeln, D. Lupo*, W. Prass, and U.Scheunemann

HOECHST AG, POSTFACH 800 320, D-6230 FRANKFURT 80, FRG

R. Keosian and G. Khanarian

HOECHST CELANESE RESEARCH COMPANY, SUMMIT, NJ, USA

1 INTRODUCTION

Since the report by Aktsipetrov et al.[1] of significant optical second harmonic generation (SHG) from a Langmuir–Blodgett film further investigations of second order nonlinear optical effects in LB films[2-8] have indicated that such films have excellent nonlinear optical properties. The high degree of chromophore orientation leads to susceptibilities which are as a class typically higher than those of inorganic materials or of organic crystals or polymers. In addition, the exact control of film thickness and the possibility of building up highly ordered planar superstructures are promising for nonlinear waveguide applications[9].

One question which has arisen is that of the effect of neighbouring molecules on the nonlinearity of a molecule with large hyperpolarizability β in an LB monolayer. Girling et al.[7] reported significant enhancement of SHG in hemicyanine monolayers via dilution with cadmium arachidate. Schildkraut et al.[5] reported a similar enhancement, and concluded that the difference in efficiency was due to a shift out of resonance in the pure dye layers due to aggregate formation, an observation which was confirmed by Marowsky[8]. They concluded as well that orientational effects and electronic coupling play a smaller role in affecting the SHG efficiency.

We have investigated the effect of dilution on spectra, nonlinearity and chromophore orientation in a series of LB monolayers of two dyes, using SHG and linear polarization spectroscopy. The hemicyanine (1) has been shown to

$$CH_3(CH_2)_{15} \diagdown N - \langle O \rangle - CH = CH - \langle O \rangle \overset{+}{N} - CH_3$$
$$CH_3(CH_2)_{15} \diagup$$

(1) I^-

348

have an exceptionally large value of $\chi^{(2)}(-2\omega;\omega,\omega)$ in monolayers[4]. The phenylhydrazone ester (2) is similar to other dyes which were seen to have

$$CH_3(CH_2)_{14}-\overset{\overset{\displaystyle O}{\|}}{C}-O-\langle\bigcirc\rangle-C{=}N-NH-\langle\bigcirc\rangle-NO_2$$

(2)

high nonlinearities via SHG.[4] This has been confirmed[10] by independent electric field induced SHG (EFISHG) measurements. The diluents used were (3) palmitic acid and (4) acrylic acid octadecylamide.

2 EXPERIMENTAL DETAILS

Isotherms of the pure dyes and mixtures were measured on a Lauda Langmuir film balance. Monolayers were deposited onto hydrophilic glass slides at a surface pressure of 30 mN m^{-1} and a speed of 1 cm min^{-1} at 20°C on a pure water subphase.

The apparatus for studies of SHG is sketched in Fig. 1. Pulses from a Q–switched Nd:YAG laser operating at 1064 nm were split into sample and reference (a detuned KDP crystal) beams. The harmonic intensities at 532 nm from the sample and the reference were selected by filters (F) and detected by photomultipliers (PM) as the sample was rotated about the incidence angle α. The α dependence of the ratio of sample and reference intensities $I_s^{2\omega}/I_r^{2\omega}$ was sampled by a boxcar. Absolute intensities were calibrated by comparison with the Maker fringes from a 2 mm thick quartz plate.[11]

Polarization spectra were measured using a modified Perkin–Elmer

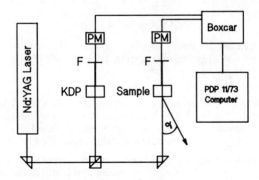

Figure 1. Schematic representation of the apparatus for SHG studies

Lambda 9 uv–vis spectrophotometer with a computer interface. The samples were mounted at 45° to the beam and the transmission change relative to an uncoated slide measured for s– and p– polarized light.

3 RESULTS AND DISCUSSION

Mixtures of dye and diluent were investigated for dye fractions of 100, 95, 90, 80, 60, and 40% in the spreading solutions. Compound (1) was diluted with palmitic acid, and compound (2) with palmitic acid and with compound (4).

<u>Hemicyanine Films</u>
For SHG studies the glass slides were coated on both sides, and the p– polarized harmonic radiation measured for both s– (s→p) and p– (p→p) polarization of the fundamental. Typical curves for hemicyanine/palmitic acid mixtures are shown in Figure 2. The envelope function for the intensity depends on the effective susceptibility as a function of incidence angle:

$$I^{2\omega}(\alpha) \approx |\chi^{(2)}_{\text{eff}}|^2 \qquad (1)$$

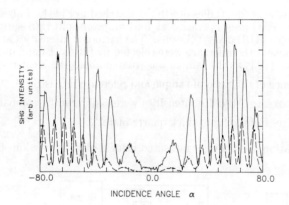

Figure 2. Second harmonic intensity as a function of incidence angle α for p→p (————) and s→p (– – –) polarization from a monolayer of 90% (1) and 10% palmitic acid

For the determination of $\chi^{(2)}$ and chromophore orientation in LB films two assumptions are typically made. First, it is assumed that the hyperpolarizability β of the individual molecule has only one component, directed along the chromophore axis, i.e. $\beta = \beta_{zzz}$. This assumption should hold for (1) and (2) as well. In addition, it is assumed that the chromophores are oriented with a narrow distribution of tilt angle θ from the surface normal z and a statistical azimuthal angle φ distribution about the z axis. In this case the only independent compo–

Table 1. Second harmonic intensities relative to quartz ($\times 10^3$) for hemicyanine/ palmitic acid mixed monolayers

%dye	(l)		(s)	
	$I^{2\omega}_{p \to p}$	$I^{2\omega}_{s \to p}$	$I^{2\omega}_{p \to p}$	$I^{2\omega}_{s \to p}$
100	2.63	——	3.1	0.5
95	2.63	0.12	1.17	0.31
90	3.47	0.39	2.47	0.21
80	2.32	0.34	1.48	0.13
60	1.14	0.12	0.46	0.03
40	0.3	0.03	0.14	0.02

nents of $\chi^{(2)}$ are $\chi^{(2)}_{zzz}$ and $\chi^{(2)}_{zxx} = \chi^{(2)}_{zyy}$ and permutations. The p→p intensity depends on $\chi^{(2)}_{zxx}$ and $\chi^{(2)}_{zzz}$ and the s→p intensity on $\chi^{(2)}_{zyy} = \chi^{(2)}_{zxx}$. From the relationship between $I^{2\omega}_{p \to p}$ and $I^{2\omega}_{s \to p}$ one can then (within the assumptions) deduce the average tilt angle θ and an effective hyperpolarizibility β_{eff}.[12]

If the second assumption is justified, one should obtain the same intensities whether the long (l) or short (s) side of the substrate is parallel to the incidence plane. As can be seen in Table 1, the intensities are quite different in the two orientations. The zero intensity at normal incidence implies a symmetry within the film plane; thus the data seem to indicate an elliptical tilt angle distribution about the z axis, i.e. that $\chi^{(2)}_{zxx} \neq \chi^{(2)}_{zyy}$. Because it was for experimental reasons not possible to probe the same spot in both orientations this observation should not be viewed as conclusive.

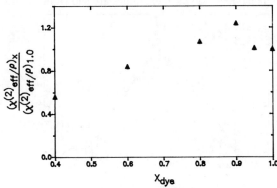

Figure 3. Effective nonlinearity per molecule, normalised to the value for a pure dye film, for mixtures of (1) and palmitic acid in monolayers

The dependence of the effective nonlinearity per molecule on concentration is shown in Figure 3. Contrary to earlier observations[5,7,8] we observed no bulk enhancement of SHG and only a small increase in the effective molecular nonlinearity at low diluent concentrations.

The absorption spectra in s− and p−polarization were measured for slides coated with a monolayer on only one side. Under the assumptions of a transition moment parallel to the chromophore axis and, as with SHG, a statistical azimuthal angle distribution and narrow tilt angle distribution,[13] one can determine the average chromophore tilt angle from the ratio of s− to p− transmission change. In all mixtures the maximum absorption was centered around 460±10 nm without major changes in band shape, and apparent tilt angles of 65–70° were found. No significant changes were observed upon going from (l) to (s) orientation of the substrate. This conflict with the asymmetry observed in SHG could imply that the SHG results were due to inhomogeneities. The zero minima in the SHG fringe patterns indicate however good sample homogeneity, and for tilt angles in the region of 60–70° the SHG intensity is more sensitive to small tilt angle changes than is linear light absorption.

The lack of spectral shifts indicates that aggregate formation is not significant with compound (1), even for pure dye monolayers. Since little enhancement of nonlinearity was seen, our results support earlier claims that SHG enhancement was primarily due to formation of spectrally shifted aggregates.

Phenylhydrazone mixtures

Monolayers of pure compound (2) or only slightly diluted with (3) or (4) showed abnormal SHG curves. There were no fringes, indicating inhomogeneity on a scale larger than a wavelength, and the maximum intensity was observed at normal incidence. Curves for mixtures with 40% or more of either diluent showed "normal" behaviour. Typical examples are shown in Figure 4.

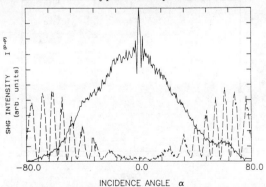

Figure 4. SHG intensity (p→p) for compound (2) pure (——) and diluted with 40% _compound (4) (− − −)

At normal incidence for p→p polarization the only component which can contribute to SHG is $\chi^{(2)}_{xxx}$, thus the large intensity at $\alpha=0$ indicates a breaking of symmetry in the film plane, which is lifted upon further dilution. Thus it is possible to alter drastically the orientation of the chromophores by varying the concentration of diluent.

Polarization spectra did not change significantly upon going from pure phenylhydrazone to mixtures. Because the analysis is dependent on the (at low concentrations clearly not justified) assumption of azimuthal symmetry about the z axis tilt angles could not be determined. The lack of dramatic spectral changes is presumably due to the fact that linear spectroscopy cannot distinguish between centrosymmetric and noncentrosymmetric orientation. SHG is, due to the stringent symmetry restrictions for second order effects, in such cases a much more sensitive probe of orientational changes.

Acknowledgement

We wish to thank Petra Ottenbreit and Irma Wöll for synthesis, Gerhard Geiß for film characterisation and Gunilla Gillberg, Dieter Neher and Isabelle Ledoux for helpful discussions. This work was financially supported by the Bundesministerium für Forschung und Technologie (Project "Ultradünne Polymerschichten).

REFERENCES

1. O.A. Aktsipetrov, N.N. Akhmediev, E.D. Mishina, and V.R. Novak, JETP Lett., 1983, 37, 207.
2. I.R. Girling, N.A. Cade, P.V. Kolinsky, and C.M. Montgomery, Electron. Lett., 1985, 21, 169.
3. I. Ledoux, D. Josse, P.V. Vidakovic, J. Zyss, R.A. Hahn, P.F. Gordon, B.D. Bothwell, S.K. Gupta, S. Allen, P. Robin, E. Chastaing, and J.–C. Dubois, Europhys. Lett., 1987, 3, 803.
4. D. Lupo, W. Praß, U. Scheunemann, A. Laschewsky, H. Ringsdorf, and I. Ledoux, J. Opt. Soc. Am. B, 1988, 5, 300.
5. J.S. Schildkraut, T.L. Penner, C.S. Willand, and A. Ulman, Optics Lett., 1988, 13, 134.
6. L.M. Hayden, S.T. Kowel, and M.P. Srinivasan, Opt. Comun., 1987, 61, 351.
7. I.R. Girling, N.A. Cade, P.V. Kolinsky, R.J. Jones, I.R. Peterson, M.M. Ahmad,D.B. Neal, M.C. Petty, G.G. Roberts, and W.J. Feast, J. Opt. Soc. Am. B, 1987, 4, 950.
8. G. Marowsky, private communication.
9. G. Stegeman, J. Appl. Phys., 1985, 58, R57.
10. G. Khanarian, private communication.
11. J. Jerphagnon and S.K. Kurtz, J. Appl. Phys., 1970, 41, 1667.
12. G. Khanarian, Thin Solid Films, 1987, 152, 265.
13. M. Orrit, D. Möbius, U. Lehmann, and H. Meyer, J. Chem. Phys., 1986, 85, 4966.

Non-linear Optical Effects in Langmuir–Blodgett Films Containing Polymers

M.C.J. Young and R.H. Tredgold

DEPARTMENT OF PHYSICS, UNIVERSITY OF LANCASTER, LANCASTER LA I 4YB, UK

P. Hodge

DEPARTMENT OF CHEMISTRY, UNIVERSITY OF LANCASTER, LANCASTER LA I 4YA, UK

1. INTRODUCTION

In order to produce opto-electronic devices which depend upon second-order non-linear optical effects, materials which can be made into non-centrosymmetric structures need to be used. The Langmuir-Blodgett, (LB), technique of fabricating thin films from organic molecules is, in principle, an ideal method for producing such structures[1]. In this technique polar molecules are spread on a water surface to form a monomolecular film which is then compressed to a close packed form. The film can then be transferred to a suitable substrate by passing the substrate slowly through the film while maintaining the film at a constant surface pressure, allowing films to be built up monolayer by monolayer in a precisely controlled way. Non-centrosymmetric multilayer samples are produced by depositing two different materials, one after the other, a monolayer at a time to produce an alternating layer structure. To produce thick films, that is from 50 to 600 layers, we have used an automated alternating layer trough[2] which allows two different materials to be spread and compressed at the same time in different compartments. A substrate can then be passed down through one monolayer and up through the second.

Although second-order optical studies have been performed on thin films i.e. monolayers and films of a few bilayers thickness[3,4], comparatively little has been done on films hundreds of bilayers thick[5,6]. If Langmuir-Blodgett films are to make an impact in the field of opto-electronics then thicker systems which produce larger effects need to be developed.

We have already shown at Lancaster that it is possible to produce thick film, 0.5 μm, waveguides[7] with acceptable attenuation. We are now

354

investigating electro-optic effects in thick LB films of chromophore- containing monomers dipped alternately with a polymer. With appropriate choice of materials and sample geometry the modulation of light can be achieved by the application of an electric field across the film. A small effect has already been shown for a monolayer coupled to a surface plasmon in a silver overlayer[8] but a much greater degree of modulation needs to be attained to produce an effective device.

The development of devices also requires the ability to produce a variety of device architectures either during production or once the non-linear system has been fabricated. Langmuir-Blodgett films of cross-linkable polymers have already been shown to act as electron beam resists[9]. Here we show that such polymers can be incorporated into an alternating layer structure along with chromophore-containing monomers while still maintaining the ability to cross-link and stabilize the film when developed. Also, the generation of a second harmonic from alternating films containing merocyanine and azo-compounds, some of which have perfluorinated chains, has been studied and is reported here.

2. EXPERIMENTAL METHOD

Materials

The compounds discussed in this paper are shown in Figure 1. The merocyanine dye, (I)[10] is shown in its more polarizable deprotonated form.

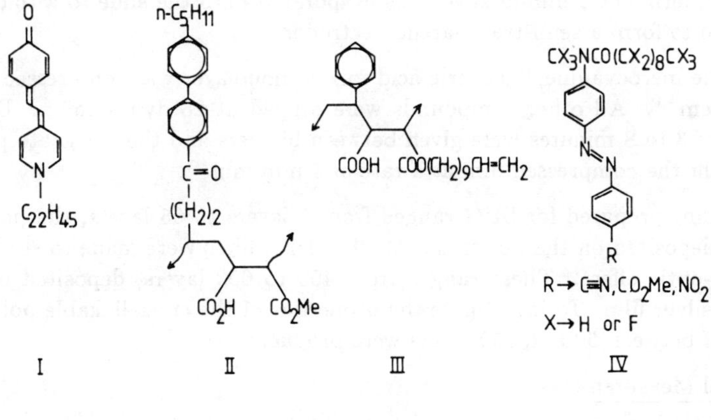

Figure 1. Materials. I Merocyanine. II Preformed polymer. III Cross-linkable polymer. IV Azobenzene compounds.

The preformed polymers (II), (III) are copolymers of vinyl derivatives and maleic anhydride, the ring of the latter being opened by reaction with an alcohol. The azobenzene compounds used for SHG generation are also shown (IV).

Preparation of Multilayers

Film deposition was performed using a double LB trough described elsewhere[12]. This was semi-automated using a BBC computer which provided control of spreading, compression of the monolayer, maintainance of surface pressure and the dipping of the substrate. Temperature and pH were monitored and controlled manually. The trough was contained within a sealed glovebox through which filtered air was circulated through a closed loop, which was passed through a domestic freezer to provide control of the water vapour content of the air.

The preformed polymer compounds and the azobenzene monomers were dissolved in ethyl acetate to a concentration of 0.2 mg/ml. The merocyanine dye was mixed with lignoceric acid to provide an equimolar mixed solution in chloroform, again at 0.2 mg/ml. Solutions were filtered through a PTFE filter of mesh 0.2 μm immediately before use. Double distilled deionized water was used for the subphase, NaOH being added to raise the pH if appropriate.

Substrates for SHG measurements were glass microscope slides, for the electro-optic experiment high refractive index glass was used while oxidized silicon provided the base for experiments on the cross-linkable polymer. The substrates were cleaned in boiling alcohol and finally placed in an alkaline etch and washed in distilled water to provide a hydrophilic surface. For the electro-optic experiment a film of silver was evaporated onto the slide to a thickness of 8 nm to form a semi-transparent electrode.

The merocyanine/lignoceric acid mixed monolayer was compressed to 35 dynes cm^{-1}. All other compounds were dipped at 28 dynes cm^{-1}. Drying times of 3 to 8 minutes were given between bilayers and the substrate passed throught the compressed film at a rate of 4 mm min^{-1}.

Films prepared for SHG ranged from 5 layers to 65 layers, the polymer being deposited on the up stroke. Much thicker films were made to study the electro-optic effect. These ranged from 400 to 600 layers, deposited on top of the silver film. To investigate the properties of the cross-linkable polymers films of between 50 and 150 layers were produced.

Optical Measurements

Measurement of second harmonic generation was made using a Nd:YAG

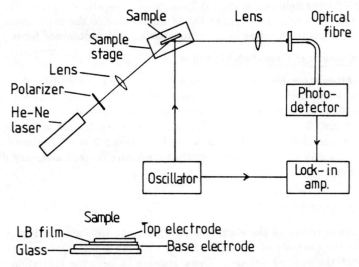

Figure 2. Schematic representation of optical equipment and sample electrodes for the electro-optic experiment.

pulsed laser operating at a wavelength of 1.064 μm and with a pulse width of 25 ns (FWHM), the light being incident at 45°. The second harmonic signal of p-polarized light was measured for transmitted light also of p-polarization.

The experimental arrangement for the measurement of the electro-optic effect is shown in Figure 2 together with an enlarged view of the structure of the sample. A 5 mW helium-neon laser operating at a wavelength of 632 nm was used as a light source. The sample was mounted on a stage allowing rotational and translational movement. Incident p-polarized light passed through the slide, the semi-transparent base electrode and the LB films and was reflected back from the thick top electrode. This top electrode was a 35 nm thick gold film evaporated onto the top of the LB film. The light reflected from the sample was detected using an optical fibre connected to a photodiode, the fibre being mounted on precision translation stages providing x-y movement. The optical equipment was mounted on an optical table with pneumatic anti-vibration mounts. An oscillator was used to provide an alternating electric field across the LB film and was also connected to a Brookdeal lock-in amplifier to provide a reference frequency for the input to the amplifier from the photo-detector. The optical fibre was positioned so that it was on the edge of the reflected spot allowing any change in the position or intensity of the spot to be more easily detected. The lock-in amplifier was used to detect any change in the intensity of the light collected by the optical fibre which was related to the frequency of the driving field across the LB films supplied by the oscillator. The sample was placed in ammonia vapour for 24 hours prior to the experiment so that the merocyanine dye was in its more polarizable deprotonated form.

Development of Cross-linkable Films

Alternating films of the cross-linkable polymer were dipped against an azo-benzene perfluorinated monomer. The samples were exposed to ultra-violet light through a mask, the peak intensity of the radiation being 9000 μW at 254 nm. The samples were exposed for between 10 and 20 minutes. The sample was then placed in chloroform for between 2 and 10 minutes. It was then rinsed in a 1:1 chloroform/methanol mixture to wash away any dissolved material.

3. RESULTS

Some results of the electro-optic experiment are shown in Figure 3. This shows the intensity of the light varying with the applied voltage, ΔI, plotted against the applied voltage. There appears to be some indication that the effect may be reaching a condition of saturation towards the larger values of applied field. This graph shows the situation for an angle of incidence of 56^0. The effect varied with angle of incidence. No effect could be detected for angles of incidence less than 40^0 and the largest signals being recorded at angles between 50^0 and 60^0.

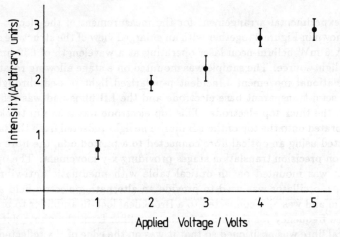

Figure 3. A graph showing the variation of the field dependent intensity with the applied voltage.

Despite the comparative thickness of the films short circuits did appear when a field has been applied for more than about 10 minutes. However, the film recovered if left for about 24 hours although, thereafter, breakdown did occur more easily.

The cross-linkable polymer, when dipped against itself, produced a well defined pattern when developed after 10 minutes exposure to the ultra-violet lamp. The unexposed film was cleaned off revealing the silicon surface. When dipped against an azobenzene monomer the polymer deposited well with a deposition ratio of ~ 1.02, if dipped on the upstroke but less well, with a deposition ratio of ~ 0.9 if dipped on the down stroke. After exposure to ultra violet light and subsequent development the alternating layer film showed a well defined pattern corresponding to the design of the mask. However, the unexposed film was not removed back to the silicon surface. There appeared to be no deterioration of the exposed film when developed in chloroform for less than five minutes. If left in the chloroform for longer than this, however, the edges of the exposed areas could be seen to have degraded when viewed under an optical microscope.

Previous studies of SHG on merocyanine dye layers have shown it to be a good producer of second harmonic light[3,6], even when many layers thick. Although one would also expect them to be highly polarizable, none of the azobenzene compounds studied here showed a comparable ability of second harmonic production.

4. CONCLUSIONS

We have shown that it is possible to incorporate cross-linkable polymers into an alternating LB structure with chromophore-containing monomers. The film is still capable of being cross-linked by ultra-violet light and maintaining its structure when developed in organic solvents. With some refinement this technique would allow waveguide or other device architectures to be drawn into non-centrosymmetric films containing optically interesting molecules.

Films made using monomers with a highly polarizable azobenzene group although showing some second harmonic generation were not as good as films of merocyanine or hemicyanine. This may, in part, be due to the packing arrangement of the non-centrosymmetric system leading to an unfavourable geometry.

Modulation of light by a bulk electro-optic effect (as opposed to surface polarons) in LB films has been demonstrated for the first time. By improving the geometry of the apparatus and use of hemicyanine instead of merocyanine we hope to produce higher levels of modulation and more reliable devices.

ACKNOWLEDGEMENTS

We would like to thank P.V. Kolinsky and R.J. Jones of G.E.C. Hirst Research Centre for performing the SHG measurements and I.R. Girling, also of G.E.C. Hirst Research Centre, for the synthesis of merocyanine. The work was funded by SERC.

REFERENCES

1. R.H. Tredgold, Reports on Progress in Physics, 1987, 50, 1609.

2. R.H. Tredgold, S. Evans, P. Hodge and A. Hoorfar, presented at the 1987 International Conference on Langmuir-Blodgett Films, to be published in Thin Solid Films.

3. I.R. Girling, P.V. Kolinsky, N.A. Cade, J.D. Earls and I.R. Peterson, Optics Communs., 1985, 55, 289.

4. L.M. Hayden, S.T. Kowel and M.P. Srinivasan, Optics Commun., 1987 61, 351.

5. P. Stroeve, M.P. Srinivasan and B.G. Higgins, Thin Solid Films, 1987, 146, 209.

6. R.H. Tredgold, M.C.J. Young, R. Jones, P. Hodge, P. Kolinsky and R.J. Jones, Electron. Lett., 1988, 24, 308.

7. R.H. Tredgold, M.C.J. Young, P. Hodge, E. Khoshdel, Thin Solid Films, 1987, 151, 441.

8. G.H. Cross, I.R. Girling, I.R. Peterson and N.A. Cade, Electron. Lett., 1986, 22, 1111.

9. R. Jones, C.S. Winter, R.H. Tredgold, P. Hodge and A. Hoorfar, Polymer, 1987, 28, 1619.

10. I.R. Girling, N.A. Cade, P.V. Kolinsky, C.M. Montgomery, Electron. Lett., 1985, 21, 169.

11. A.J. Vickers, R.H. Tredgold, P. Hodge, E. Khoshdel and I.R. Girling, Thin Solid Films, 1985, 134 43.

Second Harmonic Generation from Langmuir–Blodgett Films of Azo Dyes

L.S. Miller and P.J. Travers

APPLIED PHYSICAL SCIENCES, COVENTRY POLYTECHNIC, PRIORY STREET, COVENTRY
CV I 5FB, UK

R.S. Sethi and M.J. Goodwin

PLESSEY RESEARCH CASWELL LIMITED, CASWELL, TOWCESTER, NORTHAMPTONSHIRE
NN I2 8EQ, UK

R.M. Marsden, G.W. Gray, and R.M. Scrowston

SCHOOL OF CHEMISTRY, UNIVERSITY OF HULL, HULL HU6 7RX, UK

1. INTRODUCTION

Organic materials are attractive for nonlinear optical
devices due to their potentially large, high-frequency
second and third order optical nonlinear properties[1-3]. For
second-order nonlinear effects, such as frequency doubling
and parametric amplification, ordered (non-centrosymmetric)
material structures are required. Langmuir-Blodgett (LB)
multilayer deposition is an attractive technique, enabling
the fabrication of well controlled, ordered thin films[4,5],
potentially suitable for integrated waveguide and thin film
optical devices. However, certain molecular design criteria
have to be met to exploit this approach successfully. The
molecular structures of most of the commonly reported nonlinear organic materials[1-3] are unsuitable for deposition
by the LB technique and molecular modifications are therefore necessary before they can be used[1]. Several azo dyes
have been used to grow LB films and have been found to
exhibit relatively large optical non-linearities[6,7,8]. We
report here the results on LB films fabricated from some
new azo dyes. These were characterised using second harmonic generation to determine their second-order optical
nonlinear properties.

2. EXPERIMENTAL

Two amphiphilic amino azo dye derivatives (1) and (2)
(Figure 1) were studied as possible LB film-forming materials.

Deposition was carried out at a surface pressure of
20 mN m^{-1} and a dipping speed of 2 mm min^{-1}, using our own
design of trough[9]. For SHG measurements

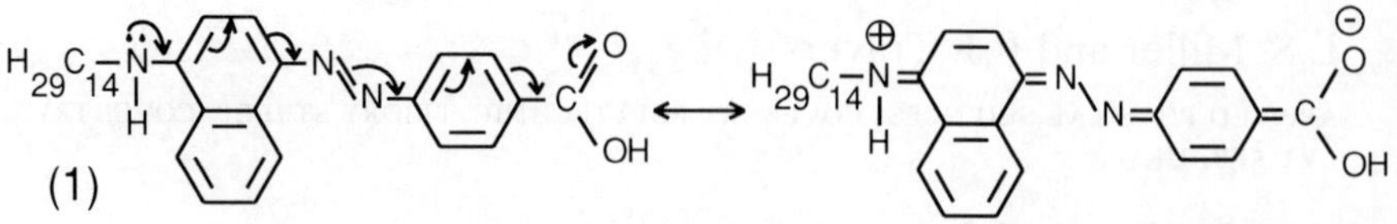

Figure 1   Ground and resonant excited states of (1) and (2)

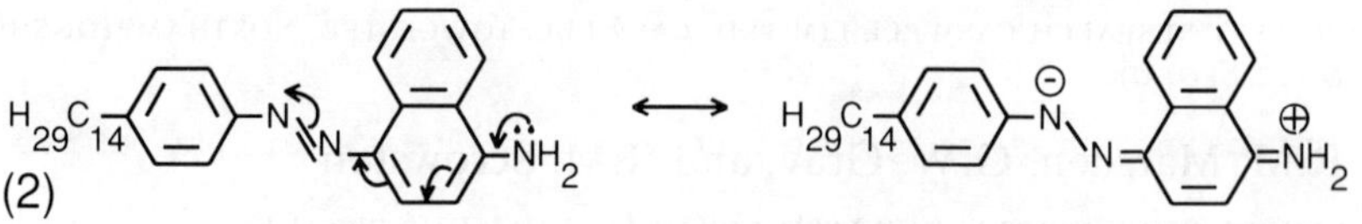

LB films were deposited onto glass microscope slides (Blue
Star, Chance Propper Ltd.) which had been rendered either
hydrophilic or hydrophobic.

The second order non-linearities were investigated by
measuring the efficiency for second harmonic generation
(SHG)[10]. The pump pulses, at a wavelength of 1.064 μm were
provided by a passively Q-switched Nd:YAG (Neodymium:
Yttrium Aluminium Garnet) laser producing temporaly smooth
pulses of 30 ns duration. The optical system enabled
plane-polarised light to be focussed on the samples; the
plane of polarisation could be selected. The samples were
held at 45° to the incident beam, and the reflected light
monitored using an S1 photocathode photomultiplier tube
(PMT). The pump wavelength was attenuated with IR absorbing
filters and a narrow band-pass dielectric filter with
maximum transmission at a wavelength of 530 nm.

3. RESULTS AND DISCUSSION

Isotherms and Langmuir monolayer stability

The isotherms of (1) and (2) were measured at a compression rate of 0.05 nm^2 mol^{-1} min^{-1} and are shown in
Figure 2(a). At surface pressures greater than 15 mN m^{-1}
the isotherms measured at a pH of 5.5 tend to be of approximately the same area per molecule (0.26 nm^2). This agrees
well with the cross sectional area of the double benzene
(naphthalene) ring (0.28 nm^2) as estimated from molecular
models, indicating that at high surface pressures the long
axis of the planar chromophore is oriented perpendicularly
or almost perpendicularly to the water surface.

The stability of the monolayers when held at a surface pressure of 20 mN m^{-1} is shown in Figure 2(b). Monolayers of (1) and (2) on a 'neutral' subphase (pH 5.5) show an initial decrease in area with time probably due to molecular rearrangement. In order that meaningful deposition ratios could be measured, films of (1) and (2) were maintained at a pressure of 20 mN m^{-1} for 10 and 90 min respectively before beginning deposition.

Deposition of (2) from a subphase acidified to a pH of 2.25 (with sulphuric acid) was started after the monolayer had been held at 20 mN m^{-1} for 20 min and terminated before reaching the 'collapse' region of the stability curve.

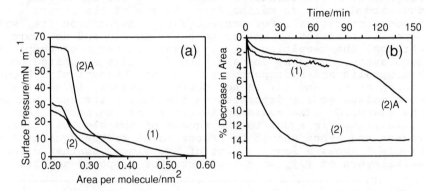

Figure 2 Isotherms (a) and surface film stability (b); (1) and (2): pH = 5.5; (2)A: dye (2) at a pH of 2.25

Multilayer deposition

Multilayers of (1) and (2) were successfully deposited onto glass substrates from a subphase with pH of 5.5. Y-type deposition was observed with typical deposition ratios of 0.9-1.0 on upward strokes and 0.7-0.8 on downward strokes. Attempts were also made to deposit (2) from a subphase acidified to a pH of 2.25 (with sulphuric acid). Monolayers were grown successfully but multilayer deposition was very poor as judged by the recorded deposition ratios and the patchy appearance of the films.

Second harmonic generation

Clear SHG signals were observed from active samples with pump pulse signal energies of a few hundred μJ. These were confirmed to be SHG by varying the centre frequency of

the narrow band filter away from 530 nm, whereupon no
signal was detected. No uncoated areas of any of the samples gave a signal; it was concluded that the SHG was a
property of the deposited films. Only an approximately
quadratic dependance of SHG intensity on incident pulse
intensity was observed for both the azo dyes and the hemicyanine reference[11].

For these thin samples, the contributions to SHG are
all essentially in phase. The resulting amplitude of the
electric vector of the second harmonic should then be the
algebraic sum of the contributions from the individual
molecules and hence be proportional to the net monolayer
alignment, m (calculated by subtracting the sum of the
downward deposition ratios from the sum of the upward
deposition ratios). For perfect Y-type deposition (ie. with
deposition ratios of one on both the upward and downward
strokes) the resulting net alignment, in terms of complete
monolayers, would be one and zero for films deposited on
hydrophilic and hydrophobic substrates respectively. Deposition of (1) and (2) is not perfect however; in general
more molecules are deposited on the upward strokes than on
the downward. The measured intensity of the second harmonic signal is equal to the square of the electric field
vector and hence should be proportional to the square of m.
Figure 3 shows a log-log plot of SHG signal against m for
multilayers of (1), which supports the proposed model.

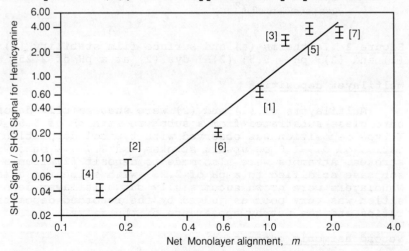

Figure 3   Log-log plot of SHG intensity for (1) relative
to hemicyanine. Line: signal $\propto m^2$.
The numbers [ ] indicate the number of layers.

The SHG efficiency of the monolayer relative to the hemicyanine standard (second order molecular polarisability, ß = 9 x 10^{-49} C^3 m^3 J^{-2}) was measured as 0.64. The ratio of the ß -values may be assumed to be in the square root of this ratio, provided the refractive indices of (1) and the hemicyanine are essentially the same[2]. This gives an approximate value of ß for (1) of 7 x 10^{-49} C^3 m^3 J^{-2}.

A study was also made of the polarisation of the second harmonic. Both s and p polarised incident radiation gave rise to predominantly p-polarised SHG signals. This indicates that the polarisation field that drives the SHG is aligned perpendicularly to the substrate[10] (i.e. along the long axis of the chromophore).

The relative SHG efficiency of a monolayer of (2) deposited at a pH of 5.5 is 0.12 which gives a value for ß of 3 x 10^{-49} C^3 m^3 J^{-2}. The results so far obtained from multilayers of (2) deposited at a pH of 5.5 do not fit the suggested model. Any signal produced by the centrosymmetric bilayer is below the sensitivity of the detection system and this is consistent with the recorded deposition ratios which give a net alignment of only 0.06. The signals produced by the five and seven layer samples however are much lower than the model predicts.

To date, almost all organics which display significant SHG possess both electron donating and electron accepting substituents separated by a conjugated system (e.g. merocyanine, 4-methylnitroanaline).

The molecule (1) is of this type and possesses a suitable resonant state. Molecule (2) has an amine group to act as the electron donor but no obvious acceptor substituent. It is possible that in this case the azo linkage is acting as the acceptor group (Figure 1).

SHG measurements were also made on a monolayer of (2) deposited from a subphase acidified to pH = 2.25 with sulphuric acid. The measured ß value was 7 x 10^{-49} C^3 m^3 J^{-2}, which means that the SHG intensity was five times greater than that for the monolayer deposited at a pH of 5.5. This may be due to protonation of the azo group which will shift the absorption band of the chromophore towards the SHG frequency, leading to resonant enhancement of the signal. In addition, protonation of the azo linkage may improve its electron accepting properties.

4. CONCLUSIONS

Both the materials reported showed strong SHG comparable to that reported for other azo dyes, with second-order polarisibilities, ß , in the range 3 to 7 x 10^{-49} C^3 m^3 J^{-2}.

The dye (2) which has no obvious acceptor substituent shows weaker SHG than dye (1) that has.

For dye (1) the increase in SHG as the number of layers was increased could be explained in terms of incomplete cancellation by alternate layers.

For dye (2), deposition from an acidic subphase resulted in an increase in SHG by a factor of 5, probably due to protonation, and making its ß-value comparable to that of (1).

REFERENCES

1. J. Zyss, J.Mol.Electron., 1985, 1, 25.
2. D.J. Williams, Agnew.Chem.Int.Ed.Engl., 1984, 23, 690.
3. 'Nonlinear Properties of Organic Polymeric Materials',
    D.J. Williams (ed), ACS Symposium Series 233,
    Washington DC, 1983.
4. G.G. Roberts, Contemp. Phys., 1984, 2, 109.
5. M. Sugi, Thin Solid Films, 1987, 152, 305.
6. O.A. Aktsipetrov, N.N. Akhemdiev E.D. Mishina and
    V.R. Novak, JETP Lett., 1983, 37, 207.
7. I. Ledoux, D. Josse, P. Vidakovic, J. Zyss, R.A. Hann,
    P.F. Gordon, B.D. Bothwell, S.K. Gupta, S. Allen,
    P. Robin, E. Chastaing, and J.C. Dubois,
    Europhys.Lett., 1987, 3, 803.
8. I. Ledoux, D. Josse, P. Fremaux, G. Post, J. Zyss,
    T. McLean, R.A. Hann, P.F. Gordon and S. Allen, Thin
    Solid Films, 1988, 158, (pages unknown). [Presented at
    the Third Int. Conf. on LB Films, Gottingen, F.R.G.,
    July 1987].
9. L.S. Miller, D.E. Hookes, P.J. Travers, and
    A.P. Murphy, J.Phys.E.:Sci.Inst., 1988, 21, 163.
10. I.R. Girling, N.A. Cade, P.V. Kolinsky and
    C.M. Montgomery, Electron.Lett., 1985, 21, 169.
11. I.R. Girling, N.A. Cade, P.V. Kolinsky, J.D. Earls,
    G.H. Cross and I.R. Peterson, Thin Solid Films, 1985,
    132, 101.

Devices

Review of Non-linear Integrated Optics Devices

G.I. Stegeman*, R. Zanoni, E.M. Wright, N. Finlayson,
K. Rochford, J. Ehrlich, and C.T. Seaton

OPTICAL SCIENCES CENTER, UNIVERSITY OF ARIZONA, TUCSON, AZ 85721, USA

Summary: A very promising application of organic materials with large nonresonant third-order nonlinearities is to nonlinear guided-wave devices. Here we review the types of devices that are possible, and summarize the progress to date with organics and other materials in implementing them.

1 INTRODUCTION

One approach to increasing the bandwidths available for the processing of serial streams of optical information is to perform the signal processing in the all-optical domain without recourse to electronics. The goal ultimately is to manipulate data on a subpicosecond time scale, and to perform various switching and logic operations. Similar functions are currently being implemented with linear guided-wave devices, usually controlled electro-optically. Essentially all of these devices can be made to operate in the all-optical domain by using waveguide materials with intensity-dependent refractive indices.

All-optical waveguide signal processing is currently evolving in two directions. Device concepts and their feasibility are being tested in a variety of configurations, both fiber[1-3] and integrated optics,[4-8] usually with non-optimum materials. Simultaneously, the figures of merit of the various devices and their material implications are being assessed.[9-11] Because of their prior history in bistable etalon devices, semiconductors were among the first materials to be used for nonlinear guided-wave devices.[4-7] Similarly, glass fibers, despite their very small nonlinearities, have also been used recently because of their availability, classical Kerr-law response, and the long low-loss propagation distances they allow.[1-3] Because nonlinear organic materials are recent entrants into this field, very few experiments have been reported to date.[10,12-16] However, the potential figures of merit for these materials appear very promising, and increased activity can be expected in the near future.[9,10]

In this paper we will summarize progress in various areas of nonlinear guided-wave optics specifically for signal processing. We start with a general summary of the possible devices and analyze one particular case to rationalize the figure of merit, which we will apply to assessing potential materials. Next, the implementation of nonlinear guided-wave devices in a number of material systems will be summarized and the prospective impact of nonlinear polymeric materials noted.

2 DEVICES AND FIGURES OF MERIT

A number of standard integrated-optics devices are summarized schematically in Figure 1. When one or more of the waveguiding media exhibit an intensity-dependent refractive index, in each case the response of the device depends on the incident beam power. Of these potential devices, the ones with the sharpest switching characteristics are the nonlinear directional coupler and its analogs, and the nonlinear distributed feedback grating. A number of these have already been implemented in material systems other than nonlinear organics,[1-8] and will be discussed later. For the moment we will concentrate on the nonlinear directional coupler to identify the figures of merit that determine desirable material characteristics.[9,10,17-19]

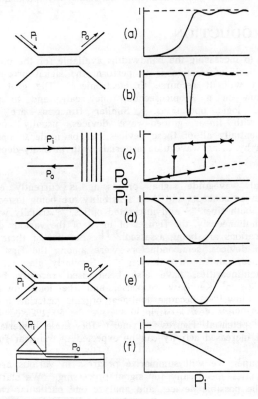

Figure 1   Standard integrated-optics devices and their response to optical power with and without nonlinearities: (a) half beat length directional coupler;   (b) 1 beat length directional coupler; (c) distributed feedback grating; (d) Mach-Zehnder interferometer; (e) mode sorter; and (f) prism coupler.

A nonlinear directional coupler consists of two parallel, identical (equal propagation wavevectors) waveguides that are weakly coupled by means of the overlap of their fields in space.[17] If light is injected into one arm of the device, it progressively couples with propagation distance across to the second channel, in close analogy to a coupled pendulum system. In a half beat length device, the device is terminated just after all of the light couples across to the cross channel at low powers. For the nonlinear version with uniform nonlinearity everywhere, injecting light into one of the channels changes the refractive index locally in that channel and breaks the degeneracy between the propagation constants of the two waveguides.[17] This decreases the coupling rate and increases the beat length, and hence the transmission in the incidence channel increases with increasing power for a half beat length device. If the nonlinearity is of the classic Kerr-law type, that is $n = n_0 + n_2 I$ where I is the local intensity, the nonlinear directional-coupler response is given in Figure 2. The critical power (at which the incident power is split 50/50 between the channels) is given by[17]

$$P_c \cong \frac{\Delta\phi^{NL} A}{k_0 L n_2} ,$$ (1)

where $\Delta\phi^{NL}$ is the minimum required phase shift available in one channel (in the absence of coupling to the second channel), L is the half beat length, and A is the effective channel area. For the nonlinear directional coupler, $\Delta\phi^{NL} = 4\pi$.[18,19] For a resonant nonlinearity, for example that associated with a two-level transition,[18]

$$\Delta n = - \frac{\Delta n_{sat}}{1 + I/I_{sat}} ,$$ (2)

where Δn_{sat} is the maximum optically induced change in refractive index, the response of the device can deviate significantly from that in shown Figure 2. We now define a dimensionless parameter

$$w = \Delta n_{sat} k_0 L / 2\pi ,$$ (3)

which, as shown in Figure 3, determines the switching behavior of this nonlinear device.[18,19] For 100% switching, w > 2, which corresponds to the $\Delta\phi^{NL} = 4\pi$ mentioned previously. For w < 1, negligible switching is obtained. Note that in principle it is possible to increase w by increasing L by reducing the coupling between the channels. However, L is usually limited by loss considerations, that is, it is desirable to have $\alpha L < 0.1$ to obtain at least 80% throughput. Taking this into account, we write[9,10]

$$W = w/\alpha L ,$$ (4)

which is now independent of device details. Although w determines device feasibility, W is a better measure of device performance, since device feasibility can always be improved at the expense of throughput. Another important parameter is n_2, since the switching power is inversely proportional to this coefficient. The relaxation (turn-off) time τ of the nonlinearity determines the maximum processing speed for "pipeline" processors since the index change attributable to one pulse must be very small by the time that the next optical pulse arrives.

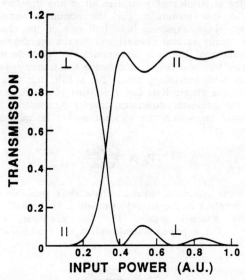

<u>Figure 2</u>   The theoretical fraction of output power in each of the output channels for a one-half beat length coupler versus incident power in the incidence (parallel) channel.

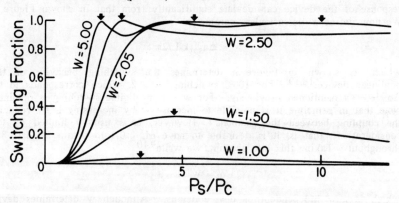

<u>Figure 3</u>   The fraction of power switched in a nonlinear directional coupler as a function of the normalized saturation parameter w.

The minimum phase shift required to produce 100% switching varies from device to device. Table I summarizes both the $\Delta\phi^{NL}$ and W required for 100% switching with >80% throughput.[9,10] Clearly the nonlinear distributed feedback grating requires the smallest nonlinear phase shift and consequently the lowest switching power. Its disadvantage, however, is that it is not a "pipeline processor," since multiple pulses cannot exist independently within the grating because of the required multireflection effects. Hence the picosecond transit time through the grating limits the processing speed.

Table I   The minimum nonlinear phase shift, $\Delta\theta^{NL}$, and the minimum dimensionless material parameter W (> 80% transmission) required for various nonlinear guided-wave devices.

Nonlinear Device	$\Delta\theta^{NL}$	W
Directional Coupler 1/2 beat length	4π	10
Directional Coupler 1 beat length	$\cong 3.3\pi$	8
Mach–Zehnder Interferometer	2π	5
Distributed-Feedback Grating	π	2.5

2.3 Nonlinear Gratings

The values of $\Delta\phi^{NL}$ required should be contrasted with those required for switching bistable etalons, typically $\pi/4$. Hence materials useful for bistable etalons may not be useful for nonlinear guided-wave applications. Because Δn_{sat} plays a very important role here, the usual nonlinear optics figure of merit $n_2/\alpha\tau$ is of limited value, where τ is the nonlinearity recovery time.

A summary of the pertinent parameters for a number of material systems is given in Table II.[20] Certain things stand out immediately. The lowest switching powers are available from semiconductors because of their large nonlinearities. However, at present only the organics and the glasses have subpicosecond response times. Furthermore, W for organics and glasses are predicted to be large enough for efficient switching with high throughputs. (Note that the material systems limits have not been reached in either semiconductors or nonlinear organics to date, as indicated in the table.) The

Table II   Typical nonlinear waveguide parameters for various classes of
materials. Determination of the parameter W requires knowledge of
all the listed values under the same operating conditions.

Waveguide material	$n_2(M^2/w)$	$\tau(sec)$	$\alpha(cm^{-1})$	Δn_{sat}	W
(a) SEMICONDUCTORS; e.g., MQW GaAs/GaAlAs					
on-resonance	-10^{-8}	10^{-8}	10^4	0.1	~0.1
off-resonance	-10^{-12}	10^{-8}	30	$\sim 2\times10^{-3}$	~0.9
theory	-10^{-13}	10^{-8}	10	0.01	~10
CdSSe-doped glasses	-10^{-14}	10^{-11}	~3	5×10^{-5}	~0.3
(b) ORGANICS; e.g., PTS					
on-resonance	2×10^{-15}	$\sim 2\times10^{-12}$	10^4	~0.1	~0.2
off-resonance	10^{-16}	$<0.03\times10^{-12}$	$<10^2$	$>10^{-3}$	>0.15
theory	10^{-16}		0.1	$>10^{-3}$	>100
(c) GLASSES;					
SiO$_2$	10^{-20}	$<0.1\times10^{-12}$	10^{-5}	$>10^{-7}$	>20
Pb-doped	10^{-18}	?	?	?	?

organics have large enough nonresonant nonlinearities to provide switching in
channel devices with watt peak powers. New glasses will have to be
developed for fibers to become serious contenders as compact all-optical
devices. Based on this table, the future for nonlinear organics in this
applications area looks very promising. Note, however, that n_2, α, and Δn_{sat}
are all equally important parameters and should all be measured in assessing
material suitability.

3 NONLINEAR DIRECTIONAL COUPLER DEVICES

This type of device has been described in some detail in the preceding section.
It has been implemented in a number of different formats and material systems.
The first demonstration was in multiple-quantum-well (MQW) GaAlAs strain-
induced waveguides,[4] sufficiently detuned from the bandgap to obtain useful
throughput with devices a few hundred micrometers long. About 10%

switching was observed with cw laser excitation and it is still not clear whether the nonlinear mechanism was thermal or electronic. More recently, there have been two reports of coupled waveguides in which the medium between the two waveguides was MQW material and the operating wavelength was near the excitonic nonlinearity.[5,6] Although impressive switching was obtained in one of the cases, again cw sources were used and it is impossible to identify whether the nonlinear mechanism was electronic or thermal.

All-optical switching in an integrated-optics nonlinear directional coupler with picosecond response time has been recently obtained in semiconductor-doped glass waveguides.[7] The nonlinear material consists of typically 100-Å crystallites of CdSSe and the nonlinearity is attributable to band-filling, that is, the excitation of carriers from the valence to conduction bands fills the bottom of the conduction band effectively shifting the band gap and causing a change in the refractive index. The measured response time is $\cong 10$ ps because the carrier recombination rate is enhanced by the small size of the crystallites. The change in the throughput of the two channels is shown in Figure 4. This

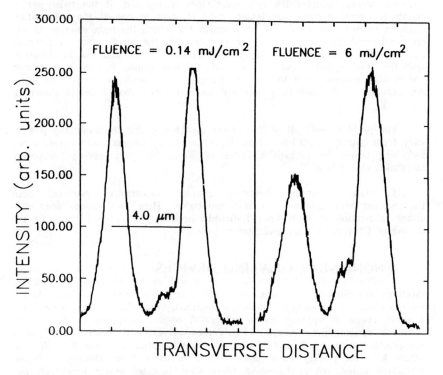

Figure 4   Light distributions at the output ports of an integrated-optics nonlinear directional coupler fabricated in semiconductor glass. Note the change in power with fluence.

switching is the result of saturation of the absorptive part of the electronic nonlinearity and is the first demonstration of all-optical switching on a picosecond time scale in an integrated-optics device.

The operating characteristics of a dual-directional coupler have also been demonstrated in specialty optical fibers. Switching of optical pulses has been measured between the two cores of a dual-core optical fiber has been reported.[2] In this case the two "channels" are the cylindrical core regions placed in close proximity so that their guided wavefields overlap.

Switching between polarizations in birefringent fibers can also be implemented all-optically with response characteristics identical to those of a nonlinear directional coupler.[1,3] In this case, the two channels correspond to the two polarization eigenmodes of the fiber. Two experiments based on this phenomenon have been reported. The first consisted of switching between right and left-circularized polarized light in a weakly birefringent fiber.[1] In the second case, a fiber filter was used.[3] Periodic twisting of the fiber during the drawing process couples the two polarization states and, if the twist period equals the birefringence beat length, light incident in one of the polarization states is rotated and there is a fiber length for which the light emerges in the cross-polarized state. If the power is increased, the birefingence beat length changes due to the intensity-dependent refractive index and the synchronism with the twist period is lost. As a result, the polarization rotation is frustrated and at high powers all of the power emerges in the incidence polarization state. An example of the switching between the two polarization states is shown in Figure 5.

The problem with all of these fiber switches is that the switching powers vary from 500 W to 50 kW. New highly nonlinear glasses could decrease the switching power by perhaps a factor of 50 to 100, still leaving the power thresholds rather large.

To date, nonlinear directional coupler phenomena have not been implemented with nonlinear organic materials. Here two options are open, either fabrication of such devices directly in integrated-optics formats, or the doping of fiber cores with nonlinear polymers.

4 NONLINEAR GRATING DEVICES

Gratings are periodic structures that reflect or deflect optical beams when a wavector-matching condition, the Bragg condition, is satisfied. They have found a wide range of applications in integrated optics, especially to distributed-feedback semiconductor lasers. For guided waves with incident and guided wavevectors given by β_i and β_r, the reflection is optimized when $\beta_r = \beta_i + K$, where K is the grating wavevector. For materials with an intensity-dependent refractive index, the guided-wave wavevector becomes power-dependent, i.e., $\beta \rightarrow \beta(P)$. Therefore the Bragg condition can be tuned optically, for example in one of the ways indicated in Figure 6.

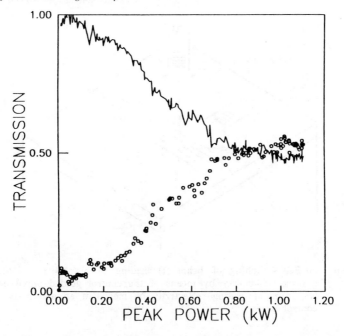

PEAK POWER (kW)

Figure 5   The dependence of measured transmission in the parallel (dots) and crossed (solid line) polarization states on pulse peak power at the resonance wavelength.

Shown in Figure 1(c) is the response of a nonlinear distributed-feedback grating, NDFBG, for which bistability as well as switching is possible.[21,22] The material requirements for this particular device are summarized in Table I. Recently a NDFBG has been implemented in InSb waveguides by means of a thermal nonlinearity excited by various lines from a CO_2 laser. A distributed feedback grating was fabricated into the film surface using a combination of holographic and ion-milling techniques. The grating transmission on resonance was essentially zero. A suitable wavelength was chosen so that the grating reflectivity was 50% at low powers. The variation in the transmission as a function of guided wave power is shown in Figure 7. Clearly the incident guided wave is tuning the reflectivity of the grating, a phenomenon that can be used for all-optial switching. This device is still under investigation.

The grating switching configuration shown in Figure 6c has also been implemented using thermal nonlinearities in a polymeric waveguide. A distributed feedback grating was first fabricated into a substrate and then a PMMA film was spun on using standard techniques.[23] The center wavelength of the grating was $\cong 0.6$ μm, and the grating response was measured as a function of wavelength using a cw dye laser, see Figure 8. When the grating was illuminated from above with a CO_2 laser, the center wavelength of the

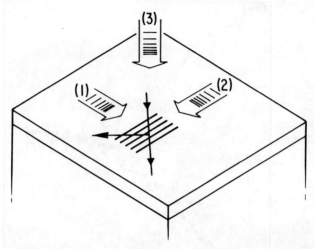

<u>Figure 6</u>   (a) Self-switching of beam (1) leading to bistability; (b) tuning of
grating reflectivity by means of the control guided wave beam (2);
(c) tuning of grating reflectivity by means of an externally incident
control beam.

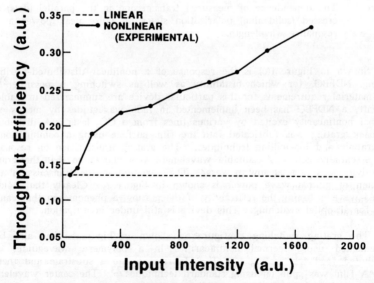

<u>Figure 7</u>   Change in the throughput of a guided wave incident onto a
distributed feedback grating fabricated on an InSb waveguide. The
geometry of Figure 6a is used.

grating response is shifted, as shown in Figure 8. By setting the guided-wave wavelength to be 50% transmitting at low powers, and zero at high power, all-optical modulation of the guided wave was demonstrated by pulsing the CO_2 laser.

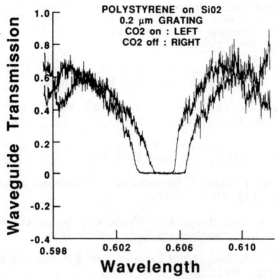

Figure 8   The guided wave power transmitted through a distributed feedback grating for a polystyrene waveguide as a function of wavelength when the CO_2 laser is off, and when the CO_2 laser is on. The grating is illuminated from above by the CO_2 beam.

5 SUMMARY

A number of all-optical guided-wave switching devices have been implemented over the last few years with a variety of nonlinear materials and nonlinear mechanisms. None of the devices have been even close to suitable for applications. Instead they can be better described as proof of principle demonstrations.

We have shown that when the figures of merit for nonlinear guided-wave devices are considered, nonlinear organic materials show considerable promise.

This research was supported by Hoechst-Celanese and by the Joint Services Optics Program of the Army Research Office and the Air Force Office of Scientific Research.

6 REFERENCES

1. S. Trillo, S. Wabnitz, R.H. Stolen, G. Assanto, C.T. Seaton, and G.I. Stegeman, Appl. Phys. Lett., 1986, 49 1224.

2. D.D. Gusovskii, E.M. Dianov, A.A. Maier, V.B. Neustreuev, E.I. Shklovsii, and I.A. Shcherbakov, Sov. J. Quant. Electron., 1985, 15, 1523 (1985); S.R. Friberg, Y. Silberberg, M.K. Oliver, M.J. Andrejco, M.A. Saifi, and P.W. Smith, Appl. Phys. Lett., 1987, 51, 1135.

3. S. Trillo, S. Wabnitz, W.C. Banyai, N. Finlayson, C.T. Seaton, G.I. Stegeman, and R.H. Stolen, Appl. Phys. Lett., in press

4. P. Li Kam Wa, J.E. Stich, N.J. Mason, J.S. Roberts, and P.N. Robson, Electron. Lett., 1985, 21, 26.

5. M. Cada, B.P. Keyworth, J.M. Glinski, A.J. SpringThorpe, and P. Mandeville, J. Opt. Soc. Am. B, 1988, 5, 462.

6. P.R. Berger, Y. Chen, P. Bhattacharya, J. Pamulapati, and G.C. Vezzoli, Appl. Phys. Lett., 1988, 52, 1125.

7. N. Finlayson, W.C. Banyai, E.M. Wright, C.T. Seaton, G.I. Stegeman, T.J. Cullen, and C.N. Ironside, Appl. Phys. Lett., in press.

8. A. Lattes, H.A. Haus, F.J. Leonberger, and E.P. Ippen, IEEE Journal of Quantum Elec., 1983, QE-19, 1718.

9. G.I. Stegeman, E.M. Wright, N. Finlayson, R. Zanoni, C.T. Seaton, L. Thylen, S. Wabnitz, S. Trillo, and Y. Silberberg, Proceedings of SPIE Conference on Electro-optics, in press.

10. G.I. Stegeman, R. Zanoni, and C.T. Seaton, Mat. Res. Soc. Symp. Proc., 1988, 109, 53.

11. C.N. Ironside, T.J. Cullen, B.S. Bhumbra, J. Bell, W.C. Banyai, N. Finlayson, C.T. Seaton, and G.I. Stegeman, J. Opt.Soc. Amer. B, 1988, 5, 492.

12. G.M. Carter, Y.J. Chen, and S.K. Tripathy, Appl. Phys. Lett., submitted.

13. M.J. Goodwin, R. Glenn, and I. Bennion, Electron. Lett., 1987, 22, 789.

14. B.P. Singh and P. Prasad, J. Opt. Soc. Am. B, 1988, 5, 453.

15. K. Sasaki, K. Fujii, T. Tomioka, and T. Kinoshita, J Opt. Soc. Am., 1988, 5, 457.

16. R. Burzynski, P. Banhu, P. Prasad, R. Zanoni, and G.I. Stegeman, Appl. Phys. Lett., submitted.

17. Stephen M. Jensen, IEEE Journ. Quantum Elec., 1982, QE-18, 1580.

18. S. Trillo, S. Wabnitz, E. Caglioti, and G.I. Stegeman, Optics Commun., 1987, 63, 281.

19. E. Gaglioti, S. Trillo, S. Wabnitz, B. Daino, and G.I. Stegeman, Appl. Phys. Lett., 1987, 51, 293.

20. G.I. Stegeman, R. Zanoni, N. Finlayson, E.M. Wright, and C.T. Seaton, J. of Lightwave Technology, 1988, 6, 953.

21. H.G. Winful, J.H. Marburge,r and E. Garmire, Appl. Phys. Lett., 1979, 35, 379.

22. C.T. Seaton, G.I. Stegeman and H.G. Winful, Opt. Engin., 1985, 24, 593; H.G. Winful and G.I. Stegeman, Proceedings of First Int. Conf. on Integrated Optical Eng., SPIE 1984, 517, 214.

23. R. Moshrefzadeh, X. Mai, C.T. Seaton, and G.I. Stegeman, Appl. Optics, 1987, 26, 2501.

Advances in Organic Integrated Optical Devices

R. Lytel*, G.F. Lipscomb, M. Stiller, J.I. Thackara, and A.J. Ticknor

RESEARCH AND DEVELOPMENT DIVISION, LOCKHEED MISSILES AND SPACE COMPANY, INC., D-9720, B-202, 3251 HANOVER ST., PALO ALTO, CA 94304, USA

1 INTRODUCTION

The synthesis of glassy polymer films containing molecular units with large nonlinear polarizabilities has led to the rapid implementation of organic integrated optics[1-5]. These films, either spun, cast, or dipped, are amorphous as produced, and can be processed to achieve a macroscopic alignment for the generation of second-order nonlinear optical effects by electric-field poling. Typically, the films are poled[6,7] by forming an electrode-polymer-electrode sandwich, and applying an electric field normal to the film surface. This produces films with their nonlinear molecular units oriented normal to the film. In this state, the electric field of an optical beam propagating through the film can be maximally modified when the field is parallel to the oriented molecular units, that is, when the propagation vector lies in the film plane. As such, the films are ideally suited for guided wave applications. The successful development of guest-host and side-chain polymer systems incorporating molecular units with large nonlinear polarizabilities in this manner has thus led to the availability of organic thin film materials for integrated optics[8-10].

Organic electro-optic materials offer a variety of potential advantages over conventional materials for integrated optical device applications. The major advantages are due to the intrinsic differences in E-O mechanisms in inorganic and organic materials[11-13]. Organics should provide flat E-O response well beyond a GHz, and, indeed, measurements of the E-O coefficients and SHG coefficients of certain poled polymer films show little or no dispersion. Second, the E-O coefficients of poled polymer films can be made nearly as large as LiNbO3. Pure MNA crystals already exhibit larger E-O coefficients[14] than LiNbO3, but it is our sense that practical, organic integrated optical devices will be made with films, not crystals. It is also true that the dielectric constant of poled polymer films is substantially lower than that of LiNbO3, implying smaller RC time constants and wider frequency bandwidths. In this connection, the bandwidth-length product for a typical poled polymer is of order 120 GHz-cm, compared to 10 GHz-cm for LiNbO3. Finally, the processing technology for integrated optical devices based on poled polymer films is relatively straightforward and fast, requiring only moderate temperatures (100-200 degrees C) for poling, and standard semiconductor fabrication equipment for fabrication of layered waveguide structures.

There are some potential drawbacks to polymers for integrated optics, as well. The microwave loss tangents in glassy poled polymers are not yet known, but are expected to be small. However, the materials have low thermal conductivity, which may lead to power dissipation problems in practical devices, as well as produce instabilities of the poled states. Further, the stability of poled states at nominal temperatures is not yet well known. These issues are of a practical nature, and are currently being addressed by us.

In light of the potential benefits of poled polymer films for integrated optical devices discussed above, many research groups in the United States, Europe, and Japan have embarked upon dedicated materials synthesis and device fabrication programs to bring this field to fruition in commercial and military products. This paper reviews our current research toward fabricating electro-optic, organic integrated optical devices. In section 2, we describe our methods for fabricating and poling slab waveguide modulators for the characterization of the glassy polymer films and the demonstration of new fabrication techniques[15]. In section 3, we describe the fabrication of several integrated optical devices, including a Y-branch interferometer, a directional coupler, and a GHz traveling-wave modulator by a new method[16,17], called the selective poling procedure (SPP), for producing active, buried-channel waveguide devices. We conclude with a summary of our work and point to new directions for future research.

We acknowledge the cooperation and support of the Hoechst Celanese Research Division (HCRD), Hoechst Celanese Corporation. The specific materials discussed in this paper were supplied to us by HCRD[8].

2 SLAB WAVEGUIDE DEVICES AND MATERIALS PROPERTIES

Our organic integrated optical devices effort initially focused on the fabrication of simple slab guided wave structures made from polymer films. These structures allow the determination of fabrication techniques for organic devices, as well as direct measurement of Kerr and electro-optic (E-O) coefficients in the waveguide configuration. Figure 1 illustrates a typical slab waveguide device and some of the components used as substrates, electrodes, buffer layers, and the polable, glassy polymers. The polymers used include: 1) guest-host MNA/PMMA films, both poled and unpoled, 2) PC6S, a yellow, pendant side-chain polymer, 3) C-22, a red, pendant side-chain polymer, and 4) HCC-1237, a more active version of C-22. Waveguide structures such as that in the figure were built up by spin-coating the various layers. The polymers were then poled to produce a non-centrosymmetric structure exhibiting a nonzero electro-optic effect.

The poling procedure typically consists of first spin-coating the electrode-coated substrate with the bottom buffer layer and then the active polymer, applying a top electrode, and applying a voltage of order 1 MV/cm to the structure. The entire procedure is monitored in real time to optimize the poling. The top electrode is then removed, and a top buffer layer and new electrode can be applied to the poled structure. Guiding of 830 nm light from a semiconductor laser over a 1-3 cm dimension with minimal loss is then achieved by locating the prism couplers directly over the ends of the poled region. Modulation to MHz frequencies is observable when the device is located on one arm of an external interferometer.

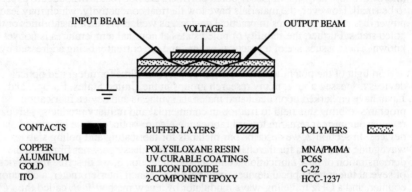

CONTACTS ■	BUFFER LAYERS ▨	POLYMERS ▨
COPPER	POLYSILOXANE RESIN	MNA/PMMA
ALUMINUM	UV CURABLE COATINGS	PC6S
GOLD	SILICON DIOXIDE	C-22
ITO	2-COMPONENT EPOXY	HCC-1237

Figure 1 Typical Slab Waveguide Modulator and Components

Table 1 summarizes the measured physical parameters for PC6S and C-22 slab waveguides. It is most significant that the linear losses in the C-22 guides are below a dB/cm. This allows path lengths of order several cm, and, consequently, half-wave voltages approaching TTL levels. Both materials exhibit even better performance at a wavelength of 1.3 μm.

Table 1 Physical Parameters Measured for PC6S and C-22 Slab Waveguides

(wavelength = 0.83 μm)	PC6S	C-22
E-O COEFFICIENT (pm/V)	2.8	16.0
TM REFRACTIVE INDEX (POLED)	1.7	1.58
TM INDEX DIFFERENCE (POLED-UNPOLED)	0.06	0.005
WG LENGTH (cm)	1.8	2.5
MEASURED LOSS (dB/cm)	2-3	0.8
HALF-WAVE VOLTAGE (volts)	48	7

3 CHANNEL WAVEGUIDE DEVICES

As discussed above, the poling process can be used to create active, electro-optic polymer slab waveguides. In addition to inducing a non-centrosymmetric structure to achieve a macroscopic electro-optic effect, a second major transformation must be engineered in the material to enable the fabrication of integrated optic circuits. Channel waveguides must be formed to confine and guide the light from one active element of the integrated optic circuit to another. The formation of channel waveguides and the poling of the material to produce an active, E-O channel must both be accomplished for device prototypes, and would usually be performed in two distinct steps: fabrication and poling.

We have developed[16] a new, powerful method, called the selective poling procedure (SPP), by which active, poled channel waveguides can be fabricated in a single fabrication step. An electrode pattern defining the channel waveguides is first deposited onto a substrate using standard photolithographic techniques. A planar buffer layer is then applied to optically isolate the active waveguide layer from the metal electrodes. The buffer material must be chosen to have an index lower than the guiding layer and to be compatible with the required processing. Thus, different buffer layers must often be used with different nonlinear polymers. A planar electrode is evaporated directly onto the nonlinear polymer for poling. The nonlinear layer is then poled by applying an electric field above the polymer glass transition and cooling the sample to room temperature under the influence of the field. The degree of alignment induced and the resultant electro-optic coefficient can be calculated based on a statistical average of the molecular susceptibilities. In this case only those regions of the material defined by the electrode pattern on the substrate are poled.

Since most organic nonlinear optical molecules also possess an anisotropic microscopic linear polarizability, the poled region becomes birefringent. The poled regions are uniaxial, with n_e oriented along the direction of the poling field. Consequently, TM and TE waves propagating in vertical and transverse device structures respectively will experience a greater refractive index in the poled regions than in the unpoled regions, and so can be confined in the lateral dimension. Thus, by applying the poling fields using electrodes patterned to define the waveguide network, including both active and passive sections, no further patterning of the organic NLO layer is required to form the channel E-O waveguide structures. The devices are then completed by etching off the planar poling electrode, applying an upper buffer layer and depositing the patterned switching electrodes. If electrode removal is not possible, the device may be finished first, and then be selectively poled through the buffer layer.

The SPP permits both vertical and transverse poling of strips. The vertical method locates the electrodes directly over the guides, implying that parallel channels can be located closer together without having an electrode in the middle of the channels. Vertically poled channels should guide TM radiation, while transverse poled channels should guide TE radiation.

Using the SPP, channel waveguides were constructed with both the PC6S and the C-22 materials, and light guiding was observed in both cases. The index change was measured by determining the prism coupling angles in poled and unpoled guides and by measuring the phase change from a double pass reflection through poled and unpoled regions. PC6S has a large index change of $\Delta n = 0.06$ for TM waves, while the C-22 material exhibited a $\Delta n = 0.005$ for TM waves. These index differences can be fine tuned by adjusting the poling field or by alterations in the nonlinear optical material. In addition to lateral confinement, no birefringence was observed outside the electrode region with a scanned, focused laser beam, indicating that the fringing fields do not cause the guiding region to spread much beyond the electrodes.

In addition to the obvious simplification of the device fabrication process, this technique has several other advantages over other methods of producing channel waveguides in poled polymer films. Fringing of the poling fields in the buffer and E-O layers acts to smooth out the edges of the guiding regions. If the edge roughness of the electrodes is small compared to the buffer layer thickness (of order 2 μm), the roughness of the waveguide boundaries should be independent of the resolution of the photolithographic process used, and should result in lower scattering losses in the waveguides. This has been confirmed by us in PC6S channel guides. Scattering losses from surface roughness are a major problem in channel guides defined by etching the nonlinear optical material or by channel filling in the substrate. Another advantage of the SPP is that the waveguides defined by the poling process have significantly different index changes for the TE and TM waves, making possible polarization selective elements.

Potential drawbacks to the SPP include partial lateral confinement of TE waves near the guide boundaries due to nonuniformly poled polymer and the fact that the guide cladding is intrinsically the unpoled polymer and cannot be arbitrarily chosen for a particular application.

The SPP has been used to fabricate three classes of integrated optical device structures: a Y-branch interferometer, a directional coupler, and a traveling wave phase modulator. These device experiments were aimed at developing the necessary processing techniques for poled polymer devices, which differ significantly form those of Ti:LiNbO$_3$, and to determine the effects of secondary, as well as primary, materials parameters on device performance. Device optimization was not carried out for the first prototypes, but is now a routine part of our work.

The prototype Y-branch interferometer was fabricated by first defining the waveguide pattern in an aluminium electrode on a glass substrate. The guides were 7 μm wide. The dimensions of these particular electrodes were chosen to facilitate tests of the poling waveguide formation process and were not scaled to optimize the completed devices made form different materials. A 3 μm lower buffer layer was deposited on the substrate using UV curing epoxy, and then a 2 μm layer of the PC6S was spun onto the substrate. A gold poling electrode was deposited directly onto the the PC6S and the material was poled at 90°C for 5 minutes with an electric field of 100V/μm. The poling electrode was etched off with dilute Aqua Regia and a glass slide containing the upper electrode over one arm of the interferometer was glued on with a 3 μm thick layer of optical epoxy. The epoxy also served as the upper buffer layer. To facilitate construction the upper electrode was much larger than necessary resulting in increased device capacitance and a reduced maximum modulation rate Figure 2 illustrates the structure of the completed device and its dimensions. Prism coupling was used to inject 780 nm light into the device and guiding was observed in each arm. Modulation to a few kHz was detected in the output beam, indicating successful confinement and poling of the Y-branch.

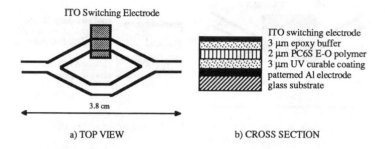

ITO Switching Electrode

ITO switching electrode
3 µm epoxy buffer
2 µm PC6S E-O polymer
3 µm UV curable coating
patterned Al electrode
glass substrate

3.8 cm

a) TOP VIEW   b) CROSS SECTION

<u>Figure 2</u> Structure of the PC6S Y-Branch Interferometer Made By SPP

A second prototype device, a directional coupler, was fabricated using the C-22 material. The electrode defining the waveguide sections consisted of two 7 µm wide sections joining together to form a 13 µm wide, 1 cm long common section at a crossing angle of 3°. Two more symmetric 7 µm sections then diverged at the same angle, resulting in a total device length of 3.8 cm. The structure consisted of the lower aluminium electrode on a glass substrate covered with a 3 µm UV curing epoxy buffer layer. A 2 µm layer of the C-22 material was then spun on and covered with a 3 µm layer of polysiloxane. A planar gold electrode was then deposited for poling and the device was poled at 105°C for 15 minutes with an electric field of 90 V/µm in the C-22 layer. The upper gold electrode was then patterned to form the switching electrode over the central section of the coupler. Figure 3 illustrates the structure of the completed device and its dimensions. Prism coupling was used to inject 830 nm light into one input arm. With an applied voltage of 125 V the output was concentrated in the arm on the same side as the input channel, producing a bar state. With an applied voltage of 65 V a significant fraction of the light was switched to the opposite output arm producing a partial crossed state[1]. Complete switching was not observed due to the large crossing angle and the non-optimum waveguide dimensions.

A third device prototype, a traveling-wave phase modulator based on C-22, was fabricated using methods similar to those described above. This device was designed to modulate light efficiently at 270 MHz. The device impedence was 9.5 Ω, and a quarter-wave transformer was used to drive the modulator with a 0.5 W electrical input. The total active length was 3 cm. The modulation achieved at 270 MHz was over 60%, almost exactly what we calculate it should have been by using the value of r_{33} measured at low frequencies, indicating little dispersion in the E-O coefficient from DC to 270 MHz.[1] The response at 1.0 GHz was small due to the impedance mismatch, but was still observable.[1] The combination of the unique features of the SPP and the properties of the C-22 and newer polymers shows some of the promise of organic materials for integrated optics.

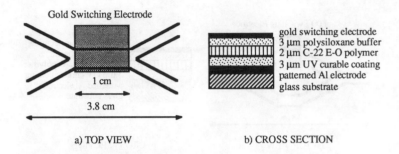

a) TOP VIEW   b) CROSS SECTION

Figure 3 Structure of the C-22 Directional Coupler Made by SPP

4 CONCLUSIONS

We have demonstrated the application of a new fabrication technique, the selective poling procedure (SPP), for producing active, buried-channel waveguide devices in poled polymer films. The SPP combines the fabrication and alignment steps required to make an active device. The method has been demonstrated in several device formats showing promise for real applications, and is under current intensive investigation at Lockheed for more advanced integrated optical structures. We have reported the achievement of a GHz response in a poled polymer channel waveguide, and have observed little or no dispersion in the E-O coefficient in our materials. The combination of unique fabrication methods and good material properties implies that the field of organic integrated optics, now in its infancy, is well on its way toward achieving much of the promise and expectations of organic electro-optic materials.

Future research on electro-optic polymer devices must address several research topics in poled polymer waveguides. From a materials standpoint, larger electro-optic coefficients and lower linear absorption are required, and higher glass transition temperatures are desirable. From a device standpoint, microwave losses must be measured and evaluated, the stability of the poled states must be determined within the operating range of the devices, and numerous other device fabrication issues, such as buffer layer selection, end-surface preparation, and fiber pigtailing must be solved. Finally, devices must eventually be integrated with semiconductor sources and detectors. Such advances are expected to occur within the next several years, and should produce a new class of high-speed, cost-effective devices for integrated optics.

REFERENCES

1.   R. Lytel, G.F. Lipscomb, M. Stiller, J.I. Thackara, and A.J. Ticknor, "Organic Integrated Optical Devices", Proc. NATO ARW on Nonlinear Optical Properties of Polymers (to be published), Nice-Sophia Antipolis, France (1988).

2. R. Lytel, G.F. Lipscomb, and J.I. Thackara, "Recent Developments in Organic Electro-optic Devices", in Nonlinear Optical Properties of Polymers, A.J. Heeger, J. Orenstein, and D.R. Ulrich , ed., Proc. Materials Research Society Vol. 109, 19 (1988).

3. R. Lytel and G.F. Lipscomb, "Nonlinear and Electro-optic Organic Devices", in Nonlinear Optical and Electro-active Polymers, P.N. Prasad and D.R. Ulrich, ed., Plenum Press, New York (1988), p. 415.

4. R. Lytel, G.F. Lipscomb, and J.I. Thackara, "Advances in Organic Electro-optic Devices", Proc. SPIE Vol. 824, pp.152-161 (1987).

5. R. Lytel, G.F. Lipscomb, P. Elizondo, B. Sullivan, and J. Thackara, "Optical Nonlinearities in Organic Materials: Fundamentals and Device Applications", Proc. SPIE 682, 125 (1986).

6. K.D. Singer, M.G. Kuzyk, and J.E. Sohn, "Second-order Nonlienar Optical Processes in Orientationally Ordered Materials: Relationships Between Molecular and Macroscopic Properties", J. Opt. Soc. Am. B4, 968 (1987).

7. C.S. Willand, S.E. Feth, M. Scozzafava, D.J. Williams, G.D. Green, J.I. Weinshenk, H.K. Hall, and J.E. Mulvaney, "Electric-Field Poling of Nonlinear Optical Polymers", p. 107; and K.D. Singer, J.E. Sohn, and M.G. Kuzyk, "Orientationally Ordered Electro-optic Materials", in Nonlinear Optical and Electro-active Polymers, P.N. Prasad and D.R. Ulrich, ed., Plenum Press, New York (1988), p. 189.

8. R.N. Demartino, E.W. Choe, G. Khanarian, D. Haas, T. Leslie, G. Nelson, J. Stamatoff, D. Stuetz, C.C. Teng, and H. Yoon, "Development of Polymeric Nonlinear Optical Materials", in Nonlinear Optical and Electro-active Polymers, P.N. Prasad and D.R. Ulrich, ed., Plenum Press, New York (1988), p. 169.

9. D.J. Williams, "Nonlinear Optical Properties of Guest-Host Polymer Structures", in Nonlinear Optical Properties of Organic Molecules and Crystals, Vol. 1, D. Chemla and J. Zyss, ed., Academic Press, FLA (1986), p. 405.

10. Nonlinear Optical Properties of Organic and Polymeric Materials , D.J. Williams ed., ACS Symposium Series 233 (American Chemical Society), 1983.

11 S.J. Lalama and A.F. Garito, "Origin of the Nonlinear Second-order Optical Susceptibilities of Organic Systems", Phys. Rev. A 20, 1179 (1979)

12. K.D. Singer and A.F. Garito, "Measurements of Molecular Second-order Optical Susceptibilities Using DC Induced Second Harmonic Generation", J. Chem. Phys. 75, 3572 (1981).

13. C.C. Teng and A.F. Garito, "Dispersion of the Nonlinear Second-order Optical Susceptibility of an Organic System: p-nitroaniline", Phys. Rev. Lett. 50, 350 (1983).

14. G.F. Lipscomb, A.F. Garito, and R.S. Narang, "An Exceptionally Large Linear Electro-optic Effect in the Organic Solid MNA", J. Chem. Phys. 75, 1509 (1981).

15. J. Thackara, M. Stiller, E. Okazaki, G.F. Lipscomb, and R. Lytel, "Optoelectronic Waveguide Devices in Thin-Film Organic Media", Conference on Lasers and Electro-optics, Baltimore, MD (1987), paper ThK29.

16. J. Thackara, M. Stiller, G.F. Lipscomb, A.J. Ticknor, and R. Lytel, "Poled Electro-optic Waveguide Formation in Thin-Film Organic Media", Appl. Phys. Lett. 52, 1031 (1988).

17. J. Thackara, M. Stiller, A.J. Ticknor, G.F. Lipscomb, and R. Lytel, "Poled Electro-optic Waveguide Devices in Thin-Film Organic Media", Conference on Lasers and Electro-optics, Anaheim, CA (1988), paper TuK4.

Properties of Non-linear Optical Organic Dopants when Incorporated into Polymer Matrices

N. Carr*, R. Cush, and M.J. Goodwin

PLESSEY RESEARCH CASWELL LTD., ALLEN CLARK RESEARCH CENTRE, CASWELL, TOWCESTER, NORTHAMPTONSHIRE NN12 8EQ, UK

1 INTRODUCTION

There is currently considerable interest in nonlinear optical waveguide devices for high speed all-optical switches[1]. Such devices are expected to find widespread use in optical signal processing. A number of material systems, including organic materials, have been investigated for such applications. Organic materials are particularly attractive because of their high nonlinear optical coefficients[2], extremely fast response times, good transmission characteristics and high laser damage threshold. In addition, molecular tailoring offers considerable flexibility to the design engineer, thereby allowing application specific materials to be produced.

For second order nonlinear processes, such as second harmonic generation or the linear electro-optic effect, it is necessary for some degree of order to be present in the material. This therefore places certain restrictions upon the material, such that centro-symmetric and amorphous structures cannot be used. Third order nonlinear processes, such as the optical Kerr effect or degenerate four-wave mixing, have no such symmetry restrictions and so amorphous materials can be used. The aim of this work was to investigate the fabrication of thin film, amorphous channel waveguide structures based upon known nonlinear optical organic materials. Although these structures were developed primarily for the investigation of third order, intensity dependent waveguide effects[3] it should be noted that poling fields can be used to orientate the dopant molecules and so enable the second order nonlinearities to be accessed[4].

A convenient method for incorporating nonlinear organic dopants into polymer matrices is by indiffusion of the dopant from solution into the host polymer [5-7]. In this work the chosen host polymer was a phenol-formaldehyde based resin (novolac) and the dopant was N,N-dimethyl-2-acetamido-4-nitroaniline (DAN)[8]. The general material requirements are for good optical properties of the films and compatibility with waveguide fabrication, as well as thermal, chemical and mechanical stability. The polymer should not be soluble in the indiffusion solvent and should resist swelling, softening, flowing, etc. The dopant should possess a large third order nonlinear optical coefficient and be soluble in the indiffusion solvent - ideally sparingly soluble to provide the driving force for penetration of the polymer. DAN was chosen for this work because of its high second order hyperpolarisability, β, since it is believed that the molecular properties resulting in a large β value are closely related to those responsible for producing a large third order hyperpolarisability, γ, i.e. an extended delocalised π-electron system that is capable of responding to an applied electric field[9]. The best third order nonlinear materials are large molecules such as β-carotene[10] or polymeric systems, notably poly-diacetylenes[11]. However, large molecules and polymers are generally unsuitable as dopants as solubility factors tend to limit their potential in most polymer/dopant systems. Whilst DAN is likely to have a lower third order nonlinearity compared with materials such as polydiacetylenes, it has been found to be an excellent dopant.

2 RESULTS AND DISCUSSION

The general structure of novolac is shown in Figure 1. This polymer has a relatively open structure which allows penetration by organic dopants when cast as a thin film. Spin cast films of a few microns thickness have been found to be amorphous in nature and of sufficiently high quality for waveguiding. Indiffusion of novolac by DAN (see Figure 1) has been monitored by measuring the absorbance spectrum in the region 200-600nm. Novolac has λ_{max} = 295nm and DAN has λ_{max} = 393nm.

A typical indiffusion profile of DAN into novolac is illustrated in Figure 2 for a series of indiffusion times. The polymer film was prepared by spin casting a solution of novolac in 2-methoxyethyl ether at 5000rpm for 40

(i) Novolac Polymer

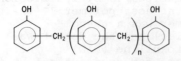

(ii) N,N - Dimethyl-2-acetamido-4-nitroaniline (DAN)

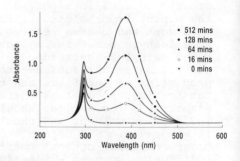

Figure 1   Figure 2

Structure of (i) host polymer,   Indiffusion profile of DAN
novolac and (ii) nonlinear   into a 1.1μm thick novolac
dopant, N,N-dimethyl-2-   film.
acetamido-4-nitroaniline
(DAN).

seconds onto a suitable glass substrate and baking the
film at 50⁰C for 18 hours to remove excess solvent.
Indiffusion of DAN was carried out at 120⁰C by immersing
the film coated substrate into a saturated solution of DAN
in a fluorocarbon solvent - an isomeric mixture of
perfluoro-1-methyl decalins.

The concentration of DAN in novolac was estimated
using equation (1) which is a standard relationship used
in spectrophotometry.

$$c = Abs/\varepsilon t \tag{1}$$

where

        c   = DAN concentration
        Abs = absorbance at 393nm
        ε   = extinction coefficient
        t   = film thickness

The validity of this expression for thin films was
assessed by measuring the film thickness and absorbance
for six different predoped DAN/novolac films covering two
known concentrations. The extinction coefficient of DAN
in 2-methoxyethyl ether (the spinning solvent) was
obtained as ε_{367} = 7098. By assuming densities of 1.2g/ml
for both DAN and novolac, the average concentration of DAN

Table 1

Percentage indiffusion of DAN into novolac for
various polymer post-baking temperatures.

		Indiffusion Time (hr)					
		0.5	2.0	8.0	14.0	34.0	54.0	100.0
Post-	50	54.1	92.1	97.6	98.8	100	100	100
bake	130	36.0	81.5	96.8	97.6	98.2	100	100
temp.	160	18.2	37.4	63.8	68.5	72.6	75.0	76.8
(^{0}C)	200	0.0	0.0	5.3	11.2	18.5	23.1	25.0

was calculated to be 3.46 and 7.43wt%. These figures
compare well with the actual pre-doped values of 3.32 and
7.45wt% DAN respectively, and confirm the usefulness of
equation (1). The red shift of 26nm in λ_{max} for DAN in
novolac relative to solution indicates a significant
polymer/dopant interaction. The choice of the spinning
solvent for determination of the DAN extinction
coefficient in solution was considered appropriate as spun
cast thin films are generally known to contain a small
amount of trapped solvent, even when post-baked.

The concentration of DAN indiffused into novolac for
the film shown in Figure 2 was calculated to be
approximately 43wt% from the measured thickness and
absorbance values of 1.10μm and 1.788 respectively. This
concentration is therefore much higher than the maximum
concentration of approximately 12wt% DAN that can normally
be achieved by co-mixing DAN and novolac in 2-methoxyethyl
ether. This fact is further illustrated in Figure 3 for a
novolac solution pre-doped with 11.5wt% DAN cast in thin
film form and post-baked as before. Upon indiffusion for
256 minutes the DAN absorbance has risen from 0.637 to
2.600. This latter value is equivalent to a DAN
concentration of 42wt% after accounting for a small amount
of polymer swelling. Consequently, by indiffusing DAN
into novolac it is possible to attain a high concentration
of nonlinear dopant in the polymer matrix.

Figure 4 shows how the DAN absorbance varies with
indiffusion time for a typical novolac film. However,
the film baking temperature after spin coating has been
found to significantly affect the rate of indiffusion.
Table 1 contains indiffusion rates, expressed as a
percentage, for different post-bake temperatures. The
values were calculated from absorbance measurements at
393nm and are quoted relative to the maximum value - that

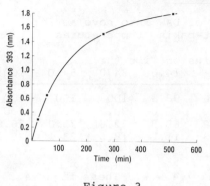

Figure 3	Figure 4

Variation in absorbance of DAN when indiffused into novolac.

Indiffusion profile of DAN into a 1.8μm thick novolac film predoped with 11.5wt% DAN.

of the 50^0C sample when indiffused for 100 hours. As the post-bake temperature is increased, the subsequent rate of indiffusion is found to decrease. The indiffusion rate for the 130^0C processed film is lower to begin with than that for the 50^0C baked film, but as the indiffusion was carried out at 120^0C, differences would only be expected during the initial stages of indiffusion. The effect of post-bake temperature on the indiffusion rate becomes particularly marked as the novolac cure temperature ($130-200^0$C) is reached. A post-bake temperature of 160^0C significantly reduces the indiffusion rate, whilst baking at 200^0C makes the polymer extremely resistant to indiffusion. The values in Table 1 for the percentage DAN indiffusion when the film is post-baked at 200^0C are estimated maximum values as the polymer itself now has a significant absorbance at 393nm. These general observations are consistent with polymer cross-linking at elevated temperatures which prevents dopant penetration.

For the fabrication of channel waveguides a two stage baking process was used. The polymer was firstly soft baked (50^0C for 18 hours) to slowly remove residual solvent, followed by a shorter, hard bake (125^0C for 30 mins) to increase the polymer resistance during subsequent processing whilst retaining its capacity for dopant indiffusion.

The indiffusion technique is ideally suited to the formation of channel waveguide structures. A multi-stage process was developed for the fabrication of DAN doped channels in novolac. The channels themselves were defined photolithographically, and as novolac is UV sensitive (indeed, novolacs are sometimes used as photoresists in microelectronics) it was necessary to protect the polymer from the exposure, development and removal processing conditions of the photoresist. This was achieved by electron-beam depositing a 1000Å gold layer on the novolac followed by sputter deposition of 500Å of silica on top of the gold layer. Using this multi-layer structure it was possible to define channels in the gold layer through which the DAN was indiffused, without causing degradation of the host polymer - the novolac/DAN system was found to be stable under gold etch (aqueous KI/I_2) and silica etch (aqueous HF) conditions.

Initial waveguide fabrication trials were carried out using glass substrates (Schott Glass BK7) that had been optically polished to remove any surface roughness and so minimise scatter at the polymer glass interface. A number of planar waveguide samples were fabricated by indiffusing different concentrations of DAN into the novolac film. The waveguide modes of these samples were excited using a high index glass prism coupler (Schott Glass SF 10) and the waveguide mode propagation constants measured. A standard inverse - Wentzel-Kramers-Brillouin (WKB) analysis[12] was performed to obtain the refractive index profile of the waveguides. Using this information it was possible to evaluate the refractive index change caused by the indiffusion as a function of dopant concentration. By controlling the dopant concentration, it was found that the refractive index change could be varied up to a value of approximately 2×10^{-2}, corresponding to the maximum dopant concentration of 40wt% DAN.

Indiffusion trials through narrow apertures have been performed to evaluate the degree of lateral diffusion occurring in these films. Novolac films were indiffused with DAN through 8μm wide apertures in the gold mask. The indiffusion took place at 120⁰C for times up to 1 hour, after which the masks were removed and the samples examined. The extent of lateral diffusion was clearly visible and could be measured microscopically. For indiffusion times up to approximately 30 minutes there was negligible lateral diffusion (less than 0.5μm). For longer times, lateral diffusion became more pronounced increasing to approximately 4μm after 60 minutes. The 30 minute indiffusion time was sufficiently long to produce a usefully high dopant concentration (24wt%) of DAN, with which sufficient index change for lateral optical confinement is available. Consequently, it is not thought that lateral diffusion will cause serious problems for channel waveguide definition using this technique. Assessment of the waveguiding properties of channel waveguides formed by DAN indiffusion into novolac has been initiated and will be reported at a future date.

3 CONCLUSION

The fabrication of waveguiding structures by indiffusion of the nonlinear material DAN into thin films of novolac polymer has been described. This process has been characterised in terms of film quality, dopant concentration, refractive index change and lateral diffusion, and shows considerable potential for production of the complex channel waveguide structures required for efficient nonlinear waveguide devices.

4 ACKNOWLEDGEMENTS

The authors gratefully acknowledge the assistance of R.M. Gibbs in the fabrication of the waveguides and thank BDH Ltd for the supply of DAN. This work was supported by the Department of Trade and Industry under the Joint Opto-electronics Research Scheme (JOERS).

5 REFERENCES

1.   G.I. Stegeman and C.T. Seaton, J. App. Phys., 1985, 58, R57.
2.   'Nonlinear Properties of Organic and Polymeric Materials', D.J. Williams, ed., A.C.S. Symposium Series 233, American Chemical Society, Washington D.C., 1983.
3.   M.J. Goodwin, C. Edge, C. Trundle and I. Bennion, J. Opt. Soc. Am. B, 1988, 5, 419.
4.   K.D. Singer, J.E. Sohn and S.J. Lalama, Appl. Phys. Lett., 1986, 49, 248.
5.   M.J. Goodwin, R. Glenn and I. Bennion, Electron. Lett., 1986, 22, 789.
6.   R. Glenn, M.J. Goodwin and C. Trundle, J. Molec. Electron., 1987, 3, 59.
7.   J. Brettle, N. Carr, R. Glenn, M. Goodwin and C. Trundle, SPIE, 1987, 824, 171.

8.   P.A. Norman, D. Bloor, J.S. Obhi, S.A. Karaulov,
      M.B. Hursthouse, P.V. Kolinsky, R.J. Jones and
      S.R. Hall, J. Opt. Soc. Am. B, 1987, 4, 1013.
9.   G.R. Meredith and B. Buchalter, J. Chem. Phys., 1983,
      78, 1938.
10.   J.P. Hermann, D. Ricard and J. Ducuing, Appl. Phys.
      Lett., 1973, 23, 178.
11.   G.M. Carter. J.V. Hryniewicz, M.K. Thakur, Y.J. Chen
      and S.E. Meyler, Appl. Phys. Lett., 1986, 49, 988.
12.   J.M. White and P.F. Heidrich, Appl. Opt., 1976, 15,
      151.

Multi-layers, Electro-optic and Thermo-optic Effects in Poly-4-vinylpyridine Optical Wave-guides

P.J. Wells* and D. Bloor

DEPARTMENT OF PHYSICS, QUEEN MARY COLLEGE, MILE END ROAD, LONDON E1 4NS, UK

1 INTRODUCTION

Polymeric films, with optically non-linear molecules included in solid solution[1,2] are potentially useful for integrated optics. We describe how refractive index variations in the plane of the film may be achieved by bleaching, and how structures can be formed in which the refractive index through the thickness of the film is varied by the alternate deposition of different polymer layers. Electro-optic and thermo-optic effects in the polymer films have also been measured.

2 BLEACHING OF RHODAMINE 6G LOADED FILMS

Films of Poly-4-Vinyl Pyridine (P4VP) were produced by withdrawal[3] of fused Quartz substrates from a solution of the polymer in Iso-Propyl Alcohol. The TE refractive index of the polymer was measured by prism wave-guide coupling to be 1.5685 at 632.8nm. The dye was incorporated into the films by adding it in various concentrations to the withdrawal solution. The refractive indices of these films were measured at 632.8 and 812nm (Figure 1). The dye concentration is defined as (Wtdye in withdrawal solution/WtP4VP+Wtdye). Also shown are theoretical indices calculated using the Kramers-Kronig (K-K) relationship assuming the film index to be that of P4VP plus a contribution solely from the main dye visible absorption peak. The difference between theory and experiment could be caused by the truncation of the K-K transformation. The very high dye loading in the film implies the film index should decrease significantly on bleaching. The irradiation of a 16.4% film with a high pressure Mercury discharge lamp gave an index difference of $5.1*10^{-3}$ between bleached and unbleached parts of the film. The index change calculated using the K-K transformation is $8.5*10^{-3}$. The difference between the two values could be caused by different residual solvent loss on the two parts of the film or by the bleaching being non-uniform. Figure 2 is a photograph of a 35.3% loaded film irradiated

FILM INDEX

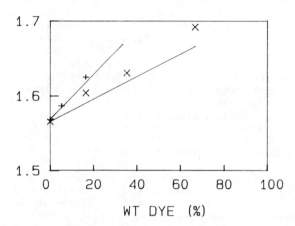

WT DYE (%)

<u>Figure 1</u> Refractive index of P4VP/dye films vs dye concentration
x = 812nm + =632.8nm.
Theoretical indices shown as straight lines for two
wave-lengths.

<u>Figure 2</u> "Y" Junction defined in P4VP with 35.3% dye (pattern
defined by partial bleaching with "Y" protected by mask).
Stripes about 8 microns wide.

through a Spectrosil mask containing fine metal lines. The unbleached
lines are 8 microns wide showing that the fine structures necessary
for horizontal confinement can be produced.

3 MULTI-LAYERS

The TE Field equations for step-index structures are formally similar
to those in 1-D quantum mechanics. This suggests that coupled
"optical quantum wells" can be made by alternate deposition of polymers of different refractive index. Figure 3(a) shows a four layer
cleaved film on a 3.5 micron Silicon Dioxide substrate on Silicon.
The polymer solutions were P4VP (index 1.5685) in IPA for the thick
layers and Poly Vinyl Alcohol (PVAL) (index 1.520) for the thin
layers. Figure 3(b) shows the experimental modes (measured using
prism coupling) and theoretical modes for a three polymer layer system
(film thicknesses estimated using single layers of each film deposited
on fused Quartz). The degeneracy of the lowest two modes is broken
by slight differences in, and tunnelling between, the two wells.

4 ELECTRO-OPTIC AND THERMO-OPTIC EFFECTS

A device was made consisting of (from Si substrate up) 0.15 microns
thermal SiO_2, 4.91 microns deposited SiO_2, 1.78 microns 10% MNA
loaded P4VP, 1.57 microns PVAL and a top evaporated silver electrode.
The cleaved film end-faces were formed by snapping the Silicon substrate which also functioned as the bottom electrode. Input light
polarised at 45 degrees was focused into one end with a microscope
objective, and collected from the other with a weak lens before
being passed through an analyser with polarisation orthogonal to the
polariser. A 102V r.m.s. 600Hz (f) sine wave voltage was applied.
The resulting modulation varied from nearly pure f (Figure 4(a)) to
2f (figure 4(b)) depending on where in the film edge the light was
input. We attribute the 2f signal to molecular reorientation in the
polymer[2] and the f signal to partially orientated thermal oxide on
the silicon. This interpretation is supported by the fact that when
f and 2f are both seen, the f signal can be increased relative to the
2f signal be moving the input beam focus from the film towards the
substrate and also by through-the-plane SHG studies which show that
although there is SHG from the loaded polymer deposited on fused
quartz (attributable to surface alignment of the dopant[4]), there is
much more SHG per micron from the SiO_2/Si substrate alone. The
maximum f modulation (pk-pk) signal/unmodulated signal) observed was
2.9% and the maximum 2f modulation 0.63%. A thermal sensor was constructed with 1.78 microns P4VP/MNA alone on a 5 micron deposited
SiO_2/0.15 micron thermal SiO_2/Si substrate. TE polarised light was
injected and collected as before through a 4.1mm sample, no analyser
was used. When the sample was warmed the TE output light was modulated (Figure 5(a)), the separation in peak maxima corresponds to a
temperature change of 1.1 degree. We attribute the effect to the
different thermal expansion coefficients of the oxide and polymer.

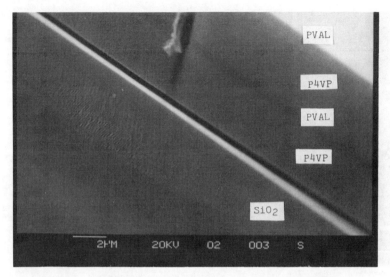

Figure 3(a) SEM of four-layer polymer film SiO_2/Si
substrate (two thin layers PVAL, two thick
layers P4VP)

REFRACTIVE INDEX

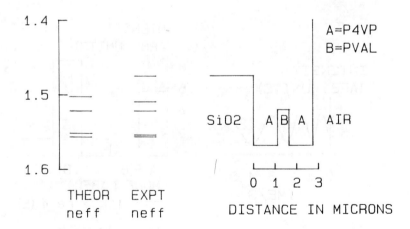

Figure 3(b) Refractive Index profile of three layer
polymer film with mode indices

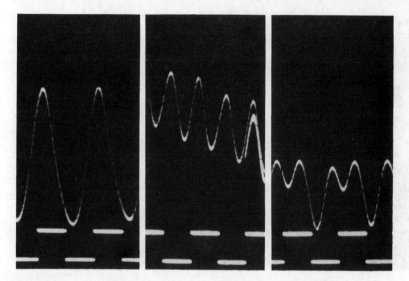

<u>Figure 4</u>   Modulation of light in P4VP/10% MNA on SiO_2
a) "f" modulation b) "2f" modulation c) mixture
Bottom trace is 600Hz TTL reference from Supply
Vertical scale A.C. signal from Photo-detector
(Arb. Units)

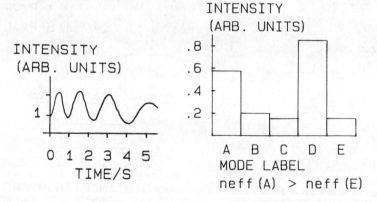

<u>Figure 5</u> (a)   TE Output Intensity of P4VP/10% MNA on SiO_2
vs time after heating lamp switched off
(b)   Intensity of modes labelled in order of neff

Modes confined mostly in the polymer (corresponding to the 2f signal) will interfere with modes with more field in the oxide (corresponding to the f signal) with the relative phase of the modes dependent on the temperature. Figure 5(b) shows the TE modes from the system collected using an output prism, we take mode A to be the lowest order "2f" mode and mode D to be a "f" mode with significant light in the oxide. The high thermal modulation achieved is probably because of the domination of the mode spectra by two modes of similar intensities.

5 CONCLUSION

We have described electro-optic and thermo-optic interference effects in polymer structures on a Si/SiO_2 substrate. We have also outlined a simple method of fabricating "vertical optical quantum wells", and bleaching "horizontal" index changes. These results suggest that by combining both types of confinement it may be possible to produce novel devices with simultaneous lateral and horizontal switching.

REFERENCES

1. 'Non-linear Optical Properties of Organic Molecules and Crystals', Ed D.S. Chemla and J. Zyss, Academic Press, 1987, Vol. 1, p. 405.
2. 'Molecular and Polymeric Opto-Electronic Materials: Fundamental and Applications', Ed G. Khanarian, SPIE proceedings, 1986, Vol. 682, p. 153.
3. C.C. Yang, et al., Thin Solid Films, 1980, 74, 117.
4. T.F. Heinz et al., Physical Review A, 1983, 28, 1883.

ADDENDUM

The index variation between bleached and unbleached parts of the 16.4% film referred to in section 2 was measured at a wave-length of 822 nm.

The PVAL solution used in section 3 used water as solvent. The thermal modulation shown in figure 5(a) was measured by collecting output diffracted light over a limited range of output diffraction angle.

Acknowledgements. The SERC are thanked for a Studentship for P.J.W. and a Fellowship for D.B. This work was also supported by the GEC Marconi Research Centre.

Synthesis and Use of Acrylate Polymers for Non-linear Optics

J.R. Hill* and P. Pantelis

BRITISH TELECOM RESEARCH LABORATORIES, MARTLESHAM HEATH, IPSWICH IP5 7RE, UK

F. Abbasi and P. Hodge

DEPARTMENT OF CHEMISTRY, UNIVERSITY OF LANCASTER, LANCASTER LA1 YA, UK

1 INTRODUCTION

Recently, there has been a significant increase in interest in the nonlinear optical properties of organic materials [1-3]. This is due to their potentially large bulk nonlinear coefficients, fast response times and high resistance to laser damage [4]. Organic polymers have the additional advantages of versatility, low cost and compatability with existing semiconductor processing technologies. Furthermore desirable optical and mechanical properties can be conveniently enhanced in suitable polymeric systems by chemical means to achieve optimisation of the materials characteristics.

In this paper we describe the synthesis and processing of an acrylate polymer to produce a demonstration linear electro-optic modulator.

For a polymeric system to exhibit second order nonlinear properties it must satisfy two essential requirements. Firstly, the polymer must possess groups having a large second order nonlinear hyperpolarizability tensor (ß) at a molecular level and secondly these groups must be organised in such a way to have no centre of symmetry in the bulk material.

Provided that the nonlinear optical groups possess a significant microscopic second order hyperpolarizability in the direction of their molecular ground state dipole moment, the required orientation may be induced by application of an external electric field. This field acts on the ground state dipole moment of the optically nonlinear groups causing them to preferentially align with the field and hence destroying the centrosymmetry of the bulk polymer[5].

Although materials having bulk second order nonlinear optical properties can be prepared by electrically ordering optically nonlinear molecules dissolved in a suitable polymeric host, the bulk nonlinearity is ultimately limited by the maximum concentration of the active guest[6,7]. By chemically incorporating the nonlinear species as part of a polymer the effective concentration of nonlinear sites may be maximised and precipitation of the active component is prevented.

We have produced a polymer where the optically nonlinear species is linked to and distanced from the backbone of the polymer by a spacer. This produces a comb like structure, with the backbone as the spine and the optically nonlinear species forming the teeth[8]. To minimise light scattering, the spacer has been kept short to prevent liquid crystalline behaviour. The polymer has pendant chromophores designed to have both high microscopic second order hyperpolarizability and a large ground state dipole moment.

The polymer was thermopoled by first heating it above its glass transition temperature (Tg) so that the pendant chromophores became mobile and then applying an electric field across the material. The polymer was then cooled to less than the Tg with the field still applied. In its glassy state the field can be removed without loss of the imposed order. Provided that the orientated polymer is not reheated above the Tg the nonlinear properties should be retained almost indefinitely.

In addition to causing the polymer to develop second order nonlinear optical properties, refractive index changes are also induced in the material which are frozen into the polymer. These refractive index changes are a result of the differences in polarizability along the axes of the pendant chromophores. For the polymer we describe, the refractive index seen orthogonal to the poling field increases while the refractive index along the poling field decreases. This change in refractive index has been recently shown to be sufficient to define waveguides by thermopoling[9].

The processibility of polymers allows them to be formed into structures other than optical waveguides and

consequently novel structures can be constructed which
would not be possible with organic or inorganic single
crystals.   In the later part of this article we present
a demonstrator modulator which illustrates this
advantage.

2 SYNTHESIS

The polymer used as the active medium in our modulator
was synthesized in three steps (Scheme 1).

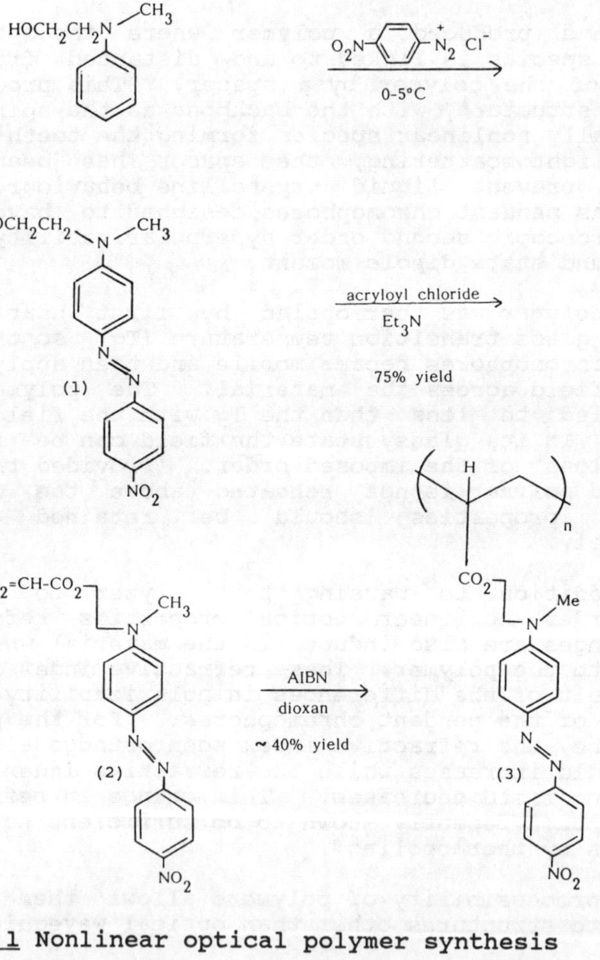

Scheme 1 Nonlinear optical polymer synthesis

4-(4'-nitrophenylazo)-N,N-methyl-2-ethanoloaniline
(1) was prepared by diazotisation of 4-nitroaniline and
subsequent diazo coupling of the product to
N-methyl-2-ethanolaniline. The product was O-acetylated
with acryloyl chloride to give the monomer,
4-(4-nitrophenylazo)-N,N-methyl-ethyl-acrylylester (2).
This was polymerized in the absence of oxygen using
azo-iso-butyronitrile as initiator to give, after
purification, the pure polymer (3). The polymer had Mn
3,400 and Mw 4,900 (as determined by a Trilab gel
permeation chromatograph calibrated with polystyrenes),
but the ^{1}H and ^{13}C-NMR spectra suggested
substantially higher absolute values. The glass
transition temperature (Tg) was found to be 82°C using
a Perkin Elmer DSC7.

3 PREPARATION OF A DEMONSTRATOR MODULATOR

A novel polymer based device has been produced for the
purpose of demonstrating the linear electro-optic effect
in a thermopoled polymer. Its simple construction and
few processing steps highlights some of the advantages
of polymers for nonlinear optical applications.

The device was an electrically modulated Fraunhofer
diffraction grating formed in the polymer via
transparent interdigital electrodes.

The electrodes were photolithographically defined
and chemically etched into indium-tin-oxide (ITO)
coated glass. They were 20 μm wide, separated by 20μm
spacing and covered a 1cm square. The electrodes were
attached to two larger supply tracks on which both
poling and signal voltages could be applied (Fig.1).

A quantity of the electro-optic polymer which was
melted into the electroded area and spread under a cover
slip at 170°C, to form a clear film a few μm thick. The
prepared slide was reheated to 110°C and a potential
difference of 200V was applied across the interdigital
electrodes while the sample was allowed to cool through
the Tg (82°C) to room temperature.

The electrical ordering of the chromophores caused
the polymer between the electrodes to become
noncentrosymmetric and to undergo a change in refractive
index which followed the pattern of the electrodes.
This periodic refractive index change in the polymer
film resulted in the formation of a diffraction grating

which has remained stable at room temperature over a
period of 6 months without measurable decay. The
thermopoled sample was positioned with the length of the
interdigital electrodes running vertically and the film
surface normal to a 1mW HeNe 633nm laser with the
electric vector horizontally polarized. The resulting
Fraunhofer diffraction pattern was viewed on a screen.
The intensity of light at one diffracted spot was
measured by a photodiode (Radio Spares No. 308-067)
having an integral amplifier(Fig.2).

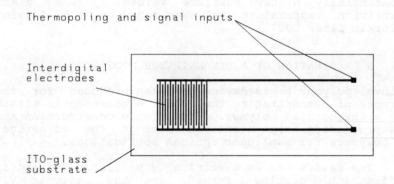

Figure 1 The electrode structure of the modulator.

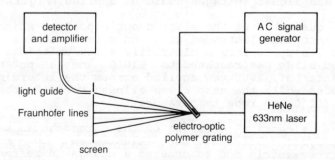

Figure 2 Schematic of the modulator in use.

When an electrical signal was applied to the transparent electrodes of the device, the frozen oriented polymer between the electrodes experienced an electric field which caused a change in refractive index owing to the polarization of the conjugated pi-electrons of the pendant chromophores. The refractive index change via the linear electro-optic effect is small, relative to the refractive index change previously frozen into the polymer on thermopoling. This caused the change in diffraction efficiency of the device to be approximately linear with the field induced index change and to be directly proportional to the applied modulating voltage. This proportionality was found to be 7.14×10^{-2} percent/V up to the 170V r.m.s. maximum output of the signal generator.

Piezoelectric deformation was eliminated as the source of the effect as mechanical vibration of the sample failed to produce any electrical signals on the electrodes.

The maximum modulation frequency of the device was limited by the capacitance of the electrodes (94pF) which restricted the rate at which the field between the electrodes could be changed. However, the ultimate response time of the material is expected to be sub-picosecond since the linear electro-optic effect arises from the polarization of molecular orbitals.

The electro-optic coefficient of the poled polymer is proportional to the electric field applied during thermopoling. Since the poling field between the electrodes of the modulator was not uniform, the value of its electro-optic coefficient varies in space. In the absence of an exact knowledge of this spatial variation we are unable even to give an average coefficient for the poled polymer. However, the nonlinear coefficient (r_{33}) of the polymer may be defined in terms of $r_{33} = Ro.Ep$ where Ep is the thermopoling field strength and Ro the electro-optic coefficient per thermopoling V/m.

To estimate the value of Ro a liquid cell having identical interdigital electrodes was prepared and filled with nitrobenzene. Unlike the thermopoled polymer, nitrobenzene is free to rotate in an electric field at room temperature. Application of a signal to the cell resulted in electro-optic modulation via molecular reorientation (the dc Kerr effect). As expected from such a system, the modulation was directly

proportional to the square of the applied voltage
(slope= 2.54×10^{-4} percent/V^2) and the frequency of the
output was double that of the applied signal[10]

Since the electrode structure of both the polymer
and nitrobenzene modulators was identical and the
dimensions of Ro and the Kerr constant (S_{33}) are both
m^2/V^2, we can directly relate Ro to S_{33}.

The value of Ro was calculated[11] to be 1.5×10^{-18}
m^2/V^2 using the established theory of phase gratings[12].
This value can be used to estimate the magnitude of the
electro-optic effect if the polymer were to be poled in
a uniform thermopoling field. For example with $20V/\mu m$
during poling the polymer should achieve an
electro-optic coefficient (r_{33}) of about 30 pm/V, which
is close to that of lithium niobate.

CONCLUSION

An acrylate polymer having pendant optically nonlinear
chromophores has been prepared. To illustrate the ease
of fabrication of potentially useful electro-optic
devices from it, the polymer has been used in the
construction of a novel linear electro-optic modulator.
By reference to a Kerr effect modulator, a value of the
electro-optic coefficient per unit thermopoling field
intensity for the polymer was calculated to be 1.5×10^{-18}
m^2/V^2 at 633nm.

ACKNOWLEDGMENT

Acknowledgment is made to the Director of British
Telecom Research Laboratories for permission to publish
this work. The authors would also like to thank their
colleagues R.J.Prodger for lithographic mask
preparation, R.Kashyap for helpful discussions and
G.J.Davies and R.Heckingbottom for support and
encouragement.

REFERENCES

1.   'Nonlinear Optics: Materials and Devices', edited
     by C.Flytzanis and J.L.Oudar, Springer-Verlag, New
     York, 1986.

2.   'Nonlinear Optical and Electroactive Polymers', edited by P.N.Prasad and D.R.Ulrich, Plenum Publishing Corp. New York, 1987.

3.   'Nonlinear Optical Properties of Polymers', edited by A.Heeger, J.Orenstein and D.R.Ulrich, Materials Research Society, Pittsburgh, 1987.

4.   A.F.Garito, K.Y.Wong, Y.M.Cai, H.T.Man and O.Zamani-Khamiri, Proc.SPIE., 1986, 682, 2.

5.   D.J.Williams, Angew.Chem.Int.Ed.Engl., 1984, 23, 690.

6.   K.D.Singer, J.E.Sohn and S.L.Lalama, Appl.Phys.Lett., 1986, 49, 248.

7.   J.R.Hill, P.L.Dunn, G.J.Davies, S.N.Oliver, P.Pantelis and J.D.Rush, Elec.Lett., 1987, 23, 700.

8.   H.Ringsdorf and H.Schmidt, Macromol.Chem., 1984, 185, 1327.

9.   J.I.Thackara, G.F.Lipscomb, M.A.Stiller, A.J.Ticknor and R.Lytel, Appl.Phys.Lett., 1988, 52, 1031.

10.   J.N.Polky and J.H.Harris, Appl.Phys.Lett., 1972, 21, 307.

11.   J.R.Hill, P.Pantelis, F.Abbasi and P.Hodge, J.Appl.Phys., 1988, 64, in press.

12.   A.Sommerfeld, 'Optics', Academic Press New York., 1954, pp 228-233.

Second Harmonic Generation in DAN Crystal Cored Fibres

G.E. Holdcroft, P.L. Dunn, and J.D. Rush

BRITISH TELECOM RESEARCH LABORATORIES, MARTLESHAM HEATH, IPSWICH IP5 7RE, UK

1 SUMMARY

The nonlinear compound DAN, 2-(N,N-dimethylamino)-5-nitroacetanilide, is the subject of much interest, particularly for growth inside fibres, as a favourable orientation for second harmonic generation can be obtained by Bridgman techniques.[1-3] In this paper we will describe the growth of DAN as long lengths of single crystal inside silica capillaries and will report on the linear and nonlinear properties of these waveguides.

2 EXPERIMENTAL

DAN was prepared by the method of Ainsworth and Suschitzky[4] and purified by repeated crystallisations from absolute ethanol containing activated charcoal to yield large sword type crystals. The crystals were stored in vacuo prior to a crystal cored fibre growth.

A silica tube was drawn down to give capillary with an external diameter of 125 μm and internal diameters of between 2 and 20 μm. The fibre was cut into 150 mm lengths and added along with a few crystals of DAN to a 6 mm id borosilicate tube. A growth ampoule was formed by evacuating this tube, together with its contents, down to below 10^{-4} torr on a vacuum line and then sealing it off to give an overall length of 200 mm.

The growth ampoule was then introduced into the upper half of a Bridgman type furnace which was heated to just above the melting point of the DAN crystals. The DAN melts and fills the fibre by capillary action. The growth ampoule was then pushed slowly through an interface into the lower half of the Bridgman type furnace which was heated to a lower temperature where the DAN crystalised inside the silica capillary. The speed at which the growth ampoule was driven through the interface was normally set at between 0.1 and 10 mm/hr.

After cooling the growth ampoule was cut open and the filled capillaries were removed and cleaned to remove excess DAN that had entrained or sublimed onto the outside of the fibres. Each capillary was then examined under a microscope, both between crossed polars and without, to

412

determine the quality and length of single crystal DAN that had grown inside. The orientation of the DAN crystals with respect to the fibre direction was ascertained qualitatively by observation of the extinction directions when viewed between crossed polars and quantitatively by rotating crystal x-ray diffraction from a Weissenberg camera.[2] The x-ray diffraction pattern obtained from a DAN crystal cored fibre was matched against predicted patterns obtained from the unit cell parameters measured on bulk crystals of DAN.[5,6]

The apparatus used to measure the linear and nonlinear properties and the scatter from the crystal cored fibres is shown schematically in figure 1. For most of the measurements the Quantronix Nd-YAG laser was operated Q-switched at 1064 nm providing 200 ns pulses at a repetition rate of 500 Hz. A fibre sample, approximately 10 to 15 mm long, was cut out of the 150 mm length and was placed on a stand such that it was supported at only two positions along its length. A combination of two Glan-Taylor polarisers and two half-wave plates was used to attenuate the power of the laser and also to vary the incident angle of polarisation. The beam was then focused by a microscope objective onto the end of crystal cored fibre with a beam diameter matched to the diameter of the core. The output face of the crystal cored fibre was brought into focus by another microscope objective and the near field image projected onto either a silicon photodetector or to a vidicon tube. An infra-red absorbing filter was placed in the path of the beam when observation of the 532 nm radiation alone was required. Sideways scatter of radiation from the crystal cored fibre was observed by having a high numerical aperture fibre mounted perpendicular to it. The light gathering fibre was traversed along the length of the crystal cored fibre and the output was fed into another silicon photodetector.

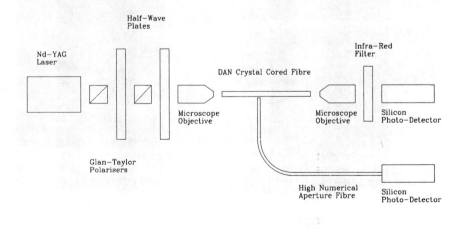

<u>Figure 1</u> Schematic of Linear and Nonlinear Optical Measurements

In addition, for a few measurements the Nd-YAG laser was operated in cw mode at 1064 nm to deliver an average power of 1 mW. The laser was also occasionally configured to give a Q-switched output at 1320 nm. Other suitable detectors and filters were required when the laser was operating in these modes.

3 RESULTS AND DISCUSSION

DAN was grown in 6 μm internal diameter hollow silica fibre up to 120 mm long, i.e almost the total length to which this diameter fibre could be filled by capillary action. Fibre with a much larger internal diameter, i.e. 20 μm, than this filled to a lesser extent and the quality of crystal grown was much poorer. When fibres of 2 μm internal diameter were used it was difficult to freeze the DAN from its melt and a super-cooled core was formed. The super-cooling phenomenon could also be induced in a 6 μm core by choosing appropriate temperature regimes for the Bridgman furnace.

Typical defects that have been observed by microscopy in crystal cored fibres were,
(i) Polycrystalline regions; always present at the beginning of a length crystal cored fibre and from which a single orientation evolves.
(ii) Transverse and longitudinal voids; caused at least in part by decomposition products.
(iii) Cracks; caused by thermal shocks to the fibre on freezing or by mechanical stress on handling.
and (iv) Spot defects; caused by particulate material in the melt.

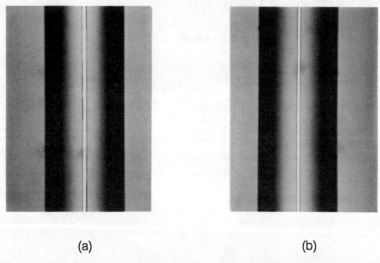

(a)   (b)

Figure 2 Photograph of DAN Crystal Cored Fibres viewed between crossed polars with tint plate, (a) Poor sample, (b) Good sample.

Some of these defects can be seen in the photograph of a poor sample of DAN crystal cored fibre shown in figure 2a. All of these gross defects have now been eliminated by various techniques and a typical photograph of a good sample, as it now can be grown, is shown in figure 2b. With the maximum resolution of our microscope the quality of the DAN crystal appears to be perfect along 100 mm of a fibre and it is from these lengths that samples are cut and submitted for linear and nonlinear optical measurements.

DAN has grown with the same orientation, with respect to the longitudinal axis of the fibre, in all our experiments. There was a strong correlation of the x-ray diffraction pattern from all the fibres examined with that predicted for the crystallographic a-axis being along the longitudinal axis of the fibre. This was also the result found by Kerkoc et al.[7]

Table 1 shows the average and best nonlinear optical results obtained from 6 μm id DAN crystal cored fibre with the laser operating Q-switched at 1064 nm. The results have all been normalised to 1 mW input power (10 W peak power) for ease of comparison.

Table 1. SHG Results for DAN in Silica Fibres

	Throughput, %	Efficiency, %	Generation Ratio
Average Sample	5 - 10	0.5 - 1	100-250
Best Sample	15	1.5	1000

Using the following definitions;

Throughput = $(P_\omega \text{ out(core)})/(P_\omega \text{ in(total)})$

SHG Efficiency = $(P_{2\omega} \text{ out(total)})/(P_\omega \text{ in(total)})$

Generation Ratio = $(\text{Max } P_{2\omega} \text{ out(total)})/(\text{Min } P_{2\omega} \text{ out(total)})$

Where, P_ω = optical power at the fundamental wavelength
and $P_{2\omega}$ = optical power at the second harmonic

A second harmonic generation (SHG) efficiency of 1.5% represents a thousand fold increase on our previously published results[2] and are a factor of 100 better than other published results[7]. This increase in crystal quality over our previously reported results is also reflected in a greater throughput and in the generation ratio. In our earlier work the maximum throughput we could obtain was in the order of a few % and the generation ratio was only about 10. The generation ratio is obtained by measuring the ratio of the maximum to minimum power of the second harmonic as the input polarisation is rotated. The ratio is a measure of the fidelity of polarisation, and thus of crystal quality, down the complete length of a sample.

Generation ratios as high as 250 are not uncommon and for the best samples it is >1000, indicating a significant improvement in crystal quality has been achieved.

In order to confirm the SHG efficiencies observed under Q-switched conditions a fibre that was 1 % efficient at 1 mW average power (10 W peak power) was tested under cw conditions. The crystal cored fibre gave an SHG efficiency of 5×10^{-4} % at 4 mW cw, in good agreement with the Q-switched data.

The maximum efficiency under Q-switched operation at 1320 nm was 0.3 %, a factor of five worse than at 1064 nm. For bulk crystals Kolinsky *et al*[8] reported comparable efficiencies at the two wavelengths. At this stage it is not clear which of many possible causes is responsible for the differing efficiencies observed in crystal cored fibres.

Figure 3a shows the scattered power, both at the fundamental and at the second harmonic, from a poor quality DAN crystal cored fibre, i.e. a sample such as shown in figure 2a and having an efficiency of 0.1 %. The peaks on the scatter plot match up with the visible defects in the fibre and there is no efficient build up of the second harmonic down the length of the fibre. Figure 3b shows the scatter from a good sample, i.e. a sample such as shown in figure 2b and having an efficiency of 1%. There is less overall scatter and there is an efficient build up of the second harmonic power towards the output end of the fibre.

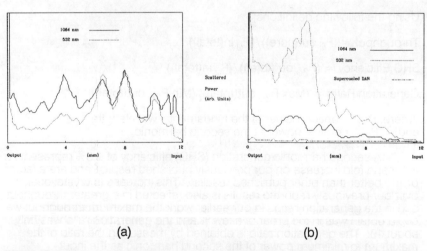

(a)   (b)

Figure 3 Sideways Scatter from DAN Crystal Cored Fibres, (a) Poor sample, (b) Good sample and a Super-Cooled sample.

As described earlier it was possible to prepare fibres in which the molten DAN did not crystalise on cooling but remained in a super-cooled liquid state. Super-cooled samples of DAN are of interest because they permit scattering from the DAN alone to be investigated without the added complications of a crystal/silica interface. Figure 3b shows a scatter plot from such a super-cooled DAN sample, only at 1064 nm, which also shows no visible defects under a microscope. The almost complete absence of scatter from the super-cooled sample with respect to a 'good' sample leads us to believe that there are still scatter centres remaining within a 'good' sample but which cannot be seen with the resolution of optical microscopy. Either the remaining scatter centres are in the bulk of the crystal core and are too small to be seen or occur at the interface between the crystal and the silica surface of the capillary and are hidden in the optical effects arising from the high refractive index change at the boundary.

4 CONCLUSION

The thousandfold increase in SHG efficiencies reported is due to a contribution of better overall crystal quality and a significant decrease in scatter centres. However even 'good' crystal cored fibres suffer from scattering far more than super-cooled samples and hence some scatter centres remain. Work is currently proceeding to identify the nature of these scatter centres and into methods of eliminating them.

ACKNOWLEDGMENTS

Acknowledgment is made to the Director of Research and Technology of British Telecom for permission to publish this paper.

REFERENCES

1.   S. Tomaru and S. Zembutsu, 'Preprints 2nd SPSJ International Polymer Conference', 1986, Japan.
2.   G.E. Holdcroft, B.K. Nayar, D. O'Brein, J.D. Rush, M.A.G. Halliwell, R. Kashyap and K.I. White, 'Proceedings Conference Lasers and Electro-optics', OSA, Washington DC, 1987, Paper THE4.
3.   G.E. Holdcroft, P.L. Dunn, J.D. Rush and M.A.G. Halliwell, 'Proceedings Conference Laser and Electro-optics, OSA, Anaheim, 1988, Paper WM55.
4.   D.P. Ainsworth and H. Suschitzky, J.Chem.Soc., 1966, 111.
5.   P.A. Norman, D. Bloor, J.S. Obhi, S.A. Karaulov, M.B. Hursthouse, P.V. Kolinsky, R.J. Jones and S.R.Hall, J.Opt.Soc.Am.B, 1987, 4, 1013.
6.   J. Baumert, R.J. Twieg, G.C. Bjorklund, J.A. Logan and C.W. Dirk, Appl.Phys.Lett., 1987, 51, 1484.
7.   P. Kerkoc, Ch. Bossard, H. Arend and P. Gunter, 'Proceedings Conference Lasers and Electro-Optics', OSA, Anaheim, 1988, Paper WM39.
8.   P.V. Kolinsky, R.J. Chad, R.J. Jones, S.R. Hall, P.A. Norman , D. Bloor and J.S. Ohbi, Electronics Lett., 1987, 23, 791.

20 μW All Optical Bistability in Nematic Liquid Crystals

A.D. Lloyd and B.S. Wherrett

DEPARTMENT OF PHYSICS, HERIOT-WATT UNIVERSITY, EDINBURGH EH14 4AS, UK

1 INTRODUCTION

To realise the parallelism that digital optics offers in principle, at a moderate total power level, requires bistable devices operating in the 1 - 100 μW regime.

Optimisation studies of nonlinear narrow pass-band interference filters[1,2] have concluded that the independent tailoring of finesse and cavity length, made possible by removal of absorption from within the cavity, can lead to a significant reduction in operating power level.

A metallic partial mirror at the rear of the cavity is thus used as both a reflector and an absorber, with thermal conduction into the spacer leading to optothermal bistability. This design allowed the construction of rigid cavities which have been used to demonstrate optical bistability with a number of liquid phase thermo-optic materials[3,4].

For a fixed cavity design of this type, the lowest power at which bistability is possible can be written[5],

$$P_c = \frac{\lambda_v}{D} \frac{1}{|\partial n/\partial T| \, \partial T/\partial P_a} \, g[R_f, R_b, \exp(-\alpha D)] \, , \qquad (1)$$

where λ_v is the wavelength of operation, D the spacer thickness, $\partial n/\partial T$ the thermo-optic coefficient, $\partial T/\partial P_a$ the temperature rise per unit power absorbed and g a cavity factor discussed in reference 5.

418

That the critical switch power is inversely proportional to the magnitude of the thermo-optic coefficient ($\partial n/\partial T$) was confirmed by experiment[3], and led to the choice of nematic phase liquid crystals as suitable, low absorption, spacer materials. Their high thermo-optic coefficient ($\partial n_e/\partial T = -2 \times 10^{-3}K^{-1}$ at room temperature) produced all optical bistability at critical power levels of 140 μW, allowing the demonstration of submilliwatt laser diode operation of these devices[6].

2 LOW POWER AND DIODE LASER OPERATION

Study of the refractive indices of the liquid crystal 4-cyano-4'-pentylbiphenyl (K15, 5CB or PCB) over its nematic range[7], have shown an increase in magnitude of the thermo-optic coefficient as the nematic-isotropic phase transition is approached. From (1) it is clear, that if $\partial T/\partial Pa$ is constant over this range, then a commensurate reduction in operating power should result.

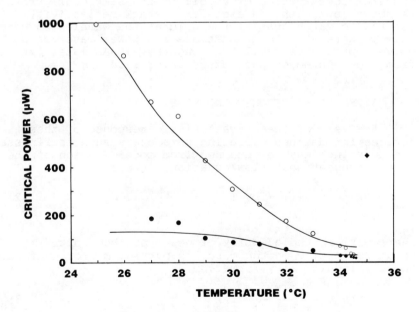

Figure 1 Comparison of theoretical and experimentally determined critical switch powers as a function of temperature, measured for a dielectric/gold 26.5 μm cell containing K15 (R_f = .85, R_b = 0.90). o = $n_\perp$ (o-ray), ● = n‖ (e-ray), ◆ = $n_\perp$ (isotropic phase).

Experimental confirmation of this thermo-optic enhancement was obtained by enclosing the entire cell within a specifically designed, controlled temperature environment. This allowed choice of ambient temperature to 1/100 of a degree and stability approaching 1/1000 of a degree[8]. Figure 1 shows the observed reduction in critical power, for both the ordinary and extraordinary refractive indices, as the phase transition is approached. This is compared with the expected reduction in operating power level due to the increasing thermo-optic coefficient, inferred from a set of high precision index measurements of K15 at 633 nm[7]. Over an ambient range of just 8°C the critical power reduces by approximately one order of magnitude to 13.3 µW at 34.72°C, corresponding to a $\partial n_e/\partial T$ of approximately $- 2 \times 10^{-2}$ K^{-1}. This low power requirement allows bistable operation well below 100 µW and, when operated in reflection, these devices display an adequate switch contrast of ∿ 2:1 (Fig. 2).

Operation of these devices with laser diodes simply requires construction of a cell of the same configuration as above, but with the dielectric stack reflectivity centred to match that of the gold at the wavelength used. Thus we were able to demonstrate low power laser diode driven optical bistability using the AlGaAs laser output from a demounted compact disc player read arm[9].

3 TEMPORAL RESPONSE AND STABILITY

Analysis of the characteristic response of these devices for defined operating parameters such as finesse and detuning, may be obtained from consideration of the time dependent heat flow equation,

$$\rho C_p \frac{\partial T}{\partial t} = Q(t) + \kappa \nabla^2 T \quad , \tag{2}$$

where Q(t) is the quantity of heat absorbed/unit volume/unit time in the metallic partial mirror, and which has the regular Airy formalism,

$$Q(t) = \frac{A \ I(t)}{1 + F \sin^2 \phi} \quad , \tag{3}$$

where A is the peak absorption coefficient, F is the coefficient of finesse, and ϕ is the single pass phase-change across the cavity.

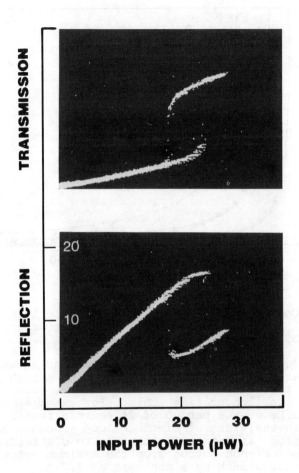

<u>Figure 2</u> Bistable operation of a dielectric/gold K15 cell observed in transmission and reflection. Spot size = 25 μm.

Making the approximation, $\kappa \nabla^2 T \cong - \Delta T/B$, where B is a diffusion related parameter, allows a numerical integration of equation (1) to be performed, yielding a family of responses dependent on detuning. Overlaying these with experimentally determined results for the same operating parameters (Fig. 3), shows a close agreement, particularly in the critical slowing down regime, with a characteristic switching time of ∿ 1.5 ms.

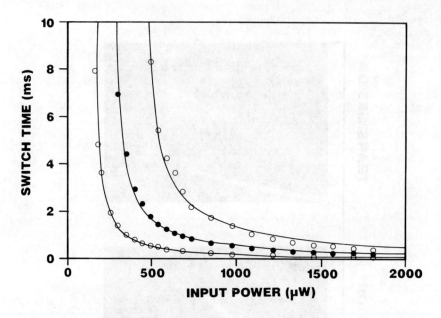

<u>Figure 3</u> Comparison of theoretical and experimentally
determined temporal responses for a dielectric/gold K15
cell with different detunings.

Extended operation of these devices has shown a
characteristic drift in detuning, and thus switch power,
similar to that experienced in ZnSe interference filters[10]
though on a far longer time scale. For example, switching
a cell at 1 Hz over a period of 24 hours effected a slow
reduction in the width of bistable loop, though, at the end
of this period, it was still possible to distinguish the
minimum upper branch output from the maximum output in the
original lower branch by a contrast of 1.7:1.

    4 DISCUSSION

To take advantage of the inherent parallelism of optics
with conventional cw laser sources requires devices with
high sensitivity. Techniques such as pixellation and
cavity/substrate optimisation will produce significant
increases in the sensitivity of all opto-thermal dispersive
bistable systems. The limits are set by sophistication of
construction techniques and the magnitude of refractive
index change with absorbed power.

Liquid crystals have presently demonstrated the greatest exploitable thermo-optic nonlinearity and have produced switching levels nearly two orders of magnitude lower than that of typical ZnSe interference filter devices[11].

Planar-aligned nematic liquid crystals also display a nonlinear birefringence which has been used to produce switches in both transmitted power and polarisation state [12]. This allied with their dielectric anisotropy and thus sensitivity to applied electric fields permits their application to a wide range of optical/ optoelectronic devices.

ACKNOWLEDGEMENTS

A. D. Lloyd is grateful for an SERC CASE award in cooperation with STL Ltd., and partial funding through the SERC/DTI JOERS programme is acknowledged. The K15 liquid crystal sample was obtained from BDH Chemicals Ltd., Poole, Dorset.

REFERENCES

1. B.S. Wherrett, D. Hutchings and D. Russell, J. Opt. Soc. Am., 1986, 3, 351.
2. A.C. Walker, Optics Comm., 1986, 59, 145.
3. A.D. Lloyd, I. Janossy, H.A. MacKenzie and B.S. Wherrett, Optics Comm., 1987, 61, 339.
4. C. Somerton and D.L. Tunnicliffe, Optics Comm., 1988, 65, 143.
5. D.C. Hutchings, A.D. Lloyd, I. Janossy and B.S. Wherrett, Optics Comm., 1987, 61, 345.
6. B.S. Wherrett, D.C. Hutchings, F.A.P. Tooley, Y.T. Chow and A.D. Lloyd, Proc. O.E. Lase '88, SPIE, 1988, 881, 2.
7. P.P. Karat and N.V. Madhusudana, Mol. Cryst. Liq. Cryst., 1976, 36, 51.
8. Model THW 200, Linkam Scientific Instruments Ltd., Tadworth, Surrey KT20 5HT, UK.
9. A.D. Lloyd and B.S. Wherrett, Proc. Optical Bistability (OBIV), Aussois, 1988, in press.
10. R.J. Campbell, J.G.H. Mathew, S.D. Smith and A.C. Walker, Submitted to Applied Optics.
11. Y.T. Chow, B.S. Wherrett, E.W. Van Stryland, B.T. McGuckin, D. Hutchings, J.G.H. Mathew, A. Miller and K. Lewis, J. Opt. Soc. Am. B, 1986, 11, 1535.
12. A.D. Lloyd, Optics Comm., 1987, 64, 302.